T0302040

ALGEBRA

...This great science which I have been calling "Characteristic," of which Algebra and Analysis are but small branches, ...is what gives words to languages, letters to words, digits to Arithmetic, notes to Music; this is what teaches us the secret of precise reasoning, requiring us to leave, as visible traces on paper, a volume for inspection at leisure: finally, this is what makes us reason, substituting characters in place of things, thereby unburdening the imagination.

Leibniz, De la Méthode de l'Universitalité (1674)

ALGEBRA
Groups, Rings, and Fields

Louis Rowen
Bar Ilan University
Ramat Gan, Israel

CRC Press is an imprint of the
Taylor & Francis Group, an **informa** business
AN A K PETERS BOOK

First published 1994 by A K Peters, Ltd.

Published 2019 by CRC Press
Taylor & Francis Group
6000 Broken Sound Parkway NW, Suite 300
Boca Raton, FL 33487-2742

ISBN 13: 978-1-56881-028-7 (hbk)

Library of Congress Cataloging-in-Publication Data

Rowen, Louis Halle.
Algebra: Groups, rings, and fields / Louis Rowen.
p. cm.
Includes index.
ISBN 1-56881-028-8
1. Algebra. I. Title.
QA152.2.R68 1993
512.9—dc20 93-39371
CIP

Typeset by AMS Tex

CONTENTS

PREFACE

Algebra is a subject with which we become acquainted during most of our education, largely in connection with the solution of equations. Some of the most famous questions in mathematical history have involved equations with coefficients in $\mathbb{Z}$, the set of integers. This course deals with their solutions. We shall see that the process of abstraction enables us to solve a variety of problems with economy of effort. This is the principle at the heart of abstract algebra, a subject that enables one to deduce sweeping conclusions from elementary premises. As such, this course can be used to initiate an intelligent student to the glorious world of mathematical discovery. At the same time, a course in abstract algebra, properly presented, could treat mathematics as an art as well as a science. In these notes I have tried to present underlying ideas, as well as the results they yield.

Abstract algebra received a major impetus toward the beginning of this century, when "intuition" in geometry began to lead to false assertions. It was seen that algebraic structure provides a firm foundation for a new subject called algebraic geometry, which enabled Zariski and others to put the latest developments of geometry an a solid footing. Indeed, it is not surprising that much of the structure theory of algebra was developed by Emmy Noether, the daughter of a geometer. Although algebraic geometry is outside the scope of this short book, we do attempt to lay the pedagogical foundations by introducing Noetherian rings and prime ideals.

These notes are far from comprehensive, and the serious student might continue with the far more thorough texts of Jacobson (*Basic Algebra*, Freeman 1985), Artin (*Algebra*, Prentice Hall 1991), and/or Cohn (*Algebra I*, Wiley 1974). Throughout, I have tried to stick to the main track of a one-year course, with the aim of touching on as many important theorems and applications as possible. The observant reader will notice the heavy debt to the classic text of Herstein (*Topics in Algebra*, Xerox, 1964), which in turn owes a great deal to Van der Waerden's pioneering work. The cursory treatment here is indicated by the division of the material into chapters, each of which is supposed to correspond to one of the 26 weeks of a year-long algebra course. (Some of these chapterss are extended a bit by appendices.)

References are normally within the same chapter, unless indicated otherwise by a decimal point; for example, "Theorem 2.15" means "Theorem 15 of Chapter 2."

My personal experience with this material is that often I am pressed towards the end and usually end a one-year course with Chapter 24, leaving most Galois theory for the next year. For those lecturers who would treat Galois theory in the first year, this raises the question of how one could push ahead in order to reach Chapters 25, 26, and 27. First of all, the addenda to Chapters 3 and 12 are not needed elsewhere in the text. Another two possible chapters that could be curtailed are Chapters 17 and 19. The second half of Chapter 17 contains a second proof of the basic theorem that Euclidean rings are Unique Factorization Domains (UFDs), by means of translating all the relevant concepts to ideals and then replacing the Euclidean degree function by considerations about ideals. Although one sees more generally that all principal ideal domains (PIDs) are UFDs, this generalization is nominal, since the motivating examples of PIDs are all Euclidean; indeed it is quite difficult to come up with an example of a UFD that is *not* Euclidean. Nevertheless, I would advocate a full presentation of this material for two reasons: It serves as an introduction to Noetherian rings and their techniques, one of the important advances in the early part of the twentieth century; and the methods are more elegant, belonging more intrinsically to the algebraic structure of rings, and indeed, ideals have taken on an even more important role in ring theory than has unique factorization.

The material of Chapter 19 might perhaps be more vulnerable to the red pen, since its interest is partly historical. However, I feel that every so often the lecturer should step back and let the students see what can be reaped from the theories that they have labored so hard to master. For many years number theory was considered the epitome of human inquiry, attracting the attention of some of the greatest intellects in history, and the theories described in this book were motivated largely by the material in Chapter 19.

Concerning the two appendices at the end of these notes, I feel uneasy relegating the transcendence of π to a corner unlikely to be reached in most courses, but classroom experience indicates that this proof is perhaps the hardest of all for the students to digest, and perhaps the effort could be put to better purpose by explaining Galois theory more completely.

Appendix B is a luxury, but it ties up several loose ends; and it intrigues me to see how far one can get in noncommutative algebra (including most of the Skolem-Noether theorem) simply by pushing forward some of the basic techniques of elementary abstract algebra.

Although several of the exercises are more or less routine applications of the theorems, most exercises are extensions of the text, which usually require substantially more time to solve. A few exercises are intended to lead the reader to anticipate an upcoming topic; these exercises are usually difficult where they are presented, but become much easier later in the course.

Of course the traditional role of exercises in a course is to provide more-or-less routine applications of the main results, for the student's edification and also as possible material for examinations. These are provided as the Review Exercises at the end of the book.

A note about proofs: In an elementary course one takes considerable care to find the "best" proof. The usual criterion in a mathematics book is the length of the proof (the shorter the better). However, several standard proofs involve sleight of hand, pulling some computation out of the thin air, so, in an effort to explain the underlying ideas, I have turned to the criterion, "Which proof is easiest to remember?" For the reader's edification and amusement, several of the magic proofs have been put into the exercises.

One dilemma in a course in algebra is deciding where to introduce matrices. Although various subsets of matrices provide some of the most important examples of groups, the full set of $n \times n$ matrices over a field has the structure of a "ring," which usually is not defined until *after* groups have been studied for several months. (Indeed, we shall define a group in the first chapter.) Thus, strictly speaking, groups arising from matrices should not be introduced until far into the course; however, I did not want to wait to bring in such an important example. As a compromise, I have assumed that the reader is familiar with the definition and basic properties of matrices.

I would like to express my gratitude to my colleagues Steve Shnider and Shalom Feigelstock for suggestions on improving the exposition, to Boaz Saban and Miriam Rosset for spotting errors in the draft version, and to my helpmate Rachel Rowen for sharing her expertise in computer typesetting.

TABLE OF PRINCIPAL NOTATION

NOTE: Similar notation might depend on the context, *e.g.* group or ring.

PREREQUISITES

Mathematics evolves around the study of various sets, so let us review briefly some foundations about sets. We write $A \subseteq B$ to denote that the set A is a *subset* of B, *i.e.*, every element of A is contained in B; if, furthermore, $A \neq B$, we may write $A \subset B$. The *empty set* is denoted as $\emptyset$. A set S can be described either by a list of its elements or by means of a larger set. For example, perhaps the most fundamental set in algebra is $\mathbb{N} = \{0, 1, 2, \dots\}$ (the set of natural numbers); the set of even natural numbers can be designated either as $\{0, 2, 4, \dots\}$ or $\{n \in \mathbb{N} : n \text{ is even}\}$ or simply as $\{2n : n \in \mathbb{N}\}$ (which will be condensed even further to $2\mathbb{N}$).

Given an arbitrary set S and $A, B \subseteq S$, we write $A \cap B$ for $\{s \in S : s \in A \text{ and } s \in B\}$; we say A and B are *disjoint* if $A \cap B = \emptyset$. Likewise we write $A \cup B$ for $\{s \in S : s \in A \text{ or } s \in B\}$. One can think of $A \cap B$ as the largest set contained in both A and B, in the sense that any set contained in both A and B must also be contained in $A \cap B$. Similarly $A \cup B$ is the smallest subset of S containing both A and B, in the sense that any subset of S containing both A and B must contain $A \cup B$. We shall use "the largest" and "the smallest" in this sense, throughout the text. (On the other hand, "maximal" is used in the slightly weaker sense that there is nothing larger. For example, if we are given the subsets $A_1 = \{0,\ 1,\ 3\}$, $A_2 = \{0,\ 2,\ 3\}$, and $A_3 = \{0,\ 3\}$ of $\mathbb{N}$, we would say A_1 and A_2 are maximal among these three subsets, although none of these is the largest.)

$\cup$ and $\cap$ satisfy the familiar axioms of associativity and distributivity for unions and intersections of subsets A, B, C of a set S:

$$
\begin{aligned}
A \cup (B \cup C) &= (A \cup B) \cup C; \\
A \cap (B \cap C) &= (A \cap B) \cap C; \\
A \cup (B \cap C) &= (A \cup B) \cap (A \cup C) \\
A \cap (B \cup C) &= (A \cap B) \cup (A \cap C).
\end{aligned}
$$

Similarly, given an arbitrary collection of subsets $A_i : i \in I$ of A, we define their *intersection* $\cap_{i \in I} A_i$ or *union* $\cup_{i \in I} A_i$.

A function f from a set A to a set B is denoted as $f: A \to B$, and is also called a *map*. (We stipulate that the image $f(a)$ is defined for each a in A.) Maps between sets turn out to be as important as the sets themselves. Given a map $f: A \to B$ we write $f(A)$ for $\{f(a) : a \in A\} \subseteq B$; we say f is *onto* if $f(A) = B$. On the other hand, we say f is *one-to-one*, written 1: 1, if for any elements $a_1 \neq a_2$ in A we have $f(a_1) \neq f(a_2)$ in B. (It is usually easier to verify the equivalent formulation that $f(a_1) = f(a_2)$ implies $a_1 = a_2$.) A map which is 1:1 and onto is called a 1:1 *correspondence*, or *bijection*.

Given a map $f: A \to B$ and $b \in B$, we define the *inverse image* f^{-1} as $\{a \in A : f(a) = b\}$. Analogously for $S \subseteq B$, we define $f^{-1}(S)$ as $\{a \in A : f(a) \in S\}$. Of course f^{-1} need not be a map, but if f is a bijection, then f^{-1} is a map and indeed is also a bijection. For maps $f: A \to B$ and $g: B \to C$, the *composite* $h = g \circ f: A \to C$ is defined by $h(a) = g(f(a))$.

The *Cartesian product* $A \times B$ of sets A, B is the set of ordered pairs $\{(a, b) : a \in A,\ b \in B\}$. A *binary relation* $\sim$ on a set A is defined formally as a subset R of $A \times A$; usually we simply write $a \sim b$ if $(a, b) \in R$. For example the relation "$a < b$" is defined on $\mathbb{N}$ as $\{(0,1), (0,2), (1,2), \ldots\}$; the relation "equality" in A is defined as $\{(a, a) : a \in A\}$. Generalizing "equality," one says a relation $\sim$ is an *equivalence* if it satisfies the following properties:

1. ("reflexivity") $a \sim a$ for all a in A;
2. ("symmetry") $a \sim b$ implies $b \sim a$;
3. ("transitivity") $a \sim b$ and $b \sim c$ imply $a \sim c$.

For example parallelism of lines in the Euclidean plane is an equivalence relation (provided one says a line is parallel to itself.)

Given an equivalence $\sim$ on A, we define the "equivalence class" $[s]$ of any element s of A to be $\{a \in A : a \sim s\}$. Note that any element belongs to its own equivalence class, by reflexivity, so A is the union of various equivalence classes. On the other hand, two equivalence classes either coincide or are disjoint; hence, A is the disjoint union of its equivalence classes. The set of equivalence classes of A is denoted as $A/\sim$.

Conversely, we define a "partition" of A to be a set of disjoint subsets whose union is A; any partition defines the equivalence relation defined by stipulating that $a \sim b$ iff they lie in the same subset in the partition. (By "iff" we mean "if and only if.")

In algebra one often introduces new algebraic structures by means of equivalence classes. The reader may have encountered the following construction of the integers $\mathbb{Z}$, from $\mathbb{N}$: Define the equivalence $\sim$ on $\mathbb{N} \times \mathbb{N}$ by

$$(a_1, a_2) \sim (b_1, b_2) \quad \text{iff} \quad a_1 + b_2 = a_2 + b_1.$$

For example, $(4,7) \sim (5,8)$. We define $\mathbb{Z}$ formally as the set $\mathbb{N} \times \mathbb{N}/\sim$. Intuitively the equivalence class of (a_1, a_2) has been identified with the integer $a_1 - a_2$, since (a posteriori)

$$(a_1, a_2) \sim (b_1, b_2) \quad \text{iff} \quad a_1 - a_2 = b_1 - b_2.$$

One can define algebraic operations on the equivalence classes, by means of representatives from the classes. For example one defines addition on $\mathbb{Z}$ by

$$[(a_1, a_2)] + [(b_1, b_2)] = [(a_1 + b_1, a_2 + b_2)].$$

The equality $-3+4 = 1$ can be expressed as $[(1,4)]+[(6,2)] = [(7,6)]$. This poses a new difficulty: one has to show that the outcome is independent of the particular representatives we chose in the equivalence classes. For example, if we used (2,5) instead of (1,4) we would wind up with (8,7) instead of (7,6), which is all right since they are equivalent. This condition is called "well-defined" and is one of the nuisances that we must contend with in many constructions.

Another example of construction by means of conjugate classes is given following Definition 2, below.

We shall assume such familiar properties of $\mathbb{N}$ as the unique factorization of any whole number > 1 into primes. In fact this follows from the theory to be developed in Chapter 16, but it is convenient to use these properties earlier, to develop examples in group theory. The reader is assumed to be familiar with the method of proof by mathematical induction, used in proving the bulk of our theorems, *cf.* Exercise 1.

One fundamental property of $\mathbb{N}$ is that every nonempty subset S of $\mathbb{N}$ contains a unique smallest element. This can be proved by mathematical induction, and actually provides an alternate version of induction that will be used quite frequently in the theory of finite groups.

Much of mathematics involves the sets $\mathbb{Q}$ (the rational numbers), $\mathbb{R}$ (the real numbers), and $\mathbb{C}$ (the complex numbers), so we are led to study the properties these sets have in common. One can check easily that in each of these sets the following axioms hold with respect to addition and multiplication, for any elements a, b, and c:

(F1) $(a+b)+c = a+(b+c)$
(F2) $a+0 = 0+a = a$
(F3) $a+b = b+a$
(F4) $a+(-a) = (-a)+a = 0$
(F5) $(ab)c = a(bc)$
(F6) $a1 = 1a = a$
(F7) $ab = ba$
(F8) $a\frac{1}{a} = \frac{1}{a}a = 1$ whenever $a \neq 0$
(F9) $a(b+c) = ab+ac$
(F9′) $(b+c)a = ba+ca$

(Of course (F9′) is superfluous here in view of (F7).) We are now in position to cross the threshold of abstract algebra — why not define an abstract entity satisfying (F1) through (F9)? Then we could examine its properties and apply the ensuing results to $\mathbb{Q}, \mathbb{R}$, and $\mathbb{C}$, thereby saving ourselves the trouble of studying each set separately. In order to do this we need more than just the set; we need to define "operations" corresponding to addition and multiplication. Accordingly, one defines a *binary operation* on a set S to be a map $S \times S \to S$, *i.e.*, it takes an ordered pair of elements of S and assigns an element of S as the answer. For example, the binary operation $+$ on $\mathbb{Q}$ takes $(4, -7.5)$ to their sum, -3.5.

Please note that each binary operation is assumed to be defined for *every* pair of elements (s_1, s_2) of $S \times S$; in some treatments this property is isolated as a separate axiom (called *closure*), but here it is subsumed in the definition.

Definition 1. A *field* is a set F, together with binary operations (denoted as $+$ and $\cdot$ and called addition and multiplication) and designated elements $0, 1 \in F$, such that:

(i) properties (F1),(F2),(F3),(F5),(F6),(F7), and (F9) hold for all a, b, c in F;
(ii) For any element a in F there is a unique element denoted $(-a)$ satisfying (F4);
(iii) For any $a \neq 0$ in F there is a unique element denoted $\frac{1}{a}$, or a^{-1}, satisfying (F8).

Having defined a field we might look for more examples. The advantage of arranging the properties as in Definition 1 is that the properties in (i) pass at once to subsets. On the other hand, $\mathbb{Z}$, with the usual addition and multiplication, satisfies (i) and (ii) but not (iii), since $\frac{1}{2}$ is not in $\mathbb{Z}$; $\mathbb{N}$ satisfies (i) but neither (ii) nor (iii). Nevertheless, one can show that $\mathbb{Q}$ as a set is in 1:1 correspondence with $\mathbb{Z}$ (as well as with $\mathbb{N}$). This points to the principle that operations accompanying a set are quintessential to the algebraic theory; the aggregate of set, operations, and designated elements, is called the *(algebraic) structure.* Here is an example that is very important in number theory, as we shall see.

Definition 2. Given $m \geq 1$, define $\mathbb{Z}_m = \{0, 1, \ldots, m-1\}$, provided with the addition and multiplication of "clock arithmetic," *i.e.*, addition and multiplication taken modulo m, also written $\pmod{m}$.

To understand $\mathbb{Z}_m$ more fully, we need a formal description. Define $m\mathbb{Z} = \{mn : n \in \mathbb{Z}\} = \{\ldots, -m, 0, m, 2m, \ldots\}$, the multiples of m. We define the equivalence $a \equiv b \pmod{m}$ (for $a, b \in \mathbb{Z}$), read "a is congruent to b modulo m," iff $a - b \in m\mathbb{Z}$. Let us write $[a]$ for the equivalence class containing a; thus $[a] = [a+m] = [a-m] = \ldots$. In this way $\mathbb{Z}$ is partitioned into m equivalence classes $[0], [1], \ldots, [m-1]$. Now define addition and multiplication by $[a] + [b] = [a+b]$ and $[a][b] = [ab]$. One must check that this definition is well-defined, *cf.* Exercise 1, but then it is easy to transfer properties from $\mathbb{Z}$ to $\mathbb{Z}_m$, merely by writing brackets wherever appropriate.

In this manner we see $\mathbb{Z}_m$ satisfies (i) and (ii) of Definition 1 (taking $-[a]$ to be $[-a] = [m-a]$), but again (iii) may fail. Indeed the reader can easily check that $[2]^{-1}$ does not exist in $\mathbb{Z}_4$, leading us to the question: "For what m is $\mathbb{Z}_m$ a field?" We shall soon see (Corollary 1.13) that $\mathbb{Z}_m$ is a field iff m is a prime number; for example $[2]^{-1} = [6]$ in $\mathbb{Z}_{11}$.

Of course the field $\mathbb{Z}_p$ (for p prime) has p elements, and the existence of finite fields will be important in our study of groups. Although $\mathbb{Z}_p$ suffices for many applications, the reader should be apprised that there is a field having precisely n elements, if and only if n is a power of a prime number; this and other facts about finite fields will be proved much later, in Chapters 24 and 26.

As mentioned earlier, the reader is assumed to be familiar with matrices over $\mathbb{R}$. Analogously, the set of $n \times n$ matrices with entries in an arbitrary field F also is endowed with the analogous matrix addition and matrix multiplication and is denoted as $M_n(F)$. Let us define the $n \times n$ *matric unit* e_{ij} of $M_n(F)$ to be the matrix with 1 in the $i-j$ position and 0 everywhere else. Then the following properties are satisfied:

(i) $e_{11} + \cdots + e_{nn} = 1$;
(ii) $e_{ij}e_{jk} = e_{ik}$;
(iii) $e_{ij}e_{uv} = 0$ if $j \neq u$.

Any element of $M_n(R)$ can be written uniquely in the form $\sum_{i,j=1}^{n} r_{ij}e_{ij}$ for r_{ij} in R. Matric units are a significant aid for computing with matrices.

Exercises

1. Prove that addition in $\mathbb{Z}$ defined formally in the text actually is well-defined. Similarly, define multiplication in $\mathbb{Z}$, and verify (i),(ii) of Definition 1, by induction. (This exercise becomes rather boring after a while, once one gets the hang of what is going on.)
2. Prove in any field F that $a, b \neq 0$ implies $ab \neq 0$.
3. Define $\mathbb{Q}(\sqrt{-1})$ to be $\{a+b\sqrt{-1} : a, b \in \mathbb{Q}\}$, which can be viewed as a subset of $\mathbb{C}$. Prove $\mathbb{Q}(\sqrt{-1})$ is a field. (*Hint*: the only challenging part is the multiplicative inverse.)
4. Define $\mathbb{Q}(\sqrt{2})$ to be $\{a + b\sqrt{2} : a, b \in \mathbb{Q}\}$, viewed as a subset of $\mathbb{R}$. Prove this is a field, under the natural addition and multiplication:

$$(a + b\sqrt{2}) + (c + d\sqrt{2}) = (a + c) + (b + d)\sqrt{2}$$
$$(a + b\sqrt{2})(c + d\sqrt{2}) = (ac + 2bd) + (ad + bc)\sqrt{2}$$

Although $\mathbb{Q}(\sqrt{2})$ is clearly in 1:1 correspondence with $\mathbb{Q}(\sqrt{-1})$, their algebraic structures as fields are different. Explain.

PART I — GROUPS

In this course, abstract algebra focuses on sets endowed with "algebraic structure," and axioms describing how the elements behave with respect to the given operations. The operations of basic concern to us are multiplication and/or addition, in various contexts. In this spirit, we should start with some observations about Definition 0.1:

1. Property (iii) (axiom (F8)) relies only on multiplication, not addition;
2. Properties (F5) through (F7) are the multiplicative analogs of (F1) through (F3), and (F8) is the analog of (F4), except that we delete $\{0\}$.

Thus it makes sense to isolate properties (F1)–(F3) with an eye to include (F4) shortly thereafter; then using the multiplicative analog (deleting $\{0\}$) we would understand (F5) through (F8). This will lead us first to the definition of "monoid" and then to "group." Abstract groups appear in almost every branch of mathematics and physics, as well as in other sciences and even in many aspects of day-to-day life (such as telling time). Our object is to develop enough of the theory of groups to enable us to answer basic questions concerning their structure and to familiarize the reader with certain groups that he or she is likely to encounter repeatedly in the future. Various interesting classes of groups are easier to study than groups in general. We shall obtain rather decisive theorems concerning Abelian groups; in Part III we shall need "solvable groups," a useful generalization of Abelian groups that is studied in Chapter 12.

Definition 1. A *monoid* $(M, \cdot, e)$ is a set M with a binary operation and a *neutral* element e (also called the *identity*) satisfying the following properties for all a, b, c in M:

(M1) (associativity) $(ab)c = a(bc)$;
(M2) $ae = ea = a$.

(Note that, as is customary in multiplicative notation, we write ab instead of $a \cdot b$. On the other hand, often $\cdot$ will be $+$, which is never suppressed in the notation.)

Note that the neutral element is uniquely determined by the operation; indeed if e and e' are neutral elements then $e' = e'e = e$. Thus we usually delete "e" from the notation; for example $(\mathbb{Z}, +)$ can only mean $(\mathbb{Z}, +, 0)$. We shall often delete the operation also, and write M for the monoid $(M, \cdot, e)$.

We say elements a and b in M *commute* if $ab = ba$. The monoid M is *commutative* if $ab = ba$ for all a, b in M. When the set M is finite, we say M is *finite* and write $|M|$ for the number of elements of M, called the *order* of M. The particular operation plays a crucial role in the structure of the monoid. For example $(\mathbb{Z}_2, +, 0)$ and $(\mathbb{Z}_2, \cdot, 1)$ are finite monoids each of order 2, but whose structures are not analogous. (In the first case the neutral element is 0 and satisfies $0 = 1 + 1$, but in the second case the neutral element 1 satisfies $1 \neq 0 \cdot 0$.)

Note. Associativity enables us to write products without parentheses, without ambiguity, *cf.* Exercise 1. Let us now turn to the key notion, bringing in (F8).

Definition 2. An element a of M is *left invertible* (resp. *right invertible*) if there is b in M such that $ba = e$ (resp. $ab = e$); a is *invertible* if a is left and right invertible. A *group* is a monoid in which every element is invertible.

Suppose a is invertible. Then there are elements b, b' such that $ba = e = ab'$. We shall now see that $b = b'$. Indeed $b = b(ab') = (ba)b' = b'$. Thus a has a unique element b such that $ba = e = ab$, which is called the *inverse* of a, denoted a^{-1}.

A commutative group is called *Abelian*, after the Norwegian mathematician Abel.

Examples of Groups and Monoids

Let us start with some basic examples.

(1) If F is a field then $(F, +)$ and $(F \setminus \{0\}, \cdot)$ are Abelian groups, seen

by axioms (F1) through (F4) and (F5) through (F8) respectively, given in the prerequisites. (See exercise 0.2 to clean up a sticky point.)

(2) $(\mathbb{Z}, +)$ is an Abelian group, but $(\mathbb{Z} \setminus \{0\}, \cdot)$ is a commutative monoid that is not a group.

(3) $(\mathbb{Z}_m, +)$ is an Abelian group of order m; $(\mathbb{Z}_m, \cdot)$ is a commutative monoid.

(4) (See prerequisites.) For any field F and any n, write $M_n(F)$ for the set of $n \times n$ matrices with entries in an arbitrary field F, endowed with the usual matrix addition and multiplication. $(M_n(F),+)$ is an Abelian group. $(M_n(F), \cdot, 1)$ is a monoid (but not a group) that is not commutative for $n > 1$, since the matrices $\begin{pmatrix} 1 & 0 \\ 0 & 0 \end{pmatrix}$ and $\begin{pmatrix} 0 & 1 \\ 0 & 0 \end{pmatrix}$ do not commute. (These matrices are not invertible, since each has rank 1. Two invertible 2×2 matrices that do not commute are $\begin{pmatrix} 1 & 1 \\ 0 & 1 \end{pmatrix}$ and $\begin{pmatrix} 0 & 1 \\ 1 & 0 \end{pmatrix}$.) One can define adjoints and determinants in the usual way; then $A \operatorname{adj}(A) = \det A$ for any A in $M_n(F)$, so $A^{-1} = \operatorname{adj}(A)/\det A$ exists whenever $\det A \neq 0$.

Aside. This argument also shows $AB = 1$ implies $BA = 1$, for any two matrices A, B, *cf.* Exercise 8.

(5) Suppose S is a set. Then $\{\text{functions } f: S \to S\}$ form a monoid which we denote as $\mathrm{Map}(S, S)$, whose operation is composition; the neutral element is the identity map 1_S defined by $1_S(s) = s$ for all s in S.

(6) The *trivial group* $\{e\}$ has multiplication given by $ee = e$.

Note. Although the operation of a group often is $+$, it is customary to choose multiplicative notation when studying groups in general, since "+" biases us toward commutativity.

One can extract a group from an arbitrary monoid. To understand this procedure let us first examine invertibility.

Remark 3. Suppose a, b are invertible. From the equation $a^{-1}a = e = aa^{-1}$ we also see $a = (a^{-1})^{-1}$. Furthermore $abb^{-1}a^{-1} = e = b^{-1}a^{-1}ab$, implying $(ab)^{-1} = b^{-1}a^{-1}$.

Given a monoid M, write $\mathrm{Unit}(M)$ for $\{\text{invertible elements of } M\}$.

PROPOSITION 4. *Unit(M) is a group.*

Proof. Associativity in $\mathrm{Unit}(M)$ follows at once from associativity in M. But $e \in \mathrm{Unit}(M)$, so Remark 3 implies $\mathrm{Unit}(M)$ is closed under multiplication and is thus a monoid. Moreover if $a \in \mathrm{Unit}(M)$, then $a^{-1} \in \mathrm{Unit}(M)$ so $\mathrm{Unit}(M)$ is a group. □

Example 5. Applying Proposition 4 to Examples (1) through (5) yield the following useful groups:

(1) $\text{Unit}(F, \cdot) = F \setminus \{0\}$ for any field F.
(2) $\text{Unit}(\mathbb{Z}, \cdot) = \{\pm 1\}$.
(3) $\text{Unit}(\mathbb{Z}_m, \cdot) = \{a : 1 \leq a < m, \text{ and } a \text{ is invertible} \mod m\}$. This is called the *Euler* group, denoted Euler(m). For example Euler(6) = {1,5}; Euler(7) = {1,2,3,4,5,6}, and Euler(8) = {1,3,5,7}. The order of Euler(m) is called the *Euler number* $\varphi(m)$. Thus $\varphi(3) = \varphi(4) = \varphi(6) = 2$, and $\varphi(5) = \varphi(8) = 4$.
(4) $\text{Unit}(M_n(F), \cdot)$ is the group of regular $n \times n$ matrices over F, called the *general linear group* and denoted $\text{GL}(n, F)$.

Let us pause for a moment, to assert that $\text{GL}(n, F)$ is perhaps the most important group in mathematics, since it can be interpreted as the group of invertible linear transformations of an n-dimensional vector space over the field F; as such, it has fundamental significance in geometry and in physics. (This identification also enables us to prove various properties of $\text{GL}(n, F)$, *cf.* Exercise 6.)

Other "geometrical" groups can be defined similarly in terms of various kinds of linear transformations, *cf.* Chapter 2 and Exercises 12.9ff, 12.15ff. Nevertheless, the focal role in these notes is played by the next example.

(5) Unit(Map(S, S)) is denoted as $A(S)$, the 1:1 onto maps from S to S. In the special case $S = \{1, \ldots, n\}$ we denote A(S) as S_n, the group of permutations of n symbols. S_n, often called the *symmetric group*, plays a special role in finite group theory and will be used throughout as our main example. The reader is urged to trace its development via the index.

When Is a Monoid a Group?

We want to explore the fundamental question of when a given monoid is already a group. One basic property of groups will become focal.

Remark 6. If $ab = ac$ with a left invertible, then $b = c$. (Indeed, multiply by the left inverse of a.)

Accordingly we call a monoid (left) *cancellative* if it satisfies the property

$$ab = ac \text{ implies } b = c$$

for any elements a, b, c. An example of a cancellative monoid which is not a group is $(\mathbb{Z} \setminus \{0\}, \cdot)$. On the other hand we have

THEOREM 7. *Any finite cancellative monoid M is a group.*

Before proving Theorem 7, let us accumulate some facts.

LEMMA 8. *If every element of a monoid M is right invertible, then M is a group.*

Proof. We need to show any element a of M is invertible. By hypothesis there is b such that $ab = e$; likewise there is c such that $bc = e$. But then b is invertible, so $a = b^{-1}$, as noted after Definition 2, implying $b = a^{-1}$. □

FACT 9. *Suppose S is a finite set. A function $f: S \to S$ is 1:1 iff f is onto.*

This fact is called the "pigeonhole principle"; for if a letter carrier is to distribute 17 letters into 17 boxes, clearly each box receives a letter iff no box receives at least two letters. Of course the pigeonhole principle fails for infinite sets, as illustrated by the map $f: \mathbb{N} \to \mathbb{N}$ given by $f(n) = n + 1$ for all n in $\mathbb{N}$ (which is 1:1 but not onto). The pigeonhole principle relates to monoids and cancellation via the following key observation.

PROPOSITION 10. *Suppose M is a monoid. Given $s \in M$ define the left multiplication map $\ell_s: M \to M$ by $\ell_s(a) = sa$ for all a in M.*

(i) *ℓ_s is onto iff s is right invertible;*
(ii) *ℓ_s is 1:1 iff $sb \neq sc$ for all b, c in M.*

Proof. (i) ($\Rightarrow$) $e = \ell_s(s')$ for some s' in M; thus $ss' = e$.
($\Leftarrow$) If $ss' = e$ then $a = ss'a = \ell_s(s'a)$ for any a in M.
(ii) Self-evident. □

Proof of Theorem 7. By Proposition 10(ii) we see for each s in M that ℓ_s is 1:1 and thus onto by the pigeonhole principle. Hence each s in M is right invertible, by Proposition 10(i), so, by Lemma 8, M is a group. □

Remark 11. The proof of Theorem 7 shows that, for any element g of a group G, the left multiplication map $\ell_g: G \to G$ is 1:1 and onto. We shall return to this fact later.

Let us apply theorem 7 to divisibility and the Euler group.

Remark 12. Writing $\gcd(a, b)$ to denote the greatest common divisor of a and b, we claim $\text{Euler}(m) = \{a \in \mathbb{N} : 1 \leq a < m \text{ and } \gcd(a, m) = 1\}$. Indeed, using unique factorization in $\mathbb{N}$, it is easy to identify the right-hand side with a cancellative monoid contained in $\mathbb{Z}_m$. (If $\gcd(a_1, m) = 1$ and $\gcd(a_2, m) = 1$, then $\gcd(a_1a_2, m) = 1$; also if $ab = ac$ in $\mathbb{Z}_m$, then $ab \equiv ac \pmod{m}$, implying m divides $ab - ac = a(b - c)$, so m divides $b - c$.) Hence the right-hand side is a group so by definition is contained in $\text{Euler}(m)$.

On the other hand any invertible element a of $(\mathbb{Z}_m, \cdot)$ is relatively prime to m, for if $d \in \mathbb{N}$ divides both a and m, then $a0 = 0 = \frac{a}{d}m = a\frac{m}{d}$, implying by remark 6 that $\frac{m}{d} \equiv 0 \pmod{m}$, so $d = 1$. Thus we have proved equality. □

COROLLARY 13. *A number is invertible* $\bmod m$ *iff it is relatively prime to* m. *In particular Euler(p)* $= \{1, \ldots, p-1\}$ *iff* p *is prime.*

COROLLARY 14. $\mathbb{Z}_p$ *is a field, for any prime number* p.

COROLLARY 15. $\{i,\ i+t,\ \ldots,\ i+(p-1)t\}$ *are all distinct* $\pmod{p}$, *for* p *prime, whenever* $t \not\equiv 0 \pmod{p}$.

Proof. One may assume $i = 0$; but then the first term is 0, and the others $\{t, \ldots, (p-1)t\}$ are distinct and nonzero $\pmod{p}$, seen by canceling t in Euler(p). □

COROLLARY 16. *If* $\gcd(m, n) = 1$, *then there are* a, b *in* $\mathbb{Z}$ *with* $am + bn = 1$.

Proof. Let $b = n^{-1}$ in Euler(m). Then m divides $1 - bn$, so $1 - bn = am$ for some a in $\mathbb{Z}$. □

Exercises

1. Define $a_1a_2a_3 = (a_1a_2)a_3$ and, continuing in this way, define

$$a_1 \ldots a_n = (a_1 \ldots a_{n-1})a_n.$$

   Prove that $(a_1 \ldots a_m)(a_{m+1} \ldots a_n) = a_1 \ldots a_n$ for any n and any $m < n$. (*Hint*: Induction on m.)
2. Define $A + B = \{a + b : a \in A,\ b \in B\}$ for subsets A, B of $\mathbb{Z}$. Compute $3\mathbb{Z} + 4\mathbb{Z}$, $6\mathbb{Z} + 10\mathbb{Z}$, and in general $a\mathbb{Z} + b\mathbb{Z}$.
3. Show if $a_1 \equiv b_1$ and $a_2 \equiv b_2 \pmod{m}$ then $a_1 + a_2 \equiv b_1 + b_2$ and $a_1a_2 \equiv b_1b_2$. Conclude that addition and multiplication in $\mathbb{Z}_m$ (as defined via equivalence classes) is well-defined.
4. If $a^2 \equiv 1 \pmod{p}$ for p prime, then $a \equiv \pm 1 \pmod{p}$. (*Hint*: Show that p divides $a^2 - 1 = (a+1)(a-1)$.) What happens for p not prime, *e.g.*, $p = 15$? $p = 21$?
5. Write down the Euler groups Euler(n) for all $n \le 10$. For which values of n does Euler(n) have order 2? Show that these groups are the "same" in some sense (to be made precise in Chapter 4).
6. Suppose F is a finite field of q elements (for example, if q is prime one could have $F = \mathbb{Z}_q$). Then the group $\mathrm{GL}(n, F)$ has order

$$\begin{aligned}&(q^n - 1)(q^n - q) \ldots (q^n - q^{n-1})\\ =&q^{n(n-1)/2}(q^n - 1)(q^{n-1} - 1) \ldots (q - 1).\end{aligned}$$

(*Hint*: One can define an arbitrary invertible linear transformation by sending a given base to any other base, so $|\mathrm{GL}(n, F)|$ is the number of different possible bases of the vector space $F^{(n)}$ over F (where the order of the base elements is significant). The first vector in the base can be any nonzero vector and thus chosen $(q^n - 1)$ ways; the next vector can be chosen $(q^n - q)$ ways, and so forth.)

7. How many concrete examples of finite groups can you produce at this stage? What is the non-abelian group of smallest order?
8. Consider the property $ab = 1$ implies $ba = 1$. Show this property holds in $M_n(F)$ (F a field) and in any finite monoid. However it fails in $\mathrm{Map}(S, S)$ when S is an infinite set. (*Hint*: Define $f: \mathbb{N} \to \mathbb{N}$ by $f(n) = n + 1$.)
9. A *semigroup* is a set with a binary operation satisfying M1 of definition 1, but not necessarily M2. Any semigroup S can be made into a monoid, simply by adjoining a formal identity e and defining $ea = ae = a$ for all a in S. Nevertheless, semigroups (without 1) arise in several contexts, where the elements are not necessarily invertible. (For example, let S be the set of functions $\mathbb{R} \to \mathbb{R}$ that are continuous in the neighborhood of a point.)
10. Inspired by the example in Exercise 9, call a semigroup S *inverse* if for each a in S there is b in S such that $aba = a$ and $bab = b$. Then b is unique and is called the *inverse* of a. Prove the analog of Remark 3 for inverse semigroups. Inverse semigroups have commanded fair attention in research.
11. An inverse semigroup that is cancellable is a group.

Chapter 2. How to Divide: Lagrange's Theorem, Cosets, and an Application to Number Theory

Our goal today is to introduce subgroups and show that the order of a subgroup divides the order of the group. This raises a basic question in arithmetic: Given two positive numbers m and n, how do we verify that m divides n (written $m|n$)? Two methods are already available from elementary school:

(1) The *Euclidean Algorithm* says that $n = qm+r$ for a suitable quotient $q > 0$ and remainder r with $0 \leq r < m$, so one need merely check $r = 0$.

(2) Possibly even more fundamentally, $m|n$ iff one can partition a set of n elements into subsets each having m elements.

We shall use each of these methods here, but first let us turn to the key ingredient of our present discussion. We have defined various algebraic structures thus far – fields, groups, monoids. When a given subset acquires the same structure from the larger set, we append to it the prefix "sub." For example:

Definition 1. *A* submonoid *of a monoid* $(M, \cdot, e)$ *is a subset* N *which becomes a monoid under the same operation* $\cdot$ *and neutral element* e.

A subgroup *of a monoid is a submonoid which is a group.*

Sometimes a group can be described most easily as a subgroup of another group. For example, one famous subgroup of $(\mathbb{C} \setminus \{0\}, \cdot)$ is the "unit circle," defined as $\{$complex numbers having absolute value $1\}$.

Remark 2. To show that the subset N is a submonoid of M, it is enough to check that $e \in N$ and N is closed under $\cdot$; axioms (M1) and (M2) of Definition 1.1 automatically hold in N (since they hold in the larger set M).

We shall be particularly interested in subgroups. One sees easily that $\text{Unit}(M)$ is the largest subgroup of M, in the sense that it contains every subgroup of M. Thus M and $\text{Unit}(M)$ have the same subgroups, so our attention focuses naturally on subgroups of *groups*.

A submonoid of a group need not be a group; *e.g.*, $(\mathbb{N}, +)$ is a submonoid of $(\mathbb{Z}, +)$. Nevertheless a submonoid of a group is cancellative by Remark 1.6; consequently Theorem 1.7 yields

Remark 3. Any finite submonoid of a group is a subgroup.

We write $H \leq G$ to denote that H is a subgroup of a group G. Obviously $\{e\}$ and G are subgroups of G. A subgroup H of G is *proper* when $H \neq G$; in this case we write $H < G$. Much of the theory of groups relies on studying subgroups, so we would like an easy criterion to verify whether a given subset is a group.

PROPOSITION 4. *Suppose G is a group and H is a nonempty subset. Then $H \leq G$ if either of the following two criteria is satisfied:*

(i) $h_1 h_2^{-1} \in H$ *for any* h_1, h_2 *in* H;

(ii) H *is finite and is closed under the group operation.*

Proof. (i) Take any h in H. Then $e = hh^{-1} \in H$. Now $h^{-1} = eh^{-1} \in H$ for any h in H. Finally, $h_1 h_2 = h_1(h_2^{-1})^{-1} \in H$. We have just proved H is a submonoid closed under inverses and so is a subgroup.

(ii) In view of Remark 3 we need merely check that $e \in H$. Take any h in H; noting $\{h^i : i > 0\}$ is finite, we must have $h^i = h^j$ for some $i > j$. But then $e = h^{i-j} \in H$. □

Any group G has the *trivial* subgroups $\{e\}$ and G itself. All other subgroups (if they exist) are called *nontrivial.*

Example 4′: Some examples of nontrivial subgroups.

(i) Suppose $F \subset E$ are fields. Then $F \setminus \{0\}$ is a subgroup of $(E \setminus \{0\}, \cdot)$. On the other hand, $(E \setminus \{0\}, \cdot)$ has other subgroups. One famous subgroup of $(\mathbb{C}\setminus\{0\}, \cdot)$ is the "unit circle," the set of complex numbers having absolute value 1.

(ii) The nontrivial subgroups of Euler(7) $= \{1, 2, 3, 4, 5, 6\}$ are $\{1, 2, 4\}$ and $\{1, 6\}$. In general, $\{1,\ n-1\}$ is a subgroup of Euler(n).

(iii) $m\mathbb{Z} = \{mt : t \in \mathbb{Z}\} \leq \mathbb{Z}$. We shall see shortly that these are the only subgroups of $\mathbb{Z}$.

(iv) If $m|n$ then $\{[0], [m], [2m], \ldots, [n-m]\} \leq (\mathbb{Z}_n, +)$. We shall see later that these are the only subgroups of $\mathbb{Z}_n$.

(v) GL(n, F) has many important subgroups, some of which are given in example 3.15. Other subgroups of GL(n, F) include the subset of all invertible diagonal matrices, and the subset of upper triangular matrices having nonzero diagonal entries.

The "smallest" subgroup containing a given invertible element a is called the *cyclic subgroup* generated by a and is rather easy to construct. Namely, define $a^0 = e$ and $a^i = a \ldots a$ (where a occurs i times in the product) for any $i > 0$; define $a^{-i} = (a^i)^{-1}$. Then $\{a^i : i \in \mathbb{Z}\}$ is a subgroup that is a subset of every subgroup containing a and thus is the cyclic subgroup generated by a.

Since our ultimate concern is groups, we work now in a group G and write $\langle a \rangle$ for the cyclic subgroup generated by $a \in G$. Note that if $\{a^i : i \geq 1\}$ is finite, then this is already a group by Proposition 4, so equals $\langle a \rangle$. In particular $a^n = e$ for some $n \geq 1$; the smallest such n is called the *order* of a, denoted $o(a)$. Note that $o(a) = 1$ iff $a = e$.

PROPOSITION 5. *If $o(a) = n$, then $\langle a \rangle = \{e, a, \dots, a^{n-1}\}$, and $a^n = e$.*

Proof. It is easy to see that the finite set $\{e, a, \dots, a^{n-1}\}$ is closed under multiplication (since $a^n = e$) and thus is a group containing a. It remains to show that $e, a, \dots, a^{n-1}$ are distinct. But if $a^i = a^j$ for $0 \leq i < j < n$, then $a^{j-i} = e$, contrary to the minimality of n. □

COROLLARY 5′. *(for G finite) $\langle a \rangle = G$ iff $o(a) = |G|$.*

If $o(a)$ divides a number m then $a^m = e$, indeed, writing $m = o(a)t$ we see $a^m = a^{o(a)t} = e^t = e$. Let us examine the converse.

PROPOSITION 6. *If $a^m = e$, then $o(a)$ divides m.*

Proof. Let $n = o(a)$. Using the Euclidean algorithm, we divide n into m and check that the remainder is 0. More precisely we write $m = nq + r$ where $0 \leq r < n$. Then

$$a^r = a^{m-nq} = a^m (a^n)^{-q} = ee = e.$$

But $r < n = o(a)$, contrary to definition, unless $r = 0$. □

COROLLARY. *If m divides $o(a)$ then $o(a^m) = \frac{o(a)}{m}$.*

Here is another instance of the same argument.

Example 6′: Every subgroup of $\mathbb{Z}$ is of the form $m\mathbb{Z}$ for suitable $m > 0$. Indeed if $0 \neq H < \mathbb{Z}$ take $m > 0$ minimal in H. Clearly $m\mathbb{Z} \leq H$, and we claim $m\mathbb{Z} = H$. Indeed take any $h \in H$. Then $h = mq + r$ for suitable $0 \leq r < m$. But $r = h - mq \in H$ since $mq \in m\mathbb{Z} \subseteq H$. By minimality of m we conclude $r = 0$, *i.e.*, $h = mq \in m\mathbb{Z}$. □

Note that $n\mathbb{Z} \leq m\mathbb{Z}$ iff $m|n$; in particular $n\mathbb{Z} = m\mathbb{Z}$ iff $n = \pm m$.

Cosets

Usually it is rather difficult to determine precisely the subgroups of a given group G, so we would like at least to describe certain properties of subgroups of G. The most basic property perhaps is the order, at least when G is finite. Note that any subgroup H of G contains $\{e\}$. Thus $|H| = 1$ iff $H = \{e\}$. On the other hand $|H| = |G|$ iff $H = G$.

Thus $|H|$ provides information about the subgroup H itself, and our goal is to see how $|H|$ is related to $|G|$. We shall prove the startling theorem, due to Lagrange, that $|H|$ divides $|G|$, for any subgroup H of G. This time the idea is to partition G into subsets each having order $|H|$. Obviously H itself is such a subset, and we look now for others.

Definition 7. A (right) *coset* of H in G is a set $Hg = \{hg : h \in H\}$ where $g \in G$ is fixed.

Remark 7'. Right multiplication by g provides a 1:1 correspondence from H to Hg, so $|Hg| = |H|$.

It remains to show that the cosets of H comprise a partition of G. (This could be done at once, using equivalence classes, *cf.* Exercise 1, but at this stage I prefer a more explicit approach.) Clearly for any $g \in G$, we have $g = eg \in Hg$. It follows at once that $G = \cup_{g \in G} Hg$, so it remains for us to show that distinct cosets are disjoint.

Remark 8. The following are equivalent for g, g' in G and $H < G$:

(i) $Hg' \subseteq Hg$;
(ii) $g' \in Hg$;
(iii) $g'g^{-1} \in H$.

(Indeed, one verifies $(i) \Leftrightarrow (ii) \Leftrightarrow (iii)$ directly.)

LEMMA 9. *If $H < G$ and $Hg_1 \subseteq Hg_2$, then $Hg_1 = Hg_2$.*

Proof. Write $g_1 = hg_2$. Then $g_2 = h^{-1}g_1$, so $Hg_2 \subseteq Hg_1$, by Remark 8, implying $Hg_1 = Hg_2$ as desired. □

Remark 9'. $Ha = Hb$ iff $ab^{-1} \in H$, by Remark 8 and Lemma 9. In particular, $Ha = He$ iff $a \in H$.

PROPOSITION 10. *If $Hg_1 \neq Hg_2$, then $Hg_1 \cap Hg_2 = \emptyset$. Thus the cosets of H comprise a partition of G.*

Proof. We show the contrapositive. Assume $g' \in Hg_1 \cap Hg_2$; then $Hg' \subseteq Hg_1 \cap Hg_2$, so by Lemma 9, $Hg_1 = Hg' = Hg_2$, as desired. □

THEOREM 11 (LAGRANGE'S THEOREM). *If H is a subgroup of a finite group G, then $|H|$ divides $|G|$; in fact $\frac{|G|}{|H|}$ is the number of cosets of H in G.*

Proof. We have just seen that G is a disjoint union of its cosets, each of which has the same number of elements as H. □

Motivated by this proof, we define the *index* of a subgroup H in G, written $[G : H]$, to be the number of (right) cosets of H in G. Note this could make sense even for G infinite; for example $[\mathbb{Z} : 2\mathbb{Z}] = 2$.

Remark 11′. When G is finite, Lagrange's theorem says that $[G:H] = \frac{|G|}{|H|}$. On the other hand, we could have developed the same theory by using *left* cosets gH instead of right cosets Hg. Hence, by symmetry, $\frac{|G|}{|H|}$ is also the number of left cosets of H in G. This idea is illustrated later, in the proof of Proposition 5.9 below. See Exercises 10.11ff for a discussion of what happens when we combine left and right cosets.

Example 12: Let $G = (\mathbb{Z}_8, +)$ and

$$H = \{[a] \in G : a \text{ is even}\} = \{[0], [2], [4], [6]\}.$$

H has two cosets: H itself and $H+[1] = \{[1], [3], [5], [7]\}$. Note that $|G| = 8$ and $|H| = 4$.

The converse to Lagrange's theorem is false; there is a group G of order 12 that fails to have any subgroup of order 6, *cf.* after Remark 5.23.

Returning to cyclic subgroups we recall $o(g) = |\langle g \rangle|$, and thus have

COROLLARY 13. *$o(g)$ divides $|G|$, for every element g of G.*

COROLLARY 14. *If $|G|$ is a prime number p, then G has no nontrivial subgroups. In particular, if $g \in G \setminus \{e\}$, then $G = \langle g \rangle$.*

Proof. Any subgroup $\neq \{e\}$ has some order $m > 1$ dividing p, implying $m = p$. □

Fermat's Little Theorem

Let us see how these ideas apply to a celebrated (although now easy) theorem of Pierre de Fermat. We shall have occasion to refer several times to Fermat's work in number theory, as a sounding board for the theory developed here. Although a jurist by profession, Fermat was one of the great mathematicians of all time, who loved to tantalize his colleagues by discovering deep results in number theory and challenging other mathematicians to reproduce them. The result discussed here is perhaps the most basic theorem concerning prime numbers.

THEOREM 15. *(Fermat's Little Theorem) If p is a prime number and $(a,p) = 1$, then $a^{p-1} \equiv 1 \pmod{p}$.*

Proof. Recall $\text{Euler}(p) = \{1, \ldots, p-1\}$ is a group under multiplication and contains a, so it suffices to prove $o(a) | p-1$. But this is true by Corollary 13, since $|\text{Euler}(p)| = p - 1$. □

Fermat's Little Theorem also helps test whether a given large number is prime, since it turns out that a^{p-1} is usually not congruent to 1 mod p when

p is not prime. Thus, given a large number p, if we compute a^{p-1} for twenty random values of $a < p$ and always get 1 (mod p), we can be virtually certain that p is prime. We shall touch on a related idea in Exercise 15. Although this probabilistic approach might seem inappropriate for an "exact" science such as mathematics, primality testing for large numbers relies on the use of computers, so it is fair to use a result whose uncertainty is less than the chance of computer error.

Exercises

1. An alternate way to introduce cosets. Given $H < G$, define the equivalence $a \sim b$ in G if $ab^{-1} \in H$. The equivalence classes are precisely the cosets of H, so some of the arguments in the text could be streamlined.
2. A group cannot be the union of two proper subgroups A and B. (*Hint*: If $a \notin B$ and $b \notin A$, where is ab?)
3. Give an example of a group that is the union of three proper subgroups.
4. If G is a finite Abelian group, then the product in G of the elements of G equals the product of those elements having order 2. (*Hint*: pair off each element with its inverse.)
5. If F is a field with $+1 \neq -1$, then -1 is the unique element of $(F \setminus \{0\}, \cdot)$ having order 2.
6. (Wilson's theorem) p is prime iff $(p-1)! \equiv -1 \pmod{p}$. (*Hint*: Exercises 4 and 5 applied to the field $\mathbb{Z}_p$. A more sophisticated proof is given in Exercise 3.5.)

Euler's Number

Recall that $\varphi(m)$ is defined as the order of Euler(m). There are two formulas to compute $\varphi(m)$, the first indicated in Exercise10 (and given explicitly in exercise 6.7), and the second given in Exercise 13.

7. (Euler's theorem) Prove $a^{\varphi(m)} \equiv 1 \pmod{m}$, for any relatively prime numbers a and m.
8. If $m = p_1p_2$ for p_1, p_2 prime, show $\varphi(m) = m - \frac{m}{p_1} - \frac{m}{p_2} + 1$. (*Hint*: We exclude multiples of p_1 and of p_2, but then have excluded m itself twice.)
9. For p prime, $\varphi(p^t) = p^t - p^{t-1} = p^t(1 - \frac{1}{p})$.
10. Generalizing Exercise 8, show if $m = p_1^i p_2^j$ for p_1, p_2 prime numbers, then $\varphi(m) = m(1-\frac{1}{p_1})(1-\frac{1}{p_2})$. Can this be generalized to arbitrary numbers? (See Exercise 6.7)
11. $\sum_{d|n} \varphi(d) = n$. (*Hint*: Define $f\colon \{1, 2, \ldots, n\} \to \cup_{d|n}\text{Euler}(\frac{n}{d})$ as follows: Given $k \leq n$, let $d = (k, n)$, and send k to $\frac{k}{d}$ in Euler$(\frac{n}{d})$.)

12. Define the *Möbius function*

$$\mu(n) = \begin{cases} 1 & n = 1, \\ 0 & n \text{ has a square factor} > 1, \\ (-1)^t & n \text{ is a product of } t \text{ distinct primes.} \end{cases}$$

Show $\mu(n_1 n_2) = \mu(n_1)\mu(n_2)$. Prove the *Möbius inversion formula* for a function $f: \mathbb{N} \to \mathbb{Z}$: If $g(n)$ is defined by $g(n) = \sum_{d|n} f(d)$, then $f(n) = \sum_{d|n} \mu(\frac{n}{d}) g(d)$. In particular, $\sum_{d|n} \mu(d) = 0$ for all $n > 1$. (*Hint*: For the last assertion take $f(1) = 1$ and $f(n) = 0$ for $n > 1$.)

13. Using Exercises 11 and 12 show $\varphi(n) = \sum_{d|n} \mu(\frac{n}{d}) d$, for every positive number n.

14. Compute $\varphi(100)$. What are the last two digits of 13^{41}? (*Hint*: Translate to Euler(100).)

15. Suppose p_1, p_2 are odd primes such that $\gcd(p_1 - 1,\ p_2 - 1) = 2$, and let $m = p_1 p_2$. If $a^{m-1} \equiv 1 \pmod{m}$, then $a^2 \equiv 1 \pmod{m}$; hence there are only 4 solutions for a. (*Hint*: Show $o(a)$ divides $p_1 + p_2 - 2$, and thus divides $(p_1 - 1)^2$ and $(p_2 - 1)^2$; hence $o(a)$ divides 4.)

16. Consider the rational number $\frac{m}{n}$, $m < n$, and its decimal expansion. This expansion must be infinite, iff n contains a prime factor other than 2, 5. Every expansion "repeats." The *period* of the expansion is the number of digits in the minimal repeating pattern. For example, $\frac{1}{14} = .0714285714285\ldots$ has period 6. Show that the period must divide $n - 1$. On the other hand, the periods of $\frac{m}{n}$ and $\frac{m}{2n}$ are the same; conclude that the period must either be 1 or even. Consider these assertions for other bases, *i.e.*, binary expansion and octal expansion, and compare the periods of the same fraction in binary and in octal expansion.

Chapter 3. Cauchy's Theorem: How to Show a Number Is Greater Than 1

In Chapter 2 we saw that Lagrange's theorem has the consequence that if $g \in G$ and $o(g) = m$, then m divides $|G|$. This raises the converse question: "If m divides $|G|$ then does $o(g) = m$ for suitable g in G?" Our goal will be to prove this for m prime; for m not prime the result is false, as evidenced by Euler(8) $= \{1, 3, 5, 7\}$, which has order 4 although each of its elements has order 2. In proving Lagrange's theorem we examined the process of division. In studying the converse we shall learn how to count.

Theorem 1 (Cauchy's theorem). *If a prime number p divides $|G|$, then G has an element of order p.*

There is an easy proof for G Abelian, given in Exercise 7.1. However, we proceed directly to the proof in general. The proof becomes rather easy for $p = 2$, as given in Exercise 1, so the reader is advised to try that first.

Proof of Theorem 1. Let $S = \{g \in G : g^p = e\}$. Clearly $e \in S$, so $|S| \geq 1$. In fact $g \in S$ iff $o(g)|p$, *i.e.*, $o(g) = 1$ or p; since e is the only element of order 1 our goal is to show $|S| \geq 2$. This is not an easy task, and the trick is to show p divides $|S|$; then $|S| \geq p \geq 2$.

Let $T = \{(g_1, \ldots, g_p) \in G \times \cdots \times G : g_1 \ldots g_p = e\}$. We shall show that p divides $|S|$, by counting $|T|$ in two different ways. On the one hand for any $g_1, \ldots, g_{p-1}$ in G we have a unique g_p in G such that $(g_1, \ldots, g_p) \in T$; namely $g_p = (g_1 \ldots g_{p-1})^{-1}$. Thus $|T|$ equals the number of ways we can choose $p-1$ elements arbitrarily from G, *i.e.*, $|T| = |G|^{p-1}$. Since p divides $|G|$, we see p divides $|T|$. On the other hand $(g, .., g) \in T$ iff $g^p = e$, iff $g \in S$. Thus, letting $T' = \{(g_1, \ldots, g_p) \in T :$ at least two g_i are distinct$\}$, we see $|T| = |S| + |T'|$. We shall prove p divides $|T'|$, which then implies p divides $|T| - |T'| = |S|$, as desired.

To prove p divides $|T'|$ it suffices to partition T' into disjoint subsets each having p elements. In principle this is easy, for if $(g_1, \ldots, g_p) \in T'$, then $(g_i, g_{i+1}, \ldots, g_p, g_1, \ldots, g_{i-1}) \in T'$ for every $1 \leq i \leq p$. (Indeed $(g_1, \ldots, g_p) \in T'$ implies $e = g_1 \ldots g_p = (g_1 \ldots g_{i-1})(g_i \ldots g_p)$, so $g_i \ldots g_p = (g_1 \ldots g_{i-1})^{-1}$; multiplying the other way shows $g_i \ldots g_p g_1 \ldots g_{i-1} = e$.)

It remains to show that the set $\{(g_i g_{i+1} \cdots g_p g_1 \cdots g_{i-1}) : 1 \leq i \leq p\}$ actually contains p distinct elements, *i.e.*, that

$$g_i g_{i+1} \cdots g_p g_1 \cdots g_{i-1}, \quad 1 \leq i \leq p$$

are all distinct. This part of the argument has nothing at all to do with the structure of G, but still utilizes some of the ideas developed earlier.

Suppose on the contrary

$$g_i g_{i+1} \cdots g_p g_1 \cdots g_{i-1} = g_j g_{j+1} \cdots g_p g_1 \cdots g_{j-1},$$

with $1 \le i < j \le p$. Let $t = j - i$. Then $g_i = g_j = g_{i+t}$. Looking now at the $(t+1)$ component of each side we see $g_j = g_{j+t} = g_{i+2t}$ (subscripts modulo p). Continuing in this way we have

$$g_i = g_{i+t} = g_{i+2t} = \cdots = g_{i+(p-1)t}.$$

By Corollary 1.15, the indices $\{i,\ i+t,\ \ldots,\ i+(p-1)t\}$ are all distinct modulo p and thus traverse $\{1,\ 2,\ \ldots,\ p\}$ (mod p); we have shown that $g_1 = g_2 = \cdots = g_p$, contrary to $(g_1, \ldots, g_p) \in T'$. □

This proof is an example par excellence of a "combinatorial proof," one that uses an ingenious counting argument, often needing few prerequisites, in order to prove a deep result. Cauchy's theorem usually is proved much later in the theory, only after a sufficient body of theorems has been obtained to erect a "structural proof." It is quite convenient to have it available at this stage.

Remark 1'. Actually we have proved the stronger result, that, for p prime, the number of elements g satisfying $g^p = e$ is divisible by p. It follows easily that the number of subgroups of order p in G is congruent to 1 (mod p), *cf.* exercise 2. This result holds for arbitrary powers of p and is due to Frobenius.The proof of Frobenius' theorem is sketched in Exercise 11.12.

Cauchy's theorem has many important applications, but first we use it to define a number related to $|G|$. In what follows lcm denotes the *least common multiple.*

The Exponent

Definition 2. The *exponent* of a finite group G, denoted as $\exp(G)$, is $\operatorname{lcm}\{o(g) : g \in G\}$.

PROPOSITION 3. *$\exp(G)$ divides $|G|$. Conversely, every prime number dividing $|G|$ also divides $\exp(G)$.*

Proof. Lagrange's theorem shows $o(g)$ divides $|G|$ for every g in G, so clearly $\exp(G)$ divides $|G|$. The second assertion is Cauchy's theorem. □

Example 4. (i) If $G = \langle g \rangle$ then $\exp(G) = |G|$, since $o(g) = |G|$.

(ii) If $|G| = p^t$ for p prime and $\exp(G) = |G|$, then $G = \langle g \rangle$ for some g in G (for otherwise each element has order dividing p^{t-1}, contradiction).

The same argument shows a more general fact: If $|G|$ is a prime power, then there is some g in G for which $o(g) = \exp(G)$. Later we shall obtain an analogous result for arbitrary Abelian groups, *cf.* Remark 7.12'.)

(iii) $\text{Exp}(\text{Euler}(8)) = 2$, whereas $|\text{Euler}(8)| = 4$.

(iv) If $|G|$ is a product of distinct prime numbers then $|G| = \exp(G)$, by Proposition 3. In particular, $\exp(S_3) = 6 = |S_3|$, although S_3 is far from cyclic (in contrast to (ii)).

PROPOSITION 5. *If $\exp(G) = 2$, then G is Abelian.*

Proof. For any a, b in G we have $a^2b^2 = e = (ab)^2 = abab$; canceling a on the left and b on the right yields $ab = ba$. □

This result cannot be generalized directly to exponent 3, since we shall see for any odd prime p there are non-Abelian groups of order p^3 and exponent p.

$\mathbf{S_n}$: Our Main Example

To gain intuition we now begin a detailed investigation of S_n, which is carried through the next few sections. Recall an element of S_n, called a *permutation*, is a 1:1 onto transformation $\pi: \{1, \dots, n\} \to \{1, \dots, n\}$. Writing πi for the image of i, we can rewrite π in terms of its action on each element, *i.e.*, as

$$\begin{pmatrix} 1 & 2 & \dots & n \\ \pi 1 & \pi 2 & \dots & \pi n \end{pmatrix}.$$

The neutral permutation is $\begin{pmatrix} 1 & 2 & \dots & n \\ 1 & 2 & \dots & n \end{pmatrix}$, which we also denote as (1). By convention $\pi\sigma$ denotes $\pi \circ \sigma$. Thus $\begin{pmatrix} 1 & 2 & 3 \\ 1 & 3 & 2 \end{pmatrix} \begin{pmatrix} 1 & 2 & 3 \\ 2 & 1 & 3 \end{pmatrix} = \begin{pmatrix} 1 & 2 & 3 \\ 3 & 1 & 2 \end{pmatrix}$.

The permutations in S_3 are

$$(1) = \begin{pmatrix} 1 & 2 & 3 \\ 1 & 2 & 3 \end{pmatrix}, \begin{pmatrix} 1 & 2 & 3 \\ 1 & 3 & 2 \end{pmatrix}, \begin{pmatrix} 1 & 2 & 3 \\ 2 & 1 & 3 \end{pmatrix}, \begin{pmatrix} 1 & 2 & 3 \\ 2 & 3 & 1 \end{pmatrix}, \begin{pmatrix} 1 & 2 & 3 \\ 3 & 1 & 2 \end{pmatrix}, \begin{pmatrix} 1 & 2 & 3 \\ 3 & 2 & 1 \end{pmatrix}.$$

One can see easily that in defining $\pi \in S_n$ there are n choices for $\pi 1$, $n - 1$ remaining choices for $\pi 2$, etc., and so $|S_n| = n(n-1) \cdots = n!$

S_3 claims the distinction of being the smallest non-Abelian group. Indeed, any group of prime order is cyclic, and any group of order 4 either has exponent 4 (and thus is cyclic by Example 4(ii)), or has exponent 2 (and thus is Abelian).

Subgroups of $\mathbf{S_n}$

For any $k < n$, S_n has a subgroup consisting of those permutations that fix $k+1$, $k+2$, ..., n, *i.e.*, all permutations of the form

$$\begin{pmatrix} 1 & 2 & \dots & k & k+1 & \dots & n \\ \pi 1 & \pi 2 & \dots & \pi k & k+1 & \dots & n \end{pmatrix}.$$

This subgroup can be identified with S_k, which has order $k!$. Of course S_n may have other subgroups. We consider small values of n.

Since $|S_2| = 2$ is prime, the only subgroups of S_2 are itself and (1).

Since $|S_3| = 6$, the proper subgroups of S_3 must have order 1,2, or 3, and thus are cyclic, by Corollary 2.14; explicitly they are

$\left\langle\begin{pmatrix}1&2&3\\1&2&3\end{pmatrix}\right\rangle = (1).$

$\left\langle\begin{pmatrix}1&2&3\\1&3&2\end{pmatrix}\right\rangle = \left\{(1), \begin{pmatrix}1&2&3\\1&3&2\end{pmatrix}\right\},$

$\left\langle\begin{pmatrix}1&2&3\\2&1&3\end{pmatrix}\right\rangle = \left\{(1), \begin{pmatrix}1&2&3\\2&1&3\end{pmatrix}\right\},$

$\left\langle\begin{pmatrix}1&2&3\\2&3&1\end{pmatrix}\right\rangle = \left\{(1), \begin{pmatrix}1&2&3\\2&3&1\end{pmatrix}, \begin{pmatrix}1&2&3\\3&1&2\end{pmatrix}\right\} = \left\langle\begin{pmatrix}1&2&3\\3&1&2\end{pmatrix}\right\rangle$, and

$\left\langle\begin{pmatrix}1&2&3\\3&2&1\end{pmatrix}\right\rangle = \left\{(1), \begin{pmatrix}1&2&3\\3&2&1\end{pmatrix}\right\},$

of order 1,2,2,3,2 respectively.

Cycles

The notation introduced above is cumbersome and, furthermore, masks some important information such as the order of the permutation. Toward this end we introduce a special kind of permutation, with a more concise notation.

Definition 6. A *cycle of length* t, written $(i_1\ i_2\ \dots\ i_t)$ for distinct $i_1, \dots, i_t$, is the permutation π defined by $\pi i_1 = i_2$, $\pi i_2 = i_3$, $\dots$, $\pi i_{t-1} = i_t$, $\pi i_t = i_1$, and $\pi i = i$ for all other i. A cycle of length 2 is called a *transposition*

For example, each element of S_3 is a cycle, and we could rewrite the above list respectively as {(1), (23), (12), (123), (132), (13)}. Three of these are transpositions ((12), (13), and (23)).

This notation is very useful, although slightly ambiguous in several aspects:

> It suppresses "n"; indeed (12) can denote $\begin{pmatrix}1&2\\2&1\end{pmatrix}$ in S_2, $\begin{pmatrix}1&2&3\\2&1&3\end{pmatrix}$ in S_3, and so forth.
>
> $(i_1\ i_2\ \dots\ i_t) = (i_2\ \dots\ i_t\ i_1) = \dots = (i_t\ i_1\ i_2 \dots\ )$, so the notation is not quite unique even when n is given. The notation becomes unique if we choose i_1 minimal, as is the custom.

If $\pi = (i_1\ i_2\ \dots\ i_t)$ then $\pi^2 i_1 = i_3$, $\pi^3 i_1 = i_4$ and so forth, until $\pi^t i_1 = i_1$. Likewise $\pi^t i_k = i_k$ for all k, so $\pi^t = (1)$. We conclude that $o(\pi) = t$; *i.e.*, the order of a cycle is its length.

S_4 contains the permutation $\pi = \begin{pmatrix}1&2&3&4\\2&1&4&3\end{pmatrix}$ which is not a cycle since $o(\pi) = 2$. However, $\pi = (12)(34)$ is a product of cycles, and we would like to pursue this matter further. We say two cycles $(i_1\ \dots\ i_t)$ and $(j_1\ \dots\ j_u)$ are *disjoint* if $\{i_1, \dots, i_t\} \cap \{j_1, \dots, j_u\} = \emptyset$.

Remark 7. Any two disjoint cycles commute. (Indeed if $\pi = (i_1 \ \dots \ i_t)$ and $\sigma = (j_1 \ \dots \ j_u)$, then $\pi\sigma i = \sigma\pi i = i$ for $i \notin \{i_1, \dots, i_t\} \cup \{j_1, \dots, j_u\}$; $\pi\sigma i_k = \pi i_k = i_{k+1}$ (subscripts mod t) $= \sigma i_{k+1} = \sigma\pi i_k$ for $1 \leq k \leq t$, and likewise $\pi\sigma j_k = \pi j_{k+1} = j_{k+1} = \sigma j_k = \sigma\pi j_k$ for $1 \leq k \leq u$.)

THEOREM 8. *Every permutation is a product of disjoint cycles (and these commute with each other, by Remark 7).*

No proof. The proof would be to write each cycle as it comes, but we leave the details to the reader, who might find the following example instructive:

$$\begin{pmatrix} 1 & 2 & 3 & 4 & 5 & 6 & 7 & 8 & 9 & 10 & 11 & 12 & 13 & 14 & 15 \\ 3 & 6 & 7 & 10 & 14 & 1 & 2 & 13 & 15 & 4 & 11 & 5 & 8 & 12 & 9 \end{pmatrix} =$$

$$(1\ 3\ 7\ 2\ 6)(4\ 10)(5\ 14\ 12)(8\ 13)(9\ 15)(11)$$

Of course the cycle (11) at the end could be deleted.

Remark 9. If $\sigma_1, .., \sigma_v$ are disjoint cycles, then

$$o(\sigma_1 \dots \sigma_v) = lcm(\text{length}(\sigma_1), \dots, \text{length}(\sigma_v)).$$

(Indeed $(\sigma_1 \dots \sigma_v)^m = \sigma_1^m \dots \sigma_v^m$, from which the assertion is clear.)

The Product of Two Subgroups

Let us turn now to a question whose answer relies heavily on cosets and Lagrange's theorem. Given $H, K \leq G$ define

$$HK = \{hk : h \in H, \ k \in K\}.$$

The question is, "Is HK necessarily a subgroup of G?"

PROPOSITION 10. *If $H, K \leq G$ with $HK = KH$, then $HK \leq G$.*

Proof. Suppose $h_i \in H$ and $k_i \in K$; using the criterion of Proposition 2.4(i) we note $(h_1k_1)(h_2k_2)^{-1} = h_1k_1k_2^{-1}h_2^{-1} \in HKH = HHK \subseteq HK$ □.

COROLLARY 11. *If G is Abelian and $H, K \leq G$, then $HK \leq G$.*

(The idea of rearranging an expression so that like elements appear together, recurs frequently in the sequel.) On the other hand, if $HK \neq KH$, then the proof of Proposition 10 fails, and in fact HK will not be a subgroup; *cf.* Exercise 3.

Example 12. (i) $G = S_4$, $H = \langle(12)\rangle$, $K = \langle(34)\rangle$. Since $(12)(34) = (34)(12)$, we see $HK = KH$ is a subgroup of S_4.

(ii) $G = S_3$. Take distinct subgroups H, K of order 2, say $H = \langle(23)\rangle$ and $K = \langle(12)\rangle$. By direct computation $(12)(23) = (123) \notin HK$, so HK is not a subgroup of G.

There is a neater argument for (ii). One can check easily that $|HK| = 4$. Hence $HK \not\leq G$, by Lagrange's theorem. This argument will become trivial after we obtain a formula for $|HK|$, which is reminiscent of the familiar formula for the dimension of the sum of two subspaces of a vector space.

THEOREM 13. *If H and K are subgroups of G, then*

$$|HK| = \frac{|H||K|}{|H \cap K|}.$$

Proof. We shall prove $|HK||H \cap K| = |H||K|$, by showing that every element of HK can be written precisely $|H \cap K|$ distinct ways in the form hk, for $h \in H$ and $k \in K$. Indeed let $t = |H \cap K|$, and write $H \cap K = \{a_1, \ldots, a_t\}$. For any hk in HK where $h \in H$ and $k \in K$, we put $h_i = ha_i \in H$ and $k_i = a_i^{-1}k \in K$; then $h_ik_i = hk$ for $1 \leq i \leq t$. On the other hand if $h' \in H$ and $k' \in K$ satisfies $h'k' = hk$, then $h^{-1}h' = k(k')^{-1} \in H \cap K$ and so equals some a_i, proving $h' = ha_i = h_i$ and $k' = (k^{-1}a_i)^{-1} = a_i^{-1}k = k_i$, as desired. □

COROLLARY 14. *If $H_1, \ldots, H_t$ are subgroups of G with $H_iH_j = H_jH_i$ for each i, j, then $H_1 \ldots H_t \leq G$ and $|H_1 \ldots H_t|$ divides $|H_1| \ldots |H_t|$.*

Proof. Easy induction on t, the case $t = 2$ being obvious from the theorem. Let $K = H_1 \ldots H_{t-1}$. By induction $K \leq G$ and $|K|$ divides $|H_1| \ldots |H_{t-1}|$. But $H_1 \ldots H_t = KH_t$, so $|H_1 \ldots H_t|$ divides $|K||H_t|$, which in turn divides $|H_1| \ldots |H_t|$. □

The Classical Groups

Let us digress for a moment, to introduce certain examples of groups of utmost importance in mathematics, although many of them are not finite. A group is called a *linear algebraic group* if, for suitable n and a suitable field F, it is a subgroup of $\mathrm{GL}(n, F)$ that is defined in terms of (polynomial) equations in the matrix components. This definition really should be framed with more precision, but let us illustrate it with some examples.

Example 15: Some linear algebraic groups

(1) The general linear group $\mathrm{GL}(n, F)$ is itself a linear algebraic group.
(2) The *special linear group* $\mathrm{SL}(n, F)$ is defined as the set of $n \times n$ matrices with determinant 1.

(3) The *orthogonal group* $\mathrm{O}(n,F) = \{A \in \mathrm{GL}(n,F) : A^t = A^{-1}\}$, where ^{t} denotes the matrix transpose, *i.e.*, $(a_{ij})^t = (a_{ji})$. (Note that if $A \in \mathrm{O}(n,F)$, then $\det(A) = \pm 1$.)

(4) The *special orthogonal group*

$$\mathrm{SO}(n,F) = \{A \in \mathrm{O}(n,F) : \det(A) = 1\}.$$

(5) The *symplectic group*

$$\mathrm{Sp}(2n,F) = \{A \in \mathrm{GL}(2n,F) : A^t = \begin{pmatrix} 0 & I \\ -I & 0 \end{pmatrix} A^{-1} \begin{pmatrix} 0 & -I \\ I & 0 \end{pmatrix}\},$$

where 0 denotes the $n \times n$ matrix all of whose entries are 0, and I denotes the $n \times n$ identity matrix.

These groups are called *classical groups* (also *cf.* Exercises 12.9 ff), and each has special geometric significance. Recall that $\mathrm{GL}(n,F)$ is the group of all invertible linear transformations of the n-dimensional vector space F^n; $\mathrm{SL}(n,F)$ is the subgroup of transformations preserving the volume of any parallelotope defined by n given vectors.

F^n can be provided with the usual inner product:

$$(\alpha_1, \ldots, \alpha_n) \cdot (\beta_1, \ldots, \beta_n) = (\alpha_1\beta_1 + \cdots + \alpha_n\beta_n).$$

$A \in \mathrm{O}(n,F)$ iff A preserves orthogonality, *i.e.*, $v \cdot w = 0$ implies $Av \cdot Aw = 0$. $\mathrm{Sp}(2n,F)$ can be described in terms of an analogous orthogonality condition with respect to a different scalar product.

If F is a finite field, for example $\mathbb{Z}_p$, then the classical groups are important examples of finite groups.

Exercises

1. (Cauchy's theorem for $p = 2$.) Let $T = \{a \in G : a^{-1} \neq a\}$. Pairing off each a with a^{-1} in T, show $|T|$ is even; hence there are an even number of elements a in G for which $a = a^{-1}$, *i.e.*, $a^2 = e$. All of these except e itself have order 2.
2. (Special case of Frobenius' theorem, *cf.* exercise 11.12) For any prime p dividing $|G|$, the number of subgroups of order p is congruent to 1 (mod p). (*Hint*: Let m be the number of subgroups of order p, and let t be the number of elements (in G) of order p. By Lagrange, the intersection of any two subgroups of order p is trivial, so $t = (p-1)m \equiv -m \pmod p$. Conclude using Remark 1$'$.)
3. Conversely to Proposition 10, show that if $HK \leq G$, then $HK = KH$. (*Hint*: Use the inverse.)

4. Find a noncyclic Abelian subgroup of S_4.
5. Another proof of Wilson's theorem (Exercise 2.6). The number of cycles of length p in S_p is $(p-1)!$, so the number of elements of order dividing p in S_p is $(p-1)!+1$. But this is a multiple of p.
6. Generalize proposition 5, to show that if a, b are elements of order 2 whose product also has order 2, then $ab = ba$.
7. $\mathrm{Exp}(S_n) = \mathrm{lcm}\{1, 2, \ldots, n\}$. (*Hint*: It is enough to check the cycles.)
8. What is the maximal possible order of an element of S_7? of S_n for n arbitrary?

The Classical Groups

9. Display $\mathrm{SL}(n, F)$, $\mathrm{O}(n, F)$, $\mathrm{SO}(n, F)$, and $\mathrm{Sp}(2n, F)$ explicitly as subgroups of $\mathrm{GL}(n, F)$, defined by equations in the matrix entries.
10. Show $\mathrm{SL}(n, \mathbb{Z}_p)$ has order $(p^n - 1)(p^n - p)\ldots(p^n - p^{n-2})p^{n-1}$. What are the orders of $\mathrm{O}(n, \mathbb{Z}_p)$, $\mathrm{SO}(n, \mathbb{Z}_p)$, and $\mathrm{Sp}(2n, \mathbb{Z}_p)$? (*Hint*: See Exercise 1.6. For $\mathrm{O}(n, \mathbb{Z}_p)$ compute the number of orthogonal bases.)
11. (The group of symmetries of the circle) A *symmetry* of the circle is a rigid map from the circle to itself, permitting "flips." Show that the set of symmetries is a group, taking composition of maps as the group operation, and that this group is isomorphic to $\mathrm{O}(2, \mathbb{R})$. $\mathrm{SO}(2, \mathbb{R})$ is a subgroup of $\mathrm{O}(2, \mathbb{R})$, of index 2, corresponding to the group of rotations.
12. For any n, the group of rigid transformations of a regular polygon of n sides can be viewed as a subgroup of the group from the previous exercise. (*Hint*: Inscribe the polygon in the circle.)

Chapter 4. Introduction to the Classification of Groups: Homomorphisms, Isomorphisms, and Invariants

Having proved some rather deep theorems about subgroups, let us pause and take stock of what we know about specific groups. We have applied Euler's groups directly to number theory. We have also seen groups arising in other contexts, such as permutations of sets of objects, invertible linear transformations of vector spaces, and symmetries of the circle (and possibly other geometric objects). Groups thereby have acquired a special significance, and deserve to be studied in their own right – the goal of such a study being to develop the tools to answer any question posed about a given group G. The most basic question is, "Does G have the same structure as a group we have already encountered?" or, in other words, "When are two groups the same?" To make this question precise, we consider the group structure more closely. Of course, the structure of the group is determined by its "multiplication table," the list of products of all pairs of elements. Here are some examples of multiplication tables of groups:

+	0	1
0	0	1
1	1	0

$(\mathbb{Z}_2, +)$

·	1	3
1	1	3
3	3	1

Euler(4)

·	1	5
1	1	5
5	5	1

Euler(6)

+	0	1	2	3
0	0	1	2	3
1	1	2	3	0
2	2	3	0	1
3	3	0	1	2

$(\mathbb{Z}_4, +)$

·	1	3	5	7
1	1	3	5	7
3	3	1	7	5
5	5	7	1	3
7	7	5	3	1

Euler(8)

Note that the groups of the first row all have the form

·	e	a
e	e	a
a	a	e

where e is the neutral element and a is the other element; however the structures of $(\mathbb{Z}_4, +)$ and Euler(8) differ, as seen by examining their main diagonals. Thus we see that the multiplication table provides a comprehensive method of comparing structures of different groups (also *cf.* Exercise 1). But for precisely this reason, the multiplication table is far too cumbersome in most situations, and we must find a more concise method of comparing group structures. The most direct approach is to find a correspondence of elements that will respect the algebraic structure of the groups. Even when the correspondence need not even be 1:1, it can still transfer valuable information. We have already come across such an instance, namely the function $\mathbb{Z} \to \mathbb{Z}_m$, which sends a to $[a]$. Let us generalize this example.

Definition 1. A *group homomorphism* $\varphi: G_1 \to G_2$ is a function satisfying

(H1) $\varphi(ab) = \varphi(a)\varphi(b)$ for all a, b in G_1;
(H2) $\varphi(e_1) = e_2$, where e_i is the neutral element of G_i;
(H3) $\varphi(a^{-1}) = \varphi(a)^{-1}$ for all a in G_1.

Remark 2. Conditions (H2) and (H3) are superfluous! Indeed if (H1) holds then $e_2\varphi(e_1) = \varphi(e_1) = \varphi(e_1^2) = \varphi(e_1)^2$, so $e_2 = \varphi(e_1)$ (seen by canceling $\varphi(e_1)$); then $\varphi(a^{-1})\varphi(a) = \varphi(a^{-1}a) = \varphi(e_1) = e_2$, proving $\varphi(a^{-1}) = \varphi(a)^{-1}$.

From now on we write e for the neutral element both of G_1 and of G_2, since it is clear from the context.

Digression 3. For the sake of completeness let us record what happens for monoids. Monoid homomorphisms are required to satisfy only (H1) and (H2). ((H3) is no longer relevant, since inverses are not part of the defining structure.) But now (H2) is no longer superfluous! For example, we could define $\varphi: (\mathbb{Z}_2, \cdot) \to (\mathbb{Z}_6, \cdot)$ by $\varphi(a) = 3a$; then (H1) holds but (H2) fails since $\varphi(1) = 3 \neq 1$. (Where does the proof of remark 2 go awry here?)

Definition 4. An *isomorphism* is a homomorphism $\varphi: G_1 \to G_2$ for which the inverse maps exists and also is a homomorphism.

Some philosophy: Suppose $\varphi: G_1 \to G_2$ is an isomorphism. For any a in G_1 we could write $\bar{a}$ for $\varphi(a)$ in G_2, and then note that the multiplication table for G_1 could be copied for G_2 by writing bars over all the elements. Thus "isomorphism" is the key concept for identifying the structure of two groups, and our immediate goal should be to find the weakest formal conditions to guarantee that a given homomorphism φ is an isomorphism. Clearly φ must be 1:1 and onto, so we first note

Remark 5. Any 1:1, onto homomorphism $\varphi: G_1 \to G_2$ is an isomorphism. (Indeed we must show $\varphi^{-1}(ab) = \varphi^{-1}(a)\varphi^{-1}(b)$ for any a, b in G. But

$$\varphi(\varphi^{-1}(ab)) = ab = \varphi\varphi^{-1}(a)\varphi\varphi^{-1}(b) = \varphi(\varphi^{-1}(a)\varphi^{-1}(b));$$

since φ is 1:1 we see $\varphi^{-1}(ab) = \varphi^{-1}(a)\varphi^{-1}(b)$.)

This leads us to separate the key features of isomorphism into two parts:

Definition 5′. An *injection* is a 1:1 homomorphism; a *surjection* is an onto homomorphism.

How can we tell whether a homomorphism is an injection? The sign of a good theory is that it provides guidance at times of need. At this juncture we note that any given homomorphism $\varphi: G_1 \to G_2$ naturally gives rise to a certain subgroup of G_1. It is convenient to use the following set-theoretic notation: Given a function $\varphi: G_1 \to G_2$ and $S \subseteq G_2$ write $\varphi^{-1}S$ for $\{g \in G_1 : \varphi(g) \in S\}$.

Definition 6. The *kernel* of a homomorphism $\varphi: G_1 \to G_2$, denoted $\ker \varphi$, is $\{a \in G_1 : \varphi(a) = e\} = \varphi^{-1}\{e\}$, where e is the neutral element of G_2.

Remark 7. $\ker \varphi$ is a subgroup of G_1; this is left as an exercise for the reader. (This remark will be improved in chapter 5.)

Certainly if φ is 1:1, then $\ker \varphi = \{e\}$; the surprise is that the converse is also true.

Remark 8. If $\ker \varphi = \{e\}$, then φ is an injection. In fact, for any homomorphism φ, if $\varphi(g_1) = \varphi(g_2)$ then $\varphi(g_1 g_2^{-1}) = e$, so $g_1 g_2^{-1} \in \ker \varphi$, implying $g_1 \in (\ker \varphi) g_2$.

Thus we see that an isomorphism is a surjection whose kernel is trivial. The task remains of determining that two given groups G_1 and G_2 are isomorphic, by constructing the isomorphism; we shall return later to this issue. Now we want to deal with the possibility that G_1 and G_2 are not isomorphic. Obviously we do not want to have to check that every map from G_1 to G_2 is not an isomorphism, so we look for a subtler approach.

Definition 9. An *invariant* is a number associated to a group, which is the same for isomorphic groups.

Example 10. $|G|$ is an invariant. Indeed if $\varphi: G_1 \to G_2$ is an isomorphism then $|G_1| = |G_2|$ since φ is 1:1 and onto.

The exponent of G is also an invariant, as we shall see from the following fact:

Remark 11. If $\varphi: G_1 \to G_2$ is a homomorphism, then $o(\varphi(g)) | o(g)$ for all g in G_1. (Indeed if $n = o(g)$ then $\varphi(g)^n = \varphi(g^n) = e$, so $o(\varphi(g)) | n$ by Proposition 2.6.) If φ is an isomorphism then $o(\varphi(g)) = o(g)$, seen by applying this observation also to φ^{-1}.

PROPOSITION 12. *Exp(G) is an invariant.*

Proof. Suppose $\varphi: G_1 \to G_2$ is an isomorphism. Then

$$\begin{aligned}\exp(G_1) = \operatorname{lcm}\{o(g) : g \in G_1\} &= \operatorname{lcm}\{o(\varphi(g)) : g \in G_1\} \\ &= \exp(G_2). \quad \square\end{aligned}$$

For example Euler(8) has exponent 2 and thus cannot be isomorphic to $(\mathbb{Z}_4, +)$. We shall introduce new invariants as the occasion requires.

Homomorphic Images

Recalling that homomorphisms can relate two groups that are not necessarily isomorphic, we are interested in information concerning arbitrary homomorphisms, including what happens to subgroups. First we introduce an important subgroup of G_2 arising from a homomorphism $\varphi: G_1 \to G_2$, namely the *image* $\varphi(G_1)$, and obtain a key generalization of Example 10.

PROPOSITION 13. *If $\varphi: G_1 \to G_2$ is a homomorphism of finite groups, then $\varphi(G_1)$ is a subgroup of G_2, and $|\varphi(G_1)| = [G_1 : \ker \varphi]$.*

Proof. $\varphi(g_1)\varphi(g_2)^{-1} = \varphi(g_1 g_2^{-1}) \in \varphi(G_1)$ so $\varphi(G_1) \leq G_2$, proving the first assertion. Recall that $[G_1 : \ker \varphi]$ is the number of cosets of $\ker \varphi$, which it remains to prove equals $|\varphi(G_1)|$. Letting $K = \ker \varphi$ we claim that φ induces a 1:1 correspondence $\Phi: \{\text{cosets of } K\} \to \varphi(G_1)$, given by $Kg \mapsto \varphi(g)$. Indeed Φ is well-defined and onto, since $\varphi(ag) = \varphi(a)\varphi(g) = \varphi(g)$ for any a in K; also, Φ is 1:1 by Remark 8. $\square$

This result justifies our intuition that $\ker \varphi$ is the information "lost" in applying the homomorphism φ. Let us use this approach to determine the subgroups of $(\mathbb{Z}_m, +)$. We start with a key observation. If $\varphi: G_1 \to G_2$ is a group homomorphism and $H \leq G_1$, then, viewing the action of φ on H only, we get a homomorphism $\varphi: H \to G_2$, called the *restriction* of φ to H, denoted $\varphi|_H$.

PROPOSITION 14. *Suppose $\varphi: G_1 \to G_2$ is a group homomorphism.*

(i) *If $H \leq G_1$, then $\varphi(H) \leq G_2$.*

(ii) *Conversely if $H_2 \leq G_2$, then $\varphi^{-1}(H_2)$ is a subgroup of G_1 containing $\ker \varphi$.*

(iii) *If φ is onto, then $|H_2| = [\varphi^{-1}(H_2) : \ker \varphi]$ for all $H_2 \leq G_2$.*

Proof. (i) By Proposition 13 applied to $\varphi|_H$.

(ii) $\ker \varphi = \varphi^{-1}\{e\} \subseteq \varphi^{-1}(H_2)$. To see that $\varphi^{-1}(H_2) \leq G_1$, take $a, b \in \varphi^{-1}(H_2)$; then $\varphi(ab^{-1}) = \varphi(a)\varphi(b)^{-1} \in H_2$, so $ab^{-1} \in \varphi^{-1}(H_2)$.

(iii) Apply Proposition 13 to the restriction of φ to $\varphi^{-1}(H_2)$. $\square$

This useful result can be reformulated more concisely.

PROPOSITION 15. *Suppose $\varphi: G_1 \to G_2$ is a surjection. Then φ induces a 1:1 correspondence*

$$\Phi: \{\textit{subgroups of } G_1 \textit{ containing } \ker\varphi\} \to \{\textit{subgroups of } G_2\},$$

and Φ^{-1} is induced by φ^{-1}.

Proof. One must show that $\varphi^{-1}\varphi(H_1) = H_1$ for any $H_1 \leq G_1$ containing $\ker\varphi$, and $\varphi\varphi^{-1}(H_2) = H_2$ for any $H_2 \leq G_2$. The second assertion is immediate since φ is onto; the first assertion is clear, for if $a \in \varphi^{-1}\varphi(H_1)$, then $\varphi(a) = \varphi(b)$ for some b in H_1, implying $a \in (\ker\varphi)b \subseteq H_1$, by Remark 8. □

COROLLARY 16. *The subgroups of $\mathbb{Z}_m$ are all of the form $n\mathbb{Z}_m$ for suitable $n|m$.*

Proof. We apply Proposition 15 to the natural surjection $\varphi: \mathbb{Z} \to \mathbb{Z}_m$. The subgroups of $\mathbb{Z}$ containing $\ker\varphi = m\mathbb{Z}$ have the form $n\mathbb{Z}$ for some $n|m$ in $\mathbb{N}$, and the image is $n\mathbb{Z}_m$. □

COROLLARY 17. *For any m, $\mathbb{Z}_m$ has a unique subgroup of order n, for each n dividing m.*

Proof. It must be $\frac{m}{n}\mathbb{Z}_m$, in view of Proposition 14(iii). □

Exercises

1. Each row (and each column) of the multiplication table of a group is in 1:1 correspondence with the set of elements of the group.
2. Write a computer program to find all groups of a given order n, but throwing away all "duplications," *i.e.*, when two groups are isomorphic. Use this program to construct all groups of order ≤ 8. What interesting facts can you conjecture from this list? As the course progresses, improve this program to take into account the various theorems.
3. Define a homomorphism $\mathrm{GL}(n, F) \to F\setminus\{0\}$, given by "det." What is the kernel?
4. Define a homomorphism $\mathrm{O}(n, F) \to \{\pm 1\}$, given by "det." What is the kernel?
5. Use Exercises 3,4 to recalculate the orders of $\mathrm{SL}(n, \mathbb{Z}_p)$ and $\mathrm{SO}(n, \mathbb{Z}_p)$.
6. (*cf.* example 2.4′) Construct a natural homorphism from the group of invertible upper triangular matrices to the group of diagonal matrices. What is its kernel?
7. Any monoid homomorphism $f: M_1 \to M_2$ restricts to a group homomorphism $\bar{f}: \mathrm{Unit}(M_1) \to \mathrm{Unit}(M_2)$.

8. If $m|n$, then there is a group homomorphism $\mathrm{Euler}(n) \to \mathrm{Euler}(m)$ given by $[a] \mapsto [a]$. What is the kernel?
9. Suppose there is a surjection $f: G_1 \to G_2$. If G_2 has an element of order m, then so does G_1. (*Hint*: Apply remark 11.)

Chapter 5. Normal Subgroups – The Building Blocks of the Structure Theory

We have seen that the kernel of a group homomorphism $\varphi: G \to H$ is a subgroup of G. Our object here is to study a special property of this kind of subgroup, which turns out to be the key to the whole structure theory.

Remark 1. If $a \in \ker\varphi$ and $g \in G$, then $gag^{-1} \in \ker\varphi$. Indeed,

$$\varphi(gag^{-1}) = \varphi(g)e\varphi(g^{-1}) = e.$$

Let us formalize this property.

Definition 2. A subgroup N of G is *normal*, written $N \triangleleft G$, if $gag^{-1} \in N$ for all $a \in N,\ g \in G$.

We call an element of the form gag^{-1} the *conjugate of a by g*; thus N is normal iff N contains every conjugate of each element of N. This property can be written more suggestively as $gNg^{-1} \subseteq N$ for all g in G, where gNg^{-1} denotes $\{gag^{-1} : a \in N\}$, the set of conjugates of elements of N by g.

Proposition 3. *The following conditions are equivalent:*

(i) $N \triangleleft G$;
(ii) $gNg^{-1} \subseteq N$ *for all g in G;*
(iii) $gNg^{-1} = N$ *for all g in G;*
(iv) $gN = Ng$ *for all g in G.*

Proof. (i) $\Leftrightarrow$ (ii) by definition.

(ii) $\Rightarrow$ (iii) We need the opposite inclusion to (ii); using g^{-1} instead of g in (ii) we get $g^{-1}Ng = g^{-1}N(g^{-1})^{-1} \subseteq N$, so $N = gg^{-1}Ngg^{-1} = g(g^{-1}Ng)g^{-1} \subseteq gNg^{-1}$.

(iii) $\Rightarrow$ (ii) Obvious. □

(iii) $\Leftrightarrow$ (iv) Clear (multiplying on the right by g or g^{-1}). □

Example 4. $G \triangleleft G$ and $\{e\} \triangleleft G$, for any group G. These are called the *trivial* normal subgroups of G, and we shall be interested in the nontrivial normal subgroups. On the other hand any subgroup of an Abelian group is normal, so the question of normality is pertinent only for nontrivial subgroups of non-Abelian groups.

Example 5. $G = S_3$ (the smallest nonabelian group). In Chapter 3 we saw that the nontrivial subgroups of S_3 are $\langle(12)\rangle, \langle(13)\rangle, \langle(23)\rangle$, and $\langle(123)\rangle$. In the appendix to this section we shall see by inspection that $\langle(12)\rangle, \langle(13)\rangle$, and $\langle(23)\rangle$ are not normal subgroups. One can also check that $\langle(123)\rangle$ is a normal subgroup by direct computation, but we shall now see two ways of obtaining this information with ease.

LEMMA 6. *For any $H \leq G$ and any g in G, gHg^{-1} is also a subgroup of G, and there is a group isomorphism $\varphi: H \to gHg^{-1}$ given by $\varphi(a) = gag^{-1}$.*

Proof. If $gag^{-1}, gbg^{-1} \in gHg^{-1}$ then

$$gag^{-1}(gbg^{-1})^{-1} = gag^{-1}gb^{-1}g^{-1} = gab^{-1}g^{-1} \in gHg^{-1},$$

proving $gHg^{-1} \leq G$ by our usual criterion of subgroup. Now φ is a group homomorphism, since

$$\varphi(ab) = gabg^{-1} = (gag^{-1})(gbg^{-1}) = \varphi(a)\varphi(b),$$

and φ is obviously onto. Finally $\ker\varphi = \{a \in H : gag^{-1} = e\} = \{e\}$, seen by multiplying by g^{-1} on the left and g on the right. □

Stated in words, conjugation by any element of G yields an isomorphism of subgroups.

PROPOSITION 7. *Suppose G has a unique subgroup H of order m. Then $H \triangleleft G$.*

Proof. For any g in G, gHg^{-1} is a subgroup of order m, so $gHg^{-1} = H$. □

COROLLARY 7′. *If p is prime and $|G| = pu$ for $u < p$, then G has a unique subgroup of order p, which thus is normal.*

Proof. By Cauchy's theorem G has a subgroup H of order p. H is unique (of order p), since if $K \neq H$ with $|K| = p$, then $H \cap K$ is a proper subgroup of H, which must have order 1 by Lagrange; hence $p^2 = |H||K|/|H \cap K| = |HK| \leq |G| < p^2$, contradiction. Thus $H \triangleleft G$. □

Example 8. $6 = 3 \cdot 2$. Hence $\langle(123)\rangle$ is the unique subgroup of S_3 having order 3, and so is normal.

Here is an even easier approach. Recall that $[G : H] = \frac{|G|}{|H|}$ when $|G|$ is finite.

PROPOSITION 9. *If $[G : H] = 2$, then $H \triangleleft G$.*

Proof. For any g in G we need to show $Hg = gH$. If $g \in H$ then $Hg = H = gH$, so we may assume $g \notin H$. But then H and Hg are the only two (right) cosets of H in G, by Lagrange, implying $G = H \cup Hg$; and analogously using left (instead of right) cosets we have $G = H \cup gH$, both disjoint unions. Thus $Hg = G \setminus H = gH$, as desired. $\square$

Note that $[S_3 : \langle(123)\rangle] = 2$, yielding $\langle(123)\rangle \triangleleft S_3$. This nice result will be generalized in Chapter 9.

Remark 10. If $H < G$ and $N \triangleleft G$, then $HN < G$, as a special case of Proposition 3.10. (Indeed $hN = Nh$ for all h in H, so $HN = NH$.)

Without computing, one concludes at once from this result that the subgroups of S_3 having order 2 are not normal, *cf.* Exercise 1 (which will be generalized in chapter 11).

The Residue Group

So far we have seen that $\ker \varphi \triangleleft G$, for any group homomorphism $\varphi: G \to K$. Now we want to show the converse, that every normal subgroup is the kernel of a suitable group homomorphism. In other words, given $N \triangleleft G$ we need to construct a group K and a homomorphism $\varphi: G \to K$ such that $\ker \varphi = N$. When G is finite, $|\varphi(G)| = [G : N]$ by Proposition 4.13. On the other hand, N has $[G : N]$ cosets in G, and this fact was instrumental in proving that proposition. Thus it makes sense to try to define a group structure on the cosets of N as our candidate for K. The obvious operation on cosets would be

$$NaNb = Nab \tag{1}$$

if this can be shown to be well-defined. So suppose $Na = Na'$, and $Nb = Nb'$ for a', b' in G; is $Nab = Na'b'$? Write $a' = xa$ and $b' = yb$ for x, y in N. Then $a'b' = xayb = x(aya^{-1})ab \in Nab$, since $aya^{-1} \in N$. Hence $Na'b' = Nab$ by Proposition 2.10.

Definition 11. G/N denotes {cosets of N}, endowed with the operation defined in (1).

It must be stressed that in G/N the cosets Na play the role of ordinary elements and should be considered as such.

THEOREM 12. *G/N is a group, with neutral element Ne. Furthermore, there is a surjection $\psi: G \to G/N$ given by $a \mapsto Na$, whose kernel is N.*

Proof. To prove G/N is a group, we simply transfer the analogous properties from G to G/N. For any a in G we have

$$NaNe = Nae = Na = Nea = NeNa,$$

proving Ne is the neutral element. Associativity follows from the computation

$$(NaNb)Nc = NabNc = N(ab)c = Na(bc) = NaNbc = Na(NbNc).$$

Finally, $(Na)^{-1}$ is Na^{-1} since

$$Na^{-1}Na = Na^{-1}a = Ne = Naa^{-1} = NaNa^{-1}.$$

Having proved G/N is a group, we next check $\psi: G \to G/N$ is a group homomorphism:

$$\psi(ab) = Nab = NaNb = \psi(a)\psi(b).$$

For the last assertion, $\ker\psi = \psi^{-1}(Ne) = \{a \in G : Na = Ne\} = N$, by Remark 2.9′. □

The group G/N is called the *residue group*, *factor group*, or *quotient group*. The homomorphism of Theorem 12 is called the *canonical homomorphism*. Let us now consider three instances when $N \triangleleft G$ and determine the structure of the group G/N.

Example 13.

(i) $G = (\mathbb{Z}, +)$ and $N = n\mathbb{Z}$. Then

$$G/N = \{N + 0,\ N + 1,\ \ldots,\ N + (n-1)\},$$

with the group operation given by $(N + a) + (N + b) = N + (a + b)$. This group is identified naturally with $(\mathbb{Z}_n, +)$ if we write $[a]$ in place of $N + a$.

(ii) $G = (\mathbb{Z}_6, +)$ and $N = 3\mathbb{Z}_6 = \{[3a] : a \in \mathbb{Z}\}$. Then $G/N = \{N + [0],\ N + [1],\ N + [2]\}$, which is identified naturally with $(\mathbb{Z}_3, +)$.

(iii) $G = S_3$ and $N = \langle(123)\rangle$. Then $G/N = \{Ne, Ng\}$ where $g = (12)$. Note that $NgNg = Ne$. Thus we can identify G/N with $(\{\pm 1\}, \cdot)$.

Please note that (i) provides an instance where G/N is *not* isomorphic to a subroup of G. (This observation should help one to avoid a common error.) Our next goal is to view (i), (ii) in their general contexts.

Noether's Isomorphism Theorems

Suppose $\varphi: G \to K$ is a surjection, and $N = \ker\varphi$. We have just seen that $N \triangleleft G$, and G/N is a group whose order is $[G : N] = |K|$ when K is finite. We would like to show that $G/N \approx K$. Given any $N \triangleleft G$ and $N \subseteq H < G$, write H/N for $\{Nh : h \in H\}$, easily seen to be a subgroup of G/N.

LEMMA 14. *Suppose $\varphi: G \to K$ is any group homomorphism, and N is a normal subgroup of G contained in $\ker\varphi$. Then there is a homomorphism $\bar{\varphi}: G/N \to K$ given by $\bar{\varphi}(Ng) = \varphi(g)$, and ker $\bar{\varphi} = (\ker\varphi)/N$.*

Proof. $\bar{\varphi}$ is well-defined, for if $Ng_1 = Ng_2$, then $g_1 = ag_2$ for some a in N, and thus

$$\bar{\varphi}(Ng_1) = \varphi(g_1) = \varphi(ag_2) = \varphi(a)\varphi(g_2) = \varphi(g_2) = \bar{\varphi}(Ng_2).$$

The rest is easy: φ is a homomorphism since

$$\bar{\varphi}(Ng_1Ng_2) = \bar{\varphi}(Ng_1g_2) = \varphi(g_1g_2) = \varphi(g_1)\varphi(g_2) = \bar{\varphi}(Ng_1)\bar{\varphi}(Ng_2),$$

and $\ker\bar{\varphi} = \{Ng : \varphi(g) = e\} = \{Ng : g \in \ker\varphi\} = (\ker\varphi)/N$ by definition. □

Remark 15. In Lemma 14, $\bar{\varphi}$ is onto iff φ is onto.

THEOREM 16. *(Noether I) Suppose $\varphi: G \to K$ is any surjection. Then $K \approx G/\ker\varphi$.*

Proof. Take $N = \ker\varphi$ in Lemma 14. Then $\bar{\varphi}: G/N \to K$ is onto, and $\ker\bar{\varphi} = N/N = Ne$, the neutral element of G/N, so $\bar{\varphi}$ is an isomorphism. □

Example 16′. The natural surjection $(\mathbb{Z}, +) \to (\mathbb{Z}_n, +)$ has kernel $n\mathbb{Z}$, so $(\mathbb{Z}/n\mathbb{Z}, +) \approx (\mathbb{Z}_n, +)$. (Compare with Example 13(i).)

THEOREM 17. *(Noether II) If $N \subseteq H$ are normal subgroups of G, then $G/H \approx (G/N)/(H/N)$.*

Proof. The canonical homomorphism $\varphi: G \to G/H$ is onto and has kernel H containing N, so Lemma 14 provides a surjection $\bar{\varphi}: G/N \to G/H$ with kernel H/N; we conclude with Noether I. □

Example 17′. Define $G = \mathbb{Z}$, $N = 6\mathbb{Z}$, and $H = 3\mathbb{Z}$. Then $\mathbb{Z}/3\mathbb{Z} \approx (\mathbb{Z}/6\mathbb{Z})/(3\mathbb{Z}/6\mathbb{Z}) \approx \mathbb{Z}_6/3\mathbb{Z}_6$. (Compare with Example 13(ii).)

THEOREM 18. *(Noether III) If $N \triangleleft G$ and $H < G$, then $H \cap N \triangleleft H$ and $H/\ H \cap N\ \approx NH/N$.*

Proof. Define $\varphi: H \to NH/N$ by $h \mapsto Nh$. Clearly φ is onto, and $\ker\bar{\varphi} = \{h \in H : Nh = e\} = \{h \in H : h \in N\} = H \cap N$, so we conclude with Noether I. □

For any $H < G$, the proof of Theorem 18 shows that NH/N is the image of H in G/N under the canonical homomorphism $G \to G/N$. In particular we have

COROLLARY 19. *If $N \triangleleft G$ and $H < G$ with $H \cap N = \{e\}$, then H is isomorphic to its canonical image in G/N.*

Proof. $NH/N \approx H/(H \cap N) = H/\{e\} \approx H$. □

These three theorems of Noether (especially Noether I) are powerful tools for building isomorphisms. Let us also record an application of Proposition 4.15 to G/N.

Remark 19′. Suppose $N \triangleleft G$. If $N \leq H < G$, then $H/N < G/N$; conversely every proper subgroup of G/N has the form H/N for $N \leq H < G$.

Conjugates in S_n

The computation of conjugates (used in the criterion for normality of a subgroup) is surprisingly straightforward in S_n, and we shall describe it here.

Remark 20. Suppose σ, π are arbitrary permutations, and write σ as a product of disjoint cycles

$$\sigma = (i_1 \dots i_{t_1})(i_{t_1+1} \dots i_{t_2})(i_{t_2+1} \dots i_{t_3}) \dots . \tag{2}$$

In any given cycle, we see that σi is the entry following i. Now observe what $\pi\sigma\pi^{-1}$ does to πi.

$\pi\sigma\pi^{-1}(\pi i) = \pi(\sigma i)$; *i.e.*, $\pi(\sigma i)$ follows πi the same way that σi follows i. Thus we see $\pi\sigma\pi^{-1}$ is the product of disjoint cycles

$$(\pi i_1 \dots \pi i_{t_1})(\pi i_{t_1+1} \dots \pi i_{t_2})(\pi i_{t_2+1} \dots \pi i_{t_3}) \dots . \tag{3}$$

This computation will be the key to many observations about permutations. In particular we see that each conjugate of σ can be written as a product of disjoint cycles of the same respective lengths as the cycles of σ. More colloquially, we say σ has the same *placement of parentheses* as each of its conjugates. On the other hand, if σ is as above and $\tau = (j_1 \dots j_{t_1})(j_{t_1+1} \dots j_{t_2})(j_{t_2+1} \dots j_{t_3}) \dots$, *i.e.*, τ has the same placement of parentheses as σ, then $\tau = \pi\sigma\pi^{-1}$ where $\pi = \begin{pmatrix} i_1 & i_2 & \dots & i_{t_1} & \dots \\ j_1 & j_2 & \dots & j_{t_1} & \dots \end{pmatrix}$. Therefore, the conjugates of σ are *precisely* those permutations that can be written as a product of disjoint cycles with the same placement of parentheses. We shall consider this issue in more detail in Chapter 10, when conjugacy is used to define an equivalence relation.

Now we apply this argument to determine the nontrivial normal subgroups of S_3 and S_4. $(13)(12)(13)^{-1} = (32) \notin \langle(12)\rangle$, which thus is not a normal subgroup of S_3; analogously $\langle(13)\rangle$ and $\langle(23)\rangle$ are not normal subgroups. On the other hand every conjugate of (123) is a cycle of order 3 and thus can be rewritten as (123) or $(132) = (123)^2$, and so belongs to $\langle(123)\rangle$; this proves (for the third time) $\langle(123)\rangle \triangleleft S_3$.

In S_4, every permutation is conjugate to either (1234), (123), (12), (1), or $(12)(34)$. In fact the set of conjugates of $(12)(34)$ is precisely

$$\{(12)(34), \quad (13)(24), \quad (14)(23)\}$$

(since for example $(42)(13) = (24)(13) = (13)(24)$); adjoining (1) to this set yields a subgroup K of four elements called the *Klein group*. Since every conjugate of each element of K is in K, we see in fact $K \triangleleft S_4$. Note that $\exp(K) = 2$.

S_4 has one other nontrivial proper normal subgroup, which we now consider.

The Alternating Group

We use these ideas to construct an interesting nontrivial normal subgroup of S_n, in fact the *only* one for $n > 4$. Any permutation π can be written as a product of disjoint cycles $C_1 \dots C_u$; any cycle $C = (i_1 \dots i_t)$ can be written as the product of $t-1$ transpositions $(i_1\ i_t)(i_1\ i_{t-1}) \dots (i_1\ i_2)$. Hence any permutation π can be written as a product of m transpositions, for suitable m. Note that m is not well-defined, since $(12) = (12)(12)(12)$. However, we shall see now that m is well-defined $\pmod 2$. Thus we can define the *sign* of π, denoted $\mathrm{sg}(\pi)$, to be $(-1)^m$; π is called *even* (resp. *odd*) according to whether m is *even* (resp. *odd*), *i.e.*, whether $\mathrm{sg}\,\pi = +1$ (resp. $= -1$).

THEOREM 21. *There is a (well-defined) homomorphism $S_n \to (\{\pm 1\}, \cdot)$, given by $\pi \mapsto \mathrm{sg}(\pi)$.*

Proof. First define

$$\psi(\pi) = \prod_{i>j} \frac{\pi i - \pi j}{i - j}.$$

We claim $\psi(\pi\sigma) = \psi(\pi)\psi(\sigma)$ for all permutations π, σ. Indeed,

$$\psi(\pi\sigma) = \prod_{i>j} \frac{\pi\sigma i - \pi\sigma j}{i - j} = \prod_{i>j} \frac{\pi\sigma i - \pi\sigma j}{\sigma i - \sigma j} \prod_{i>j} \frac{\sigma i - \sigma j}{i - j} = \psi(\pi)\psi(\sigma)$$

(since σ merely permutes the indices, and $\frac{\pi i - \pi j}{i-j} = \frac{\pi j - \pi i}{j - i}$).

In particular, ψ is a homomorphism, so it remains to show $\psi(\pi) = \mathrm{sg}(\pi)$. First we check it for π a transposition; to simplify notation we assume $\pi = (12)$:

$$\frac{\pi 2 - \pi 1}{2 - 1} = \frac{1 - 2}{2 - 1} = -1;$$

$$\frac{(\pi i - \pi 2)(\pi i - \pi 1)}{(i-2)(i-1)} = \frac{(i-1)(i-2)}{(i-2)(i-1)} = 1, \quad \text{for} \quad i > 2;$$

$$\frac{\pi i - \pi j}{i - j} = \frac{i - j}{i - j} = 1 \quad \text{for} \quad i, j > 2.$$

Multiplying these three equations together for all possible i, j yields

$$\psi((12)) = \prod_{i>j} \frac{\pi i - \pi j}{i - j} = (-1) \cdot 1 \cdot 1 = -1.$$

In general, writing an arbitrary permutation π as a product of m transpositions $\sigma_1 \dots \sigma_m$, yields $\psi(\pi) = \psi(\sigma_1) \dots \psi(\sigma_m) = (-1)^m = \mathrm{sg}(\pi)$, as desired. $\square$

There are proofs of Theorem 21 which are more intrinsic to the definition of permutation, *cf.* Exercises 3,4.

Definition 22. A_n, the *alternating group*, is the kernel of the group homomorphism ψ of Theorem 21, *i.e.*, $A_n = \{\text{even permutations of } S_n\}$.

Remark 23. $A_n \triangleleft S_n$, being the kernel of a homomorphism. Furthermore, $S_n/A_n \approx (\{\pm 1\}, \cdot)$ by Noether I. Proposition 4.13 implies $|A_n| = \frac{|S_n|}{2} = \frac{n!}{2}$. Thus $|A_3| = 3$, implying $A_3 = \langle (123) \rangle \approx (\mathbb{Z}_3, +)$. However A_4 is new. Let us list its 12 elements:

$$(123) \quad (132) \quad (124) \quad (142) \quad (134) \quad (143) \quad (234) \quad (243)$$

$$(1) \quad (12)(34) \quad (13)(24) \quad (14)(23)$$

Each element on the top row has order 3, whereas the bottom row comprises the elements of the Klein group, which has exponent 2. Thus $\exp(A_4) = 6$.

One interesting feature of A_4 is that it has no subgroup of 6 elements (and thus is the promised counterexample to the converse of Lagrange's theorem). Indeed suppose $N < A_4$ had order 6. Then $[A_4 : N] = 2$ so $N \triangleleft A_n$ by Proposition 9. But N would have an element of order 2, which we may assume is $a = (12)(34)$. Then $(14)(23) = (123)a(123)^{-1} \in N$ and $(13)(24) = (124)a(124)^{-1} \in N$, implying N would contain the Klein group (of order 4), which is impossible by Lagrange's theorem, since 4 does not divide 6. Other properties of A_n are given in Exercises 4 and 6, and Example 12.5. The most important property in this course is the following:

THEOREM 24. *A_n is the only nontrivial proper normal subgroup of S_n, for each $n \geq 5$.*

Proof. Suppose $\{e\} \neq N \triangleleft S_n$. We shall show $A_n \subseteq N$, and thus $N = A_n$ or S_n by Remark 19′. It suffices to show that N contains every product π of two transpositions, since every even permutation is a product of these. But any such π has the form $(i_1\ i_2\ i_3)$ or $(i_1\ i_2)(i_3\ i_4)$. In fact it is enough to show that each $(i_1\ i_2\ i_3) \in N$, since then $(i_1\ i_4\ i_3)(i_1\ i_2\ i_3) = (i_1\ i_2)(i_3\ i_4) \in N$. We shall prove that $(123) \in N$; analogously each $(i_1\ i_2\ i_3) \in N$.

Write any $\sigma \neq 0$ in N as a product of disjoint cycles $C_1C_2...C_t$, arranged in decreasing length. Write $C_1 = (i_1 i_2 \dots i_m)$.

Case I. $m \geq 3$. Let $C_1' = (i_2 i_1 i_3 \dots i_m)$. Then $C_1' C_2 \dots C_t$ has the same placement of parentheses as σ, so is a conjugate τ of σ, and thus is in N. Hence

$$\tau\sigma^{-1} = C_1' C_1^{-1} = (i_1 i_2 i_3) \in N,$$

so N contains (123), as desired.

Case II. $m = 2$. Then $t \geq 2$, and $C_1 = (i_1 i_2)$, $C_2 = (i_3 i_4)$. Let $C_1' = (i_1 i_3)$ and $C_2' = (i_2 i_4)$. Then $C_1' C_2' C_3 \dots C_t$ has the same placement of parentheses as σ, so is a conjugate τ of σ, and thus is in N. Hence

$$\tau\sigma^{-1} = C_1' C_2' C_2^{-1} C_1^{-1} = (i_1 i_4)(i_2 i_3) \in N,$$

so N contains (13)(45) and (12)(45), and thus also their product (123), as desired. □

Note that we needed $n \geq 5$ to escape from the Klein group in Case II.

Exercises

1. If G has at least two subgroups of prime order p, and p^2 does not divide $|G|$, then no subgroup of order p is normal. (*Hint:* Otherwise $HK < G$ of order p^2.) Find an analogous assertion for higher powers of p.
2. If $N \triangleleft G$ and $g \in G$ has order n, then Ng (viewed as an element of G/N) has order dividing n. Give an example where equality does not hold.

S_n and A_n

There are many proofs of Theorem 21 in the literature. Here are two good alternative approaches; Exercise 4 contains the fastest proof that I know, but is a bit tricky.

3. Any permutation π can be written as a product of disjoint cycles $C_1 \dots C_u$, whose lengths $t_1, \dots, t_u$ (when written in ascending order) are unique. Define $\ell(\pi) = t_1 + \dots + t_u - u$, which is clearly well-defined, and show $\pi \mapsto (-1)^{\ell(\pi)}$ is a homomorphism. (*Hint:* By induction, one need merely check the product of a cycle and a transposition.)
4. Show directly that no permutation is both even and odd. (*Hint:* Otherwise one could write $e = (a\ b)(c\ d)\dots$ as a product of an odd number of transpositions; of all such products beginning with a, take that one in which a appears the smallest possible number of times, and such that the appearances of a are as far to the left as possible. Since $(i\ a) = (a\ i)$ we may assume a never appears on the right side in a transposition. Clearly a appears at least twice. But $(i\ j)(a\ i) = (a\ j\ i) = (a\ j)(i\ j)$, so the product must start $(a\ i)(a\ j)$; then $(a\ i)(a\ j) = (a\ j\ i) = (a\ j)(i\ j)$, which has fewer occurrences of a, contradiction.)
5. The only nontrivial normal subgroup of A_4 is the Klein group.

6. If $(1) \neq N \triangleleft A_n$ for $n \geq 5$ then $N = A_n$. (*Hint*: Mimic the proof of Theorem 24. It is enough to show $(123) \in N$; take care to conjugate by even permutations only.)
7. If $H \leq S_n$ and H contains every transposition, then $H = S_n$.
8. If H is a subgroup of S_n containing $\sigma = (1\ 2\ \ldots\ n)$ and $\tau = (1\ 2)$, then $H = S_n$. (*Hint*: $\sigma^i \tau \sigma^{-i} = (i\ i+1) \in H$. But then one has $(i\ i+1)(i-1\ i)(i\ i+1) = (i-1\ i+1) \in H$. Continuing in this way produces all the transpositions.)
9. When do two permutations σ and τ commute? (*Hint*: When does $\sigma\tau\sigma^{-1} = \tau$? Be careful to treat cases such as $\sigma = (12)(34)$ and $\tau = (13)(24)$.)

Chapter 6. Classifying Groups – Cyclic Groups and Direct Products

The *classification problem* for finite groups is the question of how to list all finite groups "up to isomorphism," *i.e.*, one representative for each class of isomorphic groups. Having such a list, presumably we could verify assertions about groups simply by going down the list. However, there are several difficulties with this approach. First of all, the list must be infinite, since $(\mathbb{Z}_m, +)$ is a finite group for each m. Moreover, even if we had a complete list of finite groups, it might be impossible in practice to verify a given assertion. For example, $1, 2, 3, \ldots$ is a list of the natural numbers, but we do not know if there are any odd perfect numbers. Finally, the list of finite groups might be less enlightening than the body of theorems used to obtain the list. Nevertheless, the classification problem is the focal research problem in the theory of finite groups, and the ensuing results have been of great use.

Given a formidable problem such as classification, we would like to start by breaking it up into manageable parts. Thus we shall consider first only certain kinds of groups.

Cyclic Groups

Definition 1. A group G is *cyclic* if $G = \langle g \rangle$ for some element g in G.

For example, $(\mathbb{Z}, +) = \langle 1 \rangle$, since each positive number is a sum of 1's. Likewise $(\mathbb{Z}_n, +)$ is cyclic for each n. Surprisingly this gives the complete list of cyclic groups.

THEOREM 2. *(The classification of cyclic groups) Every cyclic group G is isomorphic to $(\mathbb{Z}, +)$ (for G infinite) or $(\mathbb{Z}_n, +)$ (for G finite and $n = |G|$). More precisely, writing $G = \langle g \rangle$ we have a surjection $\varphi : (\mathbb{Z}, +) \to G$ given by $\varphi(m) = g^m$. If $\ker \varphi = \{0\}$, then φ is an isomorphism; if $\ker \varphi = \langle n \rangle$ then $G \approx (\mathbb{Z}_n, +)$.*

Proof. We prove the second assertion, which in turn implies the first.

$$\varphi(m + m') = g^{m+m'} = g^m g^{m'} = \varphi(m)\varphi(m')$$

for all m, m' in $\mathbb{Z}$, proving φ is a homomorphism that clearly is onto. If $\ker \varphi = \{0\}$, then φ is an isomorphism, so assume $\ker \varphi \neq \{0\}$. Then $\ker \varphi = n\mathbb{Z}$ for some n, by Example 2.6′. But then Noether I implies $G \approx \mathbb{Z}/n\mathbb{Z} \approx (\mathbb{Z}_n, +)$. □

$(\mathbb{Z}_n, +)$ has taken on such importance that we can refer to it merely as $\mathbb{Z}_n$; the operation "+" is understood. $\mathbb{Z}_n$ also is denoted C_n in the literature.

COROLLARY 3. *If* $|G| = p$ *is prime then* $G \approx \mathbb{Z}_p$.

Proof. G is cyclic by Corollary 2.14, so apply the theorem. □

Here is an application of the classification of cyclic groups.

COROLLARY 4. *Every subgroup of a cyclic group* G *is cyclic. Furthermore, if* $G = \langle g \rangle$ *is cyclic and* m *divides* $n = |G|$, *then* G *has a unique subgroup of order* m, *namely* $\langle g^{n/m} \rangle$.

Proof. We may assume $G = (\mathbb{Z}, +)$ or $G = (\mathbb{Z}_n, +)$ (and $g = 1$), so we are done by Example 2.6′ and Corollaries 4.15 and 4.16. (The last assertion is clear since $o(\frac{n}{m}) = m$.) □

Generators of a Group

There are noncyclic Abelian groups, for example, Euler(8). Toward understanding such groups we need to generalize "cyclic."

Definition 5. A group G is *generated* by a set S, written $G = \langle S \rangle$, if G has no proper subgroups containing S. If G is generated by a finite set $S = \{g_1, \ldots, g_t\}$, we also say G is *generated by* the t elements $g_1, \ldots, g_t$. G is *finitely generated* if G is generated by a finite set.

Intuitively G is generated by S iff every element of G can be written as an expression involving the elements of S, by means of the group operations (the group product and inverse), in a manner to be made precise in Exercise 8.18. Of course G is generated by 1 element iff G is cyclic.

Remark 6. (i) If $S \subseteq \langle S' \rangle$ and $G = \langle S \rangle$, then $G = \langle S' \rangle$.

(ii) If $\varphi: G \to K$ is an onto homomorphism and G is generated by S, then $\varphi(G)$ is generated by $\varphi(S)$. (Indeed suppose $\varphi(S)$ is contained in a subgroup of $\varphi(G)$; this subgroup has the form $\varphi(H)$, where $\ker \varphi \leq H \leq G$. Clearly $S \subseteq H$ implying $H = G$, so we conclude $\varphi(H) = \varphi(G)$.)

An example of a noncyclic group generated by two elements is Euler(8), which is generated by [3] and [5], since $[1] = [3]^2$ and $[7] = [3] \cdot [5]$. On the other hand, obviously any group G generates itself, so every finite group is finitely generated. S_n is generated by two elements, by Exercise 5.8, so we may wonder how to construct a group that cannot be generated by two elements. The answer will be trivial by the end of this discussion; the solution is to find a way of piecing together small groups.

Direct Products

Definition 7. The *direct product* of two monoids H and K is the Cartesian product $H \times K$, endowed with componentwise multiplication, *i.e.*, $(h_1, k_1)(h_2, k_2) = (h_1 h_2, k_1 k_2)$.

The verifications of the following straightforward facts hinge on checking each component.

Remark 8. The direct product $H \times K$ of two monoids is a monoid, with neutral element (e_H, e_K), and has the following additional properties when H and K are groups:

(i) $H \times K$ is a group, with inverses given by $(h,k)^{-1} = (h^{-1}, k^{-1})$.
(ii) There are natural injections $H \to H \times K$ and $K \to H \times K$ given respectively by $h \mapsto (h, e)$ and $k \mapsto (e, k)$.
(iii) There are natural surjections $H \times K \to H$ and $H \times K \to K$ given respectively by $(h,k) \mapsto h$ and $(h,k) \mapsto k$. These are called the *projections* onto H and K respectively.
(iv) If $o(h) = m$ and $o(k) = n$ in H,K respectively then $o(h,k) = \text{lcm}(m,n)$. (Indeed let $t = o(h,k)$ and $t' = \text{lcm}(m,n)$. Then $(h,k)^{t'} = (h^{t'}, k^{t'}) = (e,e)$ so $t|t'$. On the other hand $(e,e) = (h,k)^t = (h^t, k^t)$ so $h^t = e$ and $k^t = e$ implying $m|t$ and $n|t$, and thus $t'|t$.)
(v) $|H \times K| = |H||K|$ since this already holds for Cartesian products;
(vi) $\exp(H \times K) = \text{lcm}(\exp(H),\exp(K))$, by (ii) and (iv);
(vii) If $H_1 \approx H_2$ and $K_1 \approx K_2$ then $H_1 \times K_1 \approx H_2 \times K_2$. More precisely, if $\varphi : H_1 \to H_2$ and $\psi : K_1 \to K_2$ are the given isomorphisms then the desired isomorphism $\varphi \times \psi : H_1 \times K_1 \to H_2 \times K_2$ is given by $(h_1, k_1) \mapsto (\varphi(h_1), \psi(k_1))$.
(viii) If H and K are Abelian then $H \times K$ is Abelian.

Example 9. $\mathbb{Z}_2 \times \mathbb{Z}_2$ is an Abelian group of order 4 and exponent 2 and obviously is not cyclic.

We would like to push Remark 8(ii) a bit further, in order to identify $\mathbb{Z}_2 \times \mathbb{Z}_2$ with Euler(8) and with the Klein group.

Remark 10. Let $\bar{H}, \bar{K}$ be the respective images of H,K under the homomorphisms of Remark 8(ii). Then $\bar{H} = H \times \{e\}$ and $\bar{K} = \{e\} \times K$ are isomorphic respectively to H and K and are each normal subgroups of $H \times K$. Furthermore the following properties hold:

(i) $\bar{H} \cap \bar{K} = \{(e,e)\}$;
(ii) $\bar{H}\bar{K} = H \times K$ (since $(h,k) = (h,e)(e,k)$);
(iii) Every element of $\bar{H}$ commutes with every element of $\bar{K}$ (since $(h,e)(e,k) = (h,k) = (e,k)(h,e)$).

In fact, assertions (i) through (iii) of Remark 10 characterize the direct product construction, as we shall see now.

Internal Direct Products

PROPOSITION 11. *Suppose a group G has subgroups H and K satisfying the conditions*

DP1 $H \cap K = \{e\}$;

DP2 $HK = G$;

DP3 *Every element of H commutes with every element of K.*

Then $G \approx H \times K$.

Proof. Define $\varphi : H \times K \to G$ by $(h,k) \mapsto hk$. φ is a homomorphism since

$$\begin{aligned}\varphi((h_1,k_1)(h_2,k_2)) &= \varphi(h_1h_2,k_1k_2) = (h_1h_2)(k_1k_2)\\ &= h_1(h_2k_1)k_2 = h_1(k_1h_2)k_2\\ &= (h_1k_1)(h_2k_2) = \varphi(h_1,k_1)\varphi(h_2,k_2).\end{aligned}$$

Also φ is onto, by DP2. Finally $\ker\varphi = \{(h,k) : hk = e\}$. But $hk = e$ implies $h = k^{-1} \in H \cap K = \{e\}$ by DP1; thus $(h,k) = (e,e)$, proving φ is 1:1. □

We shall say G is an *internal direct product* of subgroups H and K when the criteria of proposition 11 are satisfied. Note that DP3 is superfluous when G is Abelian, so in this case DP1 and DP2 suffice to prove G is isomorphic to $H \times K$. Let us find further reductions in verifying an internal direct product.

Remark 12. Suppose H and K are subgroups of a group G.

(i) If DP1 holds and $|G| = |H||K|$ then DP2 holds by Theorem 3.13, since $|HK| = \frac{|H||K|}{|H\cap K|} = \frac{|G|}{1} = |G|$.

(ii) Suppose $|G| = |H||K|$ with $|H|, |K|$ relatively prime. Then DP1 holds since $|H \cap K|$ divides $(|H|,|K|) = 1$; thus DP2 holds by (i).

There is an interesting alternative formulation of DP3.

PROPOSITION 13. *Suppose H and K are subgroups of G satisfying DP1 and DP2. Then H and K are both normal subgroups of G iff DP3 holds.*

Proof. ($\Rightarrow$) Take any h in H and k in K, and let $g = hkh^{-1}k^{-1}$. Then $g = (hkh^{-1})k^{-1} \in K$, but also $g = h(khk^{-1}) \in H$; hence $g \in H \cap K = \{e\}$, so $hk = kh$.

($\Leftarrow$) We shall show $ghg^{-1} \in H$, for any g in G and h in H. Indeed write $g = h_1k_1$ for h_1 in H and k_1 in K, and note $ghg^{-1} = h_1k_1hk_1^{-1}h_1^{-1} = h_1hh_1^{-1} \in H$, proving $H \triangleleft G$. Likewise $K \triangleleft G$. □

The idea of using $hkh^{-1}k^{-1}$ will also be used extensively in Chapter 12. Let us see some applications.

PROPOSITION 14. *Suppose G is a group of order 4 and exponent 2. Then $G \approx \mathbb{Z}_2 \times \mathbb{Z}_2$.*

Proof. G is Abelian, by Proposition 3.5. Take any two distinct elements $a, b \neq e$. By hypothesis $o(a) = o(b) = 2$, so $\langle a \rangle, \langle b \rangle$ are subgroups each of order 2, satisfying DP1 and DP3. But DP2 holds by remark 12, so $G \approx \langle a \rangle \times \langle b \rangle \approx \mathbb{Z}_2 \times \mathbb{Z}_2$ (since any group of order 2 is cyclic and thus isomorphic to $\mathbb{Z}_2$). □

Example 15. Euler(8) and the Klein group are each isomorphic to $\mathbb{Z}_2 \times \mathbb{Z}_2$.

Thus we begin to see that the direct product construction encompasses all of our examples of Abelian groups.

COROLLARY 16. *If $|G| < 6$ then G is Abelian, and isomorphic to one of the following groups: $\{e\}$, $\mathbb{Z}_2$, $\mathbb{Z}_3$, $\mathbb{Z}_4$, $\mathbb{Z}_2 \times \mathbb{Z}_2$, $\mathbb{Z}_5$.*

Proof. Let $n = |G|$. If n is prime then G is cyclic and isomorphic to $\mathbb{Z}_n$. Thus we may assume n is not prime, *i.e.*, $n = 4 = 2^2$. If $G \not\approx \mathbb{Z}_4$ then G is not cyclic, so $\exp(G) \neq 4$ by example 3.4(ii). But then $\exp(G) = 2$, so $G \approx \mathbb{Z}_2 \times \mathbb{Z}_2$ by Proposition 14. □

This result is typical of what we have in mind when we are classifying groups. Nevertheless we must be careful not to overlook hidden isomorphisms.

PROPOSITION 17. *If H and K are cyclic groups of respective orders m and n which are relatively prime, then $H \times K \approx (\mathbb{Z}_{mn}, +)$.*

Proof. In view of Theorem 2 we need only prove $(\mathbb{Z}_m, +) \times (\mathbb{Z}_n, +)$ is cyclic of order mn. But $([1], [1])$ obviously is an element of order mn, as desired. □

For example, $\mathbb{Z}_3 \times \mathbb{Z}_2 \approx \mathbb{Z}_6$.

The *direct product* of an arbitrary number of groups is defined to be the Cartesian product with the operation defined componentwise. This construction can provide groups that are generated only by large numbers of elements, *cf.* Exercise 2, thereby answering the question posed earlier.

Remark 18. $G_1 \times \ldots \times G_t \approx (G_1 \times \ldots \times G_{t-1}) \times G_t$, under the isomorphism $(g_1, .., g_t) \mapsto ((g_1, \ldots, g_{t-1}), g_t)$. $|G_1 \times \ldots \times G_t| = |G_1| \ldots |G_t|$, by induction on t.

The direct product of monoids gives information about groups of units and has a cute application to number theory, as seen in Exercises 5ff.

Exercises

1. Any group of exponent 2 and order 2^m is isomorphic to $\mathbb{Z}_2 \times \cdots \times \mathbb{Z}_2$ (taken m times).
2. $\mathbb{Z}_2 \times \mathbb{Z}_2 \times \cdots \times \mathbb{Z}_2$ (taken m times) cannot be generated by fewer than m elements.
3. $H \times K \approx K \times H$ under the isomorphism $(h, k) \mapsto (k, h)$.
4. $(G_1 \times G_2) \times G_3 \approx G_1 \times (G_2 \times G_3)$.
5. If M_1 and M_2 are monoids, then
$$\mathrm{Unit}(M_1 \times M_2) \approx \mathrm{Unit}(M_1) \times \mathrm{Unit}(M_2).$$
(*Hint*: Invertibility is by components.)
6. If m and n are relatively prime then, $(\mathbb{Z}_{mn}, \cdot) \approx (\mathbb{Z}_m, \cdot) \times (\mathbb{Z}_n, \cdot)$ as monoids. (*Hint*: $[a]_{mn} \mapsto ([a]_m, [a]_n)$ defines a 1:1 map.)
7. If m,n are relatively prime then the Euler number $\varphi(mn)$ equals $\varphi(m)\varphi(n)$, by Exercises 5,6. Using Exercise 2.9, conclude that if $p_1, \dots, p_t$ are the distinct prime numbers dividing m then
$$\varphi(m) = m(1 - \frac{1}{p_1}) \dots (1 - \frac{1}{p_t}).$$
8. If G is an internal direct product of H and K and if A, B are normal subgroups of H, K respectively, then $G/(A \times B) \approx H/A \times K/B$, viewed in the natural way as an internal direct product. (*Hint*: Noether I.)
9. (Direct product cancellation). If $G \times H_1 \approx G \times H_2$ with $|G|$ finite then $H_1 \approx H_2$. (*Extended hint*: It is the same to prove that if a group K can be written as an internal direct product $G_1 \times H_1 = G_2 \times H_2$ for subgroups G_i and H_i, $i = 1, 2$, and if $G_1 \approx G_2$ then $H_1 \approx H_2$. If $H_1 \cap G_2 = \{e\}$ then $H_1 \approx K/G_2 \approx H_2$ so one is done. Thus one can put $A = H_1 \cap G_2 \neq \{e\}$ and $B = H_2 \cap G_1 \neq \{e\}$. Use Exercise 8 repeatedly:
$$G_2/A \times H_2/B \approx G/AB = G/BA \approx G_1/B \times H_1/A \tag{1}$$
as internal direct products, so
$$\begin{aligned} G_2 \times G_2/A \times H_2/B &\approx G_2 \times G_1/B \times H_1/A \\ &\approx G_1/B \times H_1 \times G_2/A. \end{aligned} \tag{2}$$
Analogously
$$G_1 \times G_1/B \times H_1/A \approx G_2/A \times H_2 \times G_1/B. \tag{3}$$

But (1) together with $G_1 \approx G_2$ shows the left hand sides of (2) and (3) are isomorphic, so the right hand sides are isomorphic; by induction cancel G_1/B and then G_2/A.)

10. There is an injection $S_n \times S_n \to S_{2n}$, given by

$$(\sigma, \pi) \mapsto \begin{pmatrix} 1 & \dots & n & n+1 & \dots & 2n \\ \sigma 1 & \dots & \sigma n & \pi 1 & \dots & \pi n \end{pmatrix}.$$

Chapter 7. Finite Abelian Groups

Abelian groups have a much more manageable structure than groups in general. We shall see soon that Abelian groups are not much more complicated than cyclic groups, but first let us illustrate their nice structure by means of some easy facts.

Remark 0. For any Abelian group A and any positive number m, define $A(m) = \{a \in A : a^m = e\}$. Then $A(m) \leq A$. (Indeed if $a^m = e$ and $b^m = e$ then $(ab^{-1})^m = a^m(b^m)^{-1} = ee^{-1} = e$.)

Of course Remark 0 fails terribly for non-Abelian groups – S_3 has four elements of order dividing 2, so this set is too large to be a subgroup (by Lagrange's theorem).

Remark 1. For any Abelian group A and any $m > 0$, there is a homomorphism $f: A \to A$ given by $a \mapsto a^m$, and $\ker f = A(m)$.

We shall also need Cauchy's theorem, proved in general in Theorem 3.1, but the Abelian case needed here is an easy exercise, *cf.* Exercise 1.

Our present goal is to characterize finite Abelian groups. We shall do this in terms of generators. Like most general notions of group theory, generators become much easier in the context of Abelian groups; a set $\{a_1, \dots, a_t\}$ generates an Abelian group A iff every element of A can be put in the form $a_1^{i_1} \dots a_t^{i_t}$ for suitable $i_1, \dots, i_t$ in $\mathbb{Z}$; if A is finite then we can take $i_1, \dots, i_t$ in $\mathbb{N}$.

Obviously for any t the direct product $C_1 \times \cdots \times C_t$ of cyclic groups $C_1 = \langle a_1 \rangle, \dots, C_t = \langle a_t \rangle$ is Abelian, generated by the t elements

$$(a_1, e, e, \dots, e), (e, a_2, e, \dots, e), \dots, (e, e \dots, e, a_t).$$

Our main object is the following amazing converse, foreshadowed by Exercise 6.1:

Theorem 2 (Fundamental theorem of finite Abelian groups). *Every finite Abelian group generated by t elements is isomorphic to a direct product of t cyclic subgroups.*

Furthermore, we shall see that this decomposition is unique, up to isomorphism and permutation of the cyclic components. This result classifies all the finite Abelian groups, since we have already classified the cyclic groups (Theorem 6.2). However, the theorem presents a major pedagogical dilemma, since actually a much stronger result is true:

THEOREM 2′ (FUNDAMENTAL THEOREM OF FINITELY GENERATED ABELIAN GROUPS). *Every Abelian group generated by t elements is isomorphic to a direct product of t cyclic subgroups.*

One can prove Theorem 2′ directly and elegantly with the tools we have in hand, but the proof is very intricate; we follow a more straightforward approach due to C.R. MacCluer, which is more apt for finite groups and gives a strong version of Theorem 2. The proof of Theorem 2′ is sketched in Exercise 7, (also *cf.* Exercises 8ff.) in the hope that its idea of proof might actually become clearer without the distraction of certain details. Also see Exercises 14–17 for more information concerning infinite Abelian groups.

Some of the methods used in our proof are applicable to arbitrary groups, and others are suited best to Abelian groups; to keep the threads straight we shall let G denote an arbitrary group, and let A denote an Abelian group.

In Chapter 6 we showed that a group was isomorphic to a direct product of two subgroups iff it was an "internal direct product," *i.e.*, iff the two subgroups satisfied properties DP1, DP2, and DP3 of proposition 6.11.

Definition 3. A group G is called an *internal direct product* of t subgroups $H_1, \dots, H_t$ if the following properties are satisfied:

(DP1′) If $h_1 \dots h_t = e$ for h_i in H_i, then each $h_i = e$.

(DP2′) $H_1 H_2 \dots H_t = G$;

(DP3′) $h_i h_j = h_j h_i$ for any $i \neq j$, $h_i \in H_i$, and $h_j \in H_j$.

The reader should check that DP1′, DP2′, and DP3′ coincide with DP1, DP2, and DP3 respectively, when t = 2.

Remark 3′. If G is the internal direct product of H and K, and if K is the internal direct product of $K_1, \dots, K_t$, then G is the internal direct product of $H, K_1, \dots, K_t$. (Indeed, DP2′ and DP3′ are clear. To see DP1′ suppose $hk_1 \dots k_t = e$; then $k_1 \dots k_t = h^{-1} \in H \cap K = e$, so $k_1 = \dots = k_t = e$.)

PROPOSITION 4.

(i) *Given groups $H_1, \dots, H_t$, let $G = H_1 \times \dots \times H_t$, and define $\varphi_i \colon H_i \to G$ by $\varphi_i(h) = (e, \dots, e, h, e, \dots, e)$, where h appears in the i-th position. Clearly φ_i is an injection, and G is an internal direct product of its subgroups $\bar{H}_i = \varphi_i(H_i)$.*

(ii) *Conversely, if G is an internal direct product of subgroups $H_1, \dots, H_t$, then $G \approx H_1 \times \dots \times H_t$.*

Proof.

(i) For h_i in H_i write $\bar{h}_i = \varphi_i(h_i) = (e, \dots, e, h_i, e, \dots, e)$. For all $(h_1, \dots, h_t)$ in G we see $(h_1, \dots, h_t) = \bar{h}_1 \dots \bar{h}_t \in \bar{H}_1 \dots \bar{H}_t$, proving DP2′;

if $e = \bar{h}_1 \dots \bar{h}_t = (h_1, \dots, h_t)$, then each $h_i = e$, proving DP1$'$. DP3$'$ is clear.

(ii) Define $\phi: H_1 \times \dots \times H_t \to G$ by $\phi((h_1, \dots, h_t)) = h_1 \dots h_t$. ϕ is a homomorphism since (by DP3$'$)

$$\begin{aligned}\phi((h_1, \dots, h_t)(h_1', \dots, h_t')) &= h_1 h_1' \dots h_t h_t' = h_1 \dots h_t h_1' \dots h_t' \\ &= \phi((h_1, \dots, h_t))\phi((h_1', \dots, h_t')),\end{aligned}$$

and ϕ is onto by (i); finally $(h_1, \dots, h_t) \in \ker \phi$ iff $h_1 \dots h_t = e$, iff each $h_i = e$, proving ϕ is 1:1. □

Abelian p-Groups

Since we are considering Abelian groups here, DP3$'$ will be obvious. Let us start with a very important special case.

Definition 5. For p a prime number, we define a *p-group* to be a group whose order is a power of p.

Remark 5$'$. If p is the only prime dividing $\exp(G)$, then G is a p-group, by Cauchy's theorem.

By Lagrange's theorem any subgroup of a p-group is a p-group, and likewise any homomorphic image of a p-group is a p-group. Our strategy is to prove Theorem 2 first for p-groups and then find a direct product decomposition into p-groups, which enables us to write every finite Abelian group as a direct product of cyclic p-groups (albeit the number of factors may be considerably larger than the number of generators t); finally we shall show that some of the cyclic factors can be combined into larger cyclic groups, yielding a direct product of t cyclic subgroups.

We say a subgroup H of G is a *direct summand* of G if G is the internal direct product of H and a suitable subgroup K of G; K is called a *complement* of H. Recall by Example 3.4(ii) that for any p-group G there is an element h in G such that $o(h) = \exp(G)$. The next result is the key step in our treatment, since it permits one to start peeling off direct summands from A (thereby paving the way for induction).

THEOREM 6. *Suppose A is an Abelian p-group, and $h \in A$ such that $o(h) = \exp(A)$. Then $\langle h \rangle$ is a direct summand of A.*

Proof. Induction on n, where $|A| = p^n$. Clearly $\exp(A) = p^m$ for some $m \le n$. We are done if $m = n$, *i.e.*, if A is cyclic (for then $A = \langle h \rangle \times \{e\}$); thus we may assume $n > m$. Let $H = \langle h \rangle$.

We claim $A \setminus H$ has an element a of order p. Indeed, by Cauchy's theorem the residue group A/H has an element $\bar{b}$ of order p; its preimage

b in A satisfies $b^p \in H$, so $b^p = h^q$ for some q. If $p \nmid q$ then $\langle h^q \rangle = \langle h \rangle$, so $o(h^q) = o(h) = \exp(A)$, implying $o(b) = p\exp(A)$, contradiction; thus $p|q$. Writing $q = pt$ shows $b^p = (h^t)^p$; hence $a = bh^{-t}$ has order p and $a \notin H$ as desired, since $b \notin H$.

Let $L = \langle a \rangle$, a subgroup of order p. Then $L \cap H = \{e\}$ since its order properly divides p. Let $^-$ denote the canonical image in $\overline{A} = A/L$. Then $H \approx \overline{H}$ by Corollary 5.19. Hence $p^m = o(h) = o(\overline{h})$ divides $\exp(\overline{A})$, which in turn divides $\exp(A) = p^m$. Consequently, $\exp(\overline{A}) = p^m = o(\overline{h})$, and thus by induction $\overline{H}$ is a direct summand of $\overline{A}$. Let $\overline{K}$ be a complement of $\overline{H}$ in $\overline{A}$. Then $\overline{HK} = \overline{A}$ and $\overline{H} \cap \overline{K} = \{e\}$. Let K be the preimage of $\overline{K}$ in A.

Given a in A we have $\overline{h^{-i}a} \in \overline{K}$ for suitable i; then letting $k = h^{-i}a$ we see $a = h^i k$ and $k \in K$, proving $A = HK$. Furthermore, $H \cap K = H \cap (H \cap K) \subseteq H \cap L = \{e\}$, implying A is the internal direct product of H and K. □

We can find H explicitly by using any set of generators of A.

COROLLARY 7. *Suppose A is an Abelian p-group generated by $\{a_1, \ldots, a_t\}$, and $o(a_i)$ is maximal among $o(a_1), \ldots, o(a_t)$. Then $\exp(A) = o(a_i)$; thus $\langle a_i \rangle$ is a direct summand of A.*

Proof. Let $o(a_i) = m$. Clearly $m|\exp(A)$; on the other hand $o(a_j)|m$ for each j, implying $\exp(A) = m$ by Remark 6.8(vi). Hence $\langle a_i \rangle$ is a direct summand of A, by Theorem 6. □

We are ready for a strong version of Theorem 2, for p-groups.

THEOREM 8. *Every finite Abelian p-group generated by elements $a_1, \ldots, a_t$ is an internal direct product of t cyclic subgroups, one of which may be taken to be $\langle a_i \rangle$, where $o(a_i)$ is maximal among $o(a_1)$,...,$o(a_t)$.*

Proof. For convenience, take $i = 1$. Then $\langle a_1 \rangle$ is a direct summand of A by Corollary 7; take a complement K in A, and let $\varphi: A \to K$ be the projection onto K (given by $a_1{}^u k \mapsto k$). Then $\varphi(a_1) = e$, so $\varphi(a_2), \ldots, \varphi(a_t)$ generate K. By induction (on t), K is an internal direct product of cyclic p-groups $C_1, \ldots, C_{t-1}$, so A is the internal direct product of $\langle a_1 \rangle, C_1, \ldots, C_{t-1}$, by Remark 3$'$. □

Aside 9. Following the proof of Theorem 8, we see that each cyclic direct summand C_i of A actually is obtained as a projection of one of the $\langle a_i \rangle$.

Now let us make the reduction to p-groups. First we need a way of locating p-groups inside a given group.

Definition 10. A subgroup H of a group G is a *p-Sylow subgroup* of G if $|H| = p^k$ where $p^{k+1} \nmid |G|$.

Of course, if $H \leq G$ and $|H| = p^k$, then $p^k||G|$ by Lagrange's theorem; since $p^{k+1} \nmid |G|$, we conclude that a p-Sylow subgroup cannot be properly contained in any p-subgroup of G.

THEOREM 11 (SYLOW'S THEOREM FOR ABELIAN GROUPS). *Suppose A is a finite Abelian group, and p is a given prime number dividing $|A|$. Then*

(i) *A has a p-Sylow subgroup S;*

(ii) *$S = \{a \in A : o(a)$ is a power of $p\}$. S contains every p-subgroup of A, and thus is the unique p-Sylow subgroup of A.*

Proof. Write $|A| = p^k q$ for q prime to p.

(i) A contains a nontrivial subgroup N of order p, by Cauchy's theorem. $N \triangleleft A$ since A is Abelian, and $|A/N| = \frac{p^k q}{p} = p^{k-1}q$. By induction on $|A|$, we see A/N contains a p-Sylow subgroup which by Remark 5.19 has the form S/N for suitable $S \leq A$; then $|S/N| = p^{k-1}$ so $|S| = p^k$.

(ii) Let p^k be the largest power of p that divides $|G|$. Then $A(p^k) \leq A$, by Remark 5', and is a p-group, by Remark 5′. If H is any p-subgroup and $h \in H$, then $o(h)$ is a p-power so $h \in A(p^k)$, proving $H \subseteq A(p^k)$. In particular $A(p^k)$ is the unique p-Sylow subgroup of A. □

Actually, (i) also has a direct elementary proof, *cf.* Exercise 1(ii). On the other hand, one can also prove the uniqueness part of (ii) by an easy argument, as we shall see in Proposition 11.6.

COROLLARY 12. *Any finite Abelian group is the internal direct product of its p-Sylow subgroups.*

Proof. Let $p_1, \ldots, p_u$ be the distinct primes dividing $|A|$. Let H_j denote the p_j-Sylow subgroup of A, and let $A_0 = H_1 H_2 \ldots H_{u-1}$. By Corollary 3.14, $|A_0|$ divides $|H_1||H_2|| \ldots |H_{u-1}|$; hence H_j is also the p-Sylow subgroup of A_0, for $1 \leq j \leq u-1$. By induction on u, A_0 is the internal direct product of $H_1, \ldots, H_{u-1}$. Furthermore, $|A_0|$ and $|H_u|$ are relatively prime, so, by Remark 6.12(ii), A is the internal direct product of A_0 and H_u, and we are done by Remark 3′. □

Before proving Theorem 2, let us remark that Corollary 12 can be applied directly, to reduce many questions about Abelian groups to Abelian p-groups, *cf.* the next result as well as Exercise 4.

Remark 12′. If A is Abelian of exponent m, then A has some element of order m. (Indeed, write m as a product $p_1^{i_1} \ldots p_u^{i_u}$ of prime powers. Then each Sylow p_j-subgroup has some element a_j of order $p_j^{i_j}$, by Example 3.4(ii), so take $a = a_1 \ldots a_u$.)

In particular, if $\exp(A) = |A|$, then A is cyclic.

Proof of the Fundamental Theorem for Finite Abelian Groups

Proof of Theorem 2. Write an Abelian group A as the internal direct product of its p-Sylow subgroups $H_1, \dots, H_u$. For each $j \leq u$ we have the projection $\psi_j\colon A \to H_j$ given by $\psi_j(h_1 \dots h_u) = h_j$. If $a_1, \dots, a_t$ generate A then $\psi_j(a_1), \dots, \psi_j(a_t)$ generate H_j, so by Theorem 8 we can write H_j as a direct product of cyclic p_j-groups $C_{j1}, \dots, C_{jt}$. (If H_j is a direct product of fewer cyclics, just throw in several copies of the trivial group.) But letting $C_i = C_{1i} \times \cdots \times C_{ui}$, we then see (by rearranging the C_{ji}) that $A \approx C_1 \times \cdots \times C_t$, and each C_i is cyclic by Proposition 6.17. □

The Classification of Finite Abelian Groups

Our results provide a procedure for describing a finite Abelian group A generated by t elements as a direct product of cyclic p-groups (where p ranges over the prime divisors of $|A|$):

(i) Write A as the internal direct product of its p-Sylow subgroups, for those prime numbers p dividing $|A|$ (taken in decreasing order), and

(ii) Write each of these p-groups as an internal direct product of cyclic p-groups, taking in decreasing order.

To conclude the proof of Theorem 2 we merely recombined the (relatively prime) cyclic factors. Knowing this, it often is convenient to forego this step.

The reader should observe that since each p-Sylow subgroup is unique, we have uniqueness in step (i). Although step (ii) is not unique, we do have uniqueness "up to isomorphism," which we obtain by means of the following result. (Actually a more general assertion has been given in Exercise 6.9, but the proof for the Abelian case is more intuitive.)

THEOREM 13 ("CANCELLATION"). *If finite Abelian groups G, H, and K satisfy $G \times H \approx G \times K$, then $H \approx K$.*

Proof. First writing G as a direct product $G_1 \times \cdots \times G_t$ of cyclic p-groups (for varying p), we have $G_1 \times \cdots \times G_t \times H \approx G_1 \times \cdots \times G_t \times K$; peeling off $G_1, G_2, \dots$ one at a time, we see it is enough to prove the result when G is a cyclic p-group.

Write H_q, K_q for the respective q-Sylow subgroups of H and K, for an arbitrary prime number q. When $q \neq p$ we see (via the order) that $\{e\} \times H_q$ (resp. $\{e\} \times K_q$) is also the q-Sylow subgroup of $G \times H$ (resp. $G \times K$); but $G \times H \approx G \times K$, implying $H_q \approx K_q$ for $q \neq p$. By Corollary 12, it remains to show $H_p \approx K_p$. But $G \times H_p \approx G \times K_p$, since these are the respective p-Sylow subgroups of $G \times H \approx G \times K$; thus we may replace H, K by H_p, K_p. In particular we may assume G, H, K are p-groups, with G cyclic. Note $|H| = \frac{|G \times H|}{|G|} = \frac{|G \times K|}{|G|} = |K|$. We proceed inductively on $|G|$, noting the result is trivial if $|G| = 1$.

For any finite Abelian group A, define $f: A \to A$ by $a \mapsto a^p$; then f is a homomorphism by Remark 1, whose kernel is nontrivial iff p divides $|A|$. In particular, $|f(G)| < |G|$. But

$$f(G) \times f(H) \approx f(G \times H) \approx f(G \times K) \approx f(G) \times f(K),$$

implying by induction that $f(H) \approx f(K)$. Thus it remains to show that we can reconstruct H from $|H|$ and $f(H)$. This result has independent value, so we state it separately, in a slightly stronger version.

LEMMA 14. *(Notation as above.) Any finite Abelian p-group A is determined up to isomorphism by $|A|$ and $f(A)$.*

Proof. We write $A \approx C_1 \times \ldots \times C_t$, where the C_i are all cyclic groups and $|C_1| \geq \ldots \geq |C_t| \geq p$. Then $\ker f \approx C_1(p) \times \ldots, \times C_t(p)$, which has order p^t, so $p^t = \frac{|A|}{|f(A)|}$, thereby determining t. Also, taking $u \leq t$ maximal such that $|C_u| > p$, we have $f(C_i) = \{e\}$ for each $i > u$. Hence $f(A) \approx f(C_1) \times \cdots \times f(C_u)$, and, by induction on $|A|$ (since $|f(A)| < |A|$), the $f(C_i)$ are uniquely determined up to isomorphism. Thus it remains to show for each i that $f(C_i)$ determines C_i. But if $|f(C_i)| = p^m$ then $|C_i| = p^{m+1}$ implying $C_i \approx \mathbb{Z}_{p^{m+1}}$. □

In this proof we have encountered a new group invariant, the number of elements a satisfying $a^p = e$. Lemma 14 also yields at once the uniqueness of the components in the fundamental theorem of finite Abelian groups:

THEOREM 15. *Any Abelian p-group A is decomposable uniquely (up to isomorphism) as a direct product of cyclic p-groups $C_1 \times \cdots \times C_t$, where $|C_1| \geq \cdots \geq |C_t|$.*

Example 16. We shall now determine all Abelian groups of order 144, up to isomorphism. If $|A| = 144$, then $A \approx H_2 \times H_3$ where H_p denotes the p-Sylow subgroup; *i.e.*, $|H_2| = 16$ and $|H_3| = 9$. Hence H_2 could be

$$\mathbb{Z}_{16}, \quad \mathbb{Z}_8 \times \mathbb{Z}_2, \quad \mathbb{Z}_4 \times \mathbb{Z}_4, \quad \mathbb{Z}_4 \times \mathbb{Z}_2 \times \mathbb{Z}_2, \quad \text{or} \quad \mathbb{Z}_2 \times \mathbb{Z}_2 \times \mathbb{Z}_2 \times \mathbb{Z}_2,$$

which are nonisomorphic; H_3 could be $\mathbb{Z}_9$ or $\mathbb{Z}_3 \times \mathbb{Z}_3$. Therefore, A is isomorphic to one of the following ten groups:

$$\mathbb{Z}_{16} \times \mathbb{Z}_9, \quad \mathbb{Z}_8 \times \mathbb{Z}_2 \times \mathbb{Z}_9, \quad \mathbb{Z}_4 \times \mathbb{Z}_4 \times \mathbb{Z}_9, \quad \mathbb{Z}_4 \times \mathbb{Z}_2 \times \mathbb{Z}_2 \times \mathbb{Z}_9,$$

$$\mathbb{Z}_2 \times \mathbb{Z}_2 \times \mathbb{Z}_2 \times \mathbb{Z}_2 \times \mathbb{Z}_9, \quad \mathbb{Z}_{16} \times \mathbb{Z}_3 \times \mathbb{Z}_3, \quad \mathbb{Z}_8 \times \mathbb{Z}_2 \times \mathbb{Z}_3 \times \mathbb{Z}_3,$$

$$\mathbb{Z}_4 \times \mathbb{Z}_4 \times \mathbb{Z}_3 \times \mathbb{Z}_3, \quad \mathbb{Z}_4 \times \mathbb{Z}_2 \times \mathbb{Z}_2 \times \mathbb{Z}_3 \times \mathbb{Z}_3, \quad \text{or} \quad \mathbb{Z}_2 \times \mathbb{Z}_2 \times \mathbb{Z}_2 \times \mathbb{Z}_2 \times \mathbb{Z}_3 \times \mathbb{Z}_3.$$

Their respective exponents are 144, 72, 36, 36, 18, 48, 24, 12, 12, 6. Note that eight of these have different exponents and thus are nonisomorphic by inspection. The only pairs having the same exponent are $\mathbb{Z}_4 \times \mathbb{Z}_4 \times \mathbb{Z}_9$ and $\mathbb{Z}_4 \times \mathbb{Z}_2 \times \mathbb{Z}_2 \times \mathbb{Z}_9$, and $\mathbb{Z}_4 \times \mathbb{Z}_4 \times \mathbb{Z}_3 \times \mathbb{Z}_3$ and $\mathbb{Z}_4 \times \mathbb{Z}_2 \times \mathbb{Z}_2 \times \mathbb{Z}_3 \times \mathbb{Z}_3$. But canceling $\mathbb{Z}_4$ from each, we now have groups of differing exponent (36, 18, 12, 6), which are thus nonisomorphic.

In general, suppose A is an Abelian group of order p^n. Writing A as a direct product of cyclic groups $C_1 \times \cdots \times C_t$, where $|C_1| \geq |C_2| \geq \ldots$, and letting $|C_i| = p^{m_i}$, we have

$$p^n = p^{m_1} p^{m_2} \ldots p^{m_t} = p^{m_1+m_2+\cdots+m_t},$$

implying $n = m_1 + m_2 + \cdots + m_t$. Thus the number of nonisomorphic Abelian groups of order p^n is precisely the number of ways we can write $n = m_1 + m_2 + \cdots + m_t$ with $m_1 \geq m_2 \geq \cdots \geq m_t$; this is called the "partition number," as explained in Exercise 13 and Exercises 16.16 ff.

Exercises

1. (i) *Proof of Cauchy's theorem for any Abelian group A.* Induction on $n = |A|$. Take any element $a \neq e$, and let $m = o(a)$. If $p|m$ then $o(a^{m/p}) = p$. If $p \nmid m$ then by induction $A/\langle a \rangle$ has an element $\bar{b}$ of order p, so $o(b) = pt$ where t divides m, and $o(b^t) = p$.
   (ii) More generally, prove that any Abelian group A has a p-Sylow subgroup, for any p dividing $|A|$. (*Hint*: Apply induction to $A/\langle a \rangle$, where $o(a) = p$.)
2. (i) An Abelian p-group is cyclic iff it has only $p - 1$ elements of order p.
   (ii) More generally, if $A \approx \mathbb{Z}_{n_1} \times \mathbb{Z}_{n_2} \times \cdots \times \mathbb{Z}_{n_t}$ then the number of elements in A of order dividing m is $(m, n_1) \ldots (m, n_t)$. (*Hint*: Reduce to the case A is cyclic.)
   (iii) If G and H are nonisomorphic Abelian groups, then there exists some m such that G has a different number of elements of order m than H. (*Hint*: Pass to p-Sylow subgroups.) This fact does not hold for non-Abelian groups, *cf.* Exercise 10.9.
3. If a is an element of maximal order in a finite Abelian group A, then $\langle a \rangle$ is a direct summand of A.
4. Any Abelian subgroup of S_n is isomorphic to $\langle C_1 \rangle \times \cdots \times \langle C_t \rangle$, where $C_1, \ldots, C_t$ are disjoint cycles in S_n. (This is not so easy; *cf.* Exercise 5.9.)
5. Using Exercise 4, show that any Abelian subgroup of S_n has order $m = m_1 \ldots m_t$ where $m_1 + \cdots + m_t \leq n$. Show that m is maximal in the following situations:

(i) If $n = 3k$ then $m = 3^k$;
(ii) If $n = 3k+1$ then $m = 4 \cdot 3^{k-1}$;
(iii) If $n = 3k+2$ then $m = 2 \cdot 3^k$.
Thus $\frac{m}{n} \to \infty$ as $n \to \infty$.

6. Determine the analogues of Exercises 4 and 5 for A_n in place of S_n.

Finitely Generated Abelian Groups

7. In this exercise we outline the proof of Theorem 2′, leaving the reader to fill in the details. To make the idea more intuitive we write the operation of A as addition instead of multiplication; *i.e.*, the neutral element is now 0. One wants to prove that A is an internal direct product of $\langle a_1 \rangle, \dots, \langle a_t \rangle$ for suitable a_i in A. The condition DP1′ now is: If $\sum n_i a_i = 0$ for n_i in $\mathbb{Z}$ then each $n_i a_i = 0$. Accordingly we shall say $a_1, \dots, a_t$ satisfy a *nontrivial dependence relation* if $\Sigma n_i a_i = 0$ for suitable $n_1, \dots, n_t$ in $\mathbb{Z}$ such that some $n_i a_i \neq 0$.

Take t minimal such that A is generated by t elements. We are done unless each generating set $\{a_1, \dots, a_t\}$ satisfies a nontrivial dependence relation $\sum n_i a_i = 0$; by symmetry (*i.e.*, renumbering the a_i if necessary) assume $n_1 a_1 \neq 0$. Of all generating sets $\{a_1, \dots, a_t\}$ and all dependence relations $\sum n_i a_i = 0$ with $n_1 a_1 \neq 0$, choose $\{a_1, \dots, a_t\}$ and the dependence, such that $|n_1|$ is minimal. One may assume $n_1 > 0$. Note that $|n_j| \geq n_1$ for any j such that $n_j a_j \neq 0$. By creating new dependence relations with smaller values of n_1 prove the following reductions (in increasing order of difficulty):

Reduction 1. If $m a_1 = 0$ with $m > 0$, then $m > n_1$.

Reduction 2. If $\sum m_i a_i = 0$ is an arbitrary dependence relation (involving the same $a_1, \dots, a_t$), then $n_1 | m_1$.

Reduction 3. One may assume $n_1 | n_j$ for all j (seen by replacing a_1 by $a_1 + a_j$, and applying induction to n_j).

Write $n_i = n_1 q_i$ for all $1 < i \leq t$. Define $a_1' = a_1 + \sum_{i>1} q_i a_i$; then $a_1', a_2, \dots, a_t$ generate A. Furthermore $n_1 a_1' = 0$. Let H be the subgroup of A generated by $a_2, \dots, a_t$. Then $A \approx \langle a_1' \rangle \times H$; conclude by induction on t.

There is a way of bypassing these reductions, which casts light on generating sets, as we see in Exercises 8 through 11. Write $M_t(\mathbb{Z})$ for the $t \times t$ matrices with entries in $\mathbb{Z}$.

8. A matrix in $M_t(\mathbb{Z})$ is invertible if and only if its determinant is ± 1.
9. If $m_1, \dots, m_t$ are relatively prime integers, then there is an invertible matrix in $M_t(\mathbb{Z})$ in which $m_1, \dots, m_t$ constitutes the first row. (*Hint:*

Induction on t. Let $d = \gcd(m_1, \ldots, m_{t-1})$ and write $m_i = dn_i$ for $1 \leq i \leq t-1$. Let Q be an invertible $t-1 \times t-1$ matrix with first row $n_1, \ldots, n_{t-1}$, and let Q' be the $t-2 \times t-1$ matrix consisting of all but the first row of Q. Then take $Q = \begin{pmatrix} m_1 & \ldots & m_{t-1} & m_t \\ vn_1 & \ldots & vn_{t-1} & u \\ & & & 0 \\ & Q' & & \vdots \\ & & & 0 \end{pmatrix}$, where $m_t v - ud = 1$.)

10. Suppose an Abelian group A is generated by $a_1, \ldots, a_t$, and let Q be any invertible matrix in $M_t(\mathbb{Z})$. Let $(b_1, \ldots, b_t) = (a_1, \ldots, a_t)Q$. Then $b_1, \ldots, b_t$ also generate A. (*Hint*: $(a_1, \ldots, a_t) = (b_1, \ldots, b_t)Q^{-1}$.)
11. If an Abelian group A is generated by $a_1, \ldots, a_t$, and if $m_1, \ldots, m_t$ are relatively prime integers, then there is another set of t generators that includes $\sum_{i=1}^{t} m_i a_i$. (*Hint*: Use Exercises 9 and 10.)
12. Bypass reductions 1,2,3 in Exercise 7, by means of Exercise 11.
13. Define the *partition number* P_t to be the number of ways one can write t as a sum of positive integers $t_1 + t_2 + \cdots + t_k$, such that $t_1 \geq t_2 \geq \cdots \geq t_k$. For example, $P_3 = 3$, $P_4 = 5$. Let $\#(n)$ denote the number of Abelian groups of order n. Then $\#(p^t) = P_t$, for any prime number p, and in general if $n = p_1^{t_1} \ldots p_u^{t_u}$ with the p_i distinct primes, then $\#(n) = P_{t_1} \ldots P_{t_u}$. (Compare with Example 16.) The partition number is computed in Exercise 16.16ff.
14. Define the *torsion subgroup* $t(A) = \{a \in A : a \text{ has finite order}\}$. Thus $a \in t(A)$ iff $a^u = e$ for suitable $u > 0$. A is called *torsion-free* if $t(A) = \{e\}$. Show that $t(A) \leq A$, and $A/t(A)$ is torsion-free.
15. Any finitely generated torsion-free Abelian group A is isomorphic to a direct product of m copies of $\mathbb{Z}$, for suitable m, and m is uniquely determined (called the *rank* of A). (*Hint*: To show that m is uniquely determined, it suffices to show that if $m < n$ and there are $n \times m$ and $m \times n$ matrices (over $\mathbb{Z}$) whose product is the identity $n \times n$ matrix, then $n = m$. But this follows from Exercise 8.)
16. Any finitely generated Abelian group A is the internal direct product of $t(A)$ and a torsion-free Abelian group that is unique up to isomorphism.
17. Any finitely generated Abelian group is isomorphic to a finite direct product of copies of $\mathbb{Z}$ and cyclic p-groups (for suitable distinct prime numbers p.) This decomposition is unique, up to isomorphism and permutation of the cyclic factors.

In contrast to the Abelian case, generation in arbitrary groups is very complicated; we need some other ingredient to describe a group, namely the *relations* among the generators, *i.e.*, those expressions among the generators that are equal.

Example 0. Suppose A is any finite Abelian group. By the Fundamental Theorem, A is an internal direct product of subgroups $\langle a_1 \rangle, \dots, \langle a_t \rangle$. Letting $n_i = \mathrm{o}(a_i)$ we see that any element of A can be written uniquely in the form $a_1^{u_1} \dots a_t^{u_t}$, where $0 \leq u_i < n_i$ for each i. (We shall say an element described this way is in "normal form.") In particular, A is generated by elements $a_1, \dots, a_t$ which satisfy the relations

(i) $\qquad a_i^{n_i} = e \quad$ for each i, $\quad$ where $\quad n_i = \mathrm{o}(a_i)$;

(ii) $\qquad a_i a_j = a_j a_i \quad$ for each i, j.

Of course, we also have other relations such as $a_i a_j^2 = a_j^2 a_i$, but this is a consequence of (ii) and associativity. All other possible relations among the generators are extraneous. To be sure that our relations determine A, we could build the multiplication table of A, using only these relations and associativity. Indeed, we check by means of repeated applications of (ii) that

$$\begin{aligned}(a_1^{u_1} \dots a_t^{u_t})(a_1^{v_1} \dots a_t^{v_t}) &= a_1^{u_1} \dots a_t^{u_t} a_1^{v_1} \dots a_t^{v_t} \\ &= a_1^{u_1} a_1^{v_1} \dots a_t^{u_t} a_t^{v_t} \\ &= a_1^{u_1+v_1} \dots a_t^{u_t+v_t},\end{aligned}$$

and then we use (i) to reduce the powers of a_i modulo n_i, thereby arriving at an expression in normal form.

By a *presentation* of a group G, we mean the description of G in terms of generators and enough relations to reconstruct the multiplication table of G. Any group G can be presented by means of the generators $\{a : a \in G\}$ and obvious relations $\{ab = c : a, b \in G \text{ and } ab = c\}$; however, we are interested in finding efficient presentations, in terms of minimal sets of generators and relations. Actually we are treading on rather thin ice here, since it is not guaranteed that G can be described "effectively" through generators and relations, in the sense that there may be no procedure to check when two elements are equal! Nevertheless, many groups are described best in terms of generators and their relations. The notion of "generating" can be described intrinsically (*cf.* Exercises 18ff), but we shall rely mainly on Definition 6.5 and our intuition.

Generation of a non-Abelian group is often much more subtle than generation of an Abelian group; although $|S_n| = n!$, which grows rapidly as n increases, S_n is generated by two elements having respective orders 2 and n, *cf.* Exercise 5.8. Nevertheless, the non-Abelian groups G of immediate interest to us will be described explicitly in terms of two generators a, b such that $G = \langle a \rangle \langle b \rangle$, *i.e.*, $G = \{a^i b^j : 0 \le i < u,\ 0 \le j < v\}$ for suitable u, v (which is impossible for S_n, $n \ge 6$, *cf.* Exercise 11). Also these groups will be defined by rather few relations.

Remark 1. Suppose G is generated by $\{a, b\}$.

(i) If a, b commute then G is Abelian.

(ii) If $N < G$ and $aNa^{-1} = N$ and $bNb^{-1} = N$, then $N \triangleleft G$.

(iii) If $ba^ib^{-1} \in \langle a^i \rangle$, then $\langle a^i \rangle \triangleleft G$, by (ii).

(iv) If $\varphi: G \to H$ is an onto group homomorphism then $\{\varphi(a), \varphi(b)\}$ generates H.

Description of Groups of Low Order

Let us start by describing all groups of order 6.

Example 2. Suppose $|G|$ is an arbitrary group of order 6. By Cauchy's theorem, G contains an element a of order 3 and an element b of order 2. Then $G = \langle a \rangle \langle b \rangle$ by Remark 6.12 (*i.e.*, $|\langle a \rangle \langle b \rangle| = \frac{3 \cdot 2}{1} = 6$). The information obtained so far is given by the two obvious relations

$$a^3 = e; \tag{1}$$

$$b^2 = e. \tag{2}$$

In order to foster a new meaningful relation, note that $\langle a \rangle$ has index 2 in G and thus is a normal subgroup, by Proposition 5.9. Hence $bab^{-1} \in \langle a \rangle = \{e, a, a^2\}$. But $\mathrm{o}(bab^{-1}) = \mathrm{o}(a) = 3$, implying $bab^{-1} \ne e$. Thus we have two possibilities (noting $a^2 = a^{-1}$):

$$bab^{-1} = a, \text{ i.e., } ba = ab; \tag{3'}$$

$$bab^{-1} = a^{-1}, \text{ i.e., } ba = a^{-1}b. \tag{3''}$$

By Example 0, relations (1),(2), and (3′) define $\mathbb{Z}_3 \times \mathbb{Z}_2 \approx \mathbb{Z}_6$, so we need only concern ourselves with (3″), which implies

$$\begin{aligned}(a^ib^j)(a^ub^v) &= a^ib^{j-1}(ba)a^{u-1}b^v = a^ib^{j-1}a^{-1}ba^{u-1}b^v \\ &= a^ib^{j-1}(a^{-1})^ub^{1+v} = a^ib^{j-1}a^{-u}b^{1+v} \\ &= a^ib^{j-2}a^ub^{2+v} = \cdots = a^{i+(-1)^ju}b^{j+v}\end{aligned}$$

Then we can use (1) and (2) to reduce $i + (-1)^j u$ to its residue mod 3, and $j + v$ to its residue mod 2. Since we have constructed the complete multiplication table, we have proved

Remark 3. There are at most two nonisomorphic groups of order 6, each generated by two elements; one group is $\mathbb{Z}_3 \times \mathbb{Z}_2$, given by relations (1), (2), (3′), and the other is given by (1), (2), (3″).

In order to push this analysis further we need an obvious observation.

Example 3′. In Example 2, the group corresponding to (3″) is not Abelian (since $ba = a^2b \neq ab$). But S_3 is non-Abelian. We conclude:

Relations (1), (2), (3′) define $\mathbb{Z}_6 \approx \mathbb{Z}_2 \times \mathbb{Z}_3$.
Relations (1), (2), (3″) define S_3.

Thus all groups of order 6 are isomorphic to $\mathbb{Z}_2 \times \mathbb{Z}_3$ or S_3.

A general principle in mathematical discovery is that good ideas go a long way. Perhaps we can generalize Example 3″ to handle degrees other than $6 = 2 \cdot 3$. Note that the reasoning that (1), (2), and (3) determine the multiplication of G also yields the following more general observation:

Remark 4. Suppose G is generated by elements a and b such that $\langle a \rangle \triangleleft G$.

(i) $G = \langle a \rangle \langle b \rangle$. Indeed, $bab^{-1} \in \langle a \rangle$, so $bab^{-1} = a^i$ for some i. But then $ba = a^i b$, so that we can always reduce an expression in a and b to the form $a^u b^v$ for suitable u, v; namely, we replace any occurrence ba by $a^i b$, thereby eventually moving all occurrences of b to the right of all occurrences of a.

(ii) $|\langle a \rangle \cap \langle b \rangle| = \frac{o(a)o(b)}{|G|}$, seen from $|G| = \frac{|\langle a \rangle|\,|\langle b \rangle|}{|\langle a \rangle \cap \langle b \rangle|}$.

(iii) $\langle a \rangle \cap \langle b \rangle = \langle a^{|G|/o(b)} \rangle = \langle b^{|G|/o(a)} \rangle$. Indeed,

$$|G| = \frac{|\langle a \rangle||\langle b \rangle|}{|\langle a \rangle \cap \langle b \rangle|} = \frac{o(a)o(b)}{|\langle a \rangle \cap \langle b \rangle|};$$

in particular $\frac{|G|}{o(b)} = \frac{o(a)}{|\langle a \rangle \cap \langle b \rangle|}$, which divides $o(a)$. Hence, by Corollary 6.4, $\langle a \rangle \cap \langle b \rangle$ is the (unique) cyclic subgroup of $\langle a \rangle$ of order $\frac{o(a)o(b)}{|G|} = o(a^{|G|/o(b)})$, yielding the first equality; likewise, $\langle a \rangle \cap \langle b \rangle = \langle b^{|G|/o(a)} \rangle$.

Note that when $o(a)o(b) = |G|$, (iii) says $\langle a \rangle \cap \langle b \rangle = \{e\}$. Often we can make do with this easier special case.

(iv) Any element of G can be written uniquely in the form $a^i b^j$ for $0 \leq i < o(a)$, $0 \leq j < \frac{|G|}{o(b)}$. Indeed, we lower the power of b by means of (iii), since we can replace $b^{|G|/o(a)}$ by a suitable power of a. If $a^i b^j = a^{i'} b^{j'}$ for $0 \leq i, i' < o(a)$ and $0 \leq j, j' \leq \frac{|G|}{o(a)}$, then $a^{i-i'} = b^{j'-j} \in \langle a \rangle \cap \langle b \rangle = \langle b^{|G|/o(a)} \rangle$, so $j - j' = 0$ yielding $j = j'$; hence, $a^{i-i'} = e$, so $i - i' = 0$, yielding $i = i'$.

Definition 5. (n arbitrary.) The *dihedral group* D_n (if it exists) is the group generated by two elements, satisfying the relations

$$a^n = e, \quad b^2 = e, \quad bab^{-1} = a^{n-1}(= a^{-1}).$$

Of course, we have not yet proved that D_n is a group of order $2n$, but we assume it now, deferring the proof until Chapter 9. One can also prove this fact by direct verification, or geometrically (Exercise 2), or algebraically (Exercise 3).

Remark 6. $\langle a \rangle \triangleleft D_n$ by inspection, so Remark 4 shows that $D_n = \langle a \rangle \langle b \rangle$; also $\langle a \rangle \cap \langle b \rangle = \{e\}$, and every element of D_n can be written uniquely in the form $a^i b^j : 0 \leq i \leq n-1,\ 0 \leq j \leq 1$.

Now we can characterize groups of order $2p$, for any odd prime p.

Example 7. (Generalization of Example 2) If $|G| = 2p$ with p an odd prime, then $G \approx \mathbb{Z}_2 \times \mathbb{Z}_p \approx \mathbb{Z}_{2p}$ or $G \approx D_p$.

Indeed, by Cauchy's theorem, G contains an element a of order p and an element b of order 2. $[G : \langle a \rangle] = 2$, so $\langle a \rangle \triangleleft G$ by Proposition 5.9. Since p and 2 are relatively prime we see that $\langle a \rangle \cap \langle b \rangle = \{e\}$, so the discussion above yields $G = \{a^i b^j : 0 \leq i < p,\ 0 \leq j \leq 1\}$. On the other hand, a, b satisfy the relations

$$a^p = e, \tag{4}$$

$$b^2 = e, \tag{5}$$

$$bab^{-1} = a^{\mu} \quad \text{for suitable } \mu; \text{ i.e., } ba = a^{\mu} b. \tag{6}$$

As in Example 2, these relations determine the multiplication table and thus define the group up to isomorphism; furthermore we can use (5) to determine the possibilities for μ :

$$a = b^2 ab^{-2} = b(bab)^{-1} b^{-1} = ba^{\mu} b^{-1} = (bab^{-1})^{\mu} = (a^{\mu})^{\mu} = a^{\mu^2}$$

implying $a^{\mu^2 - 1} = e$, so $p = o(a) \mid \mu^2 - 1 = (\mu - 1)(\mu + 1)$. Thus p divides $\mu + 1$ or $\mu - 1$, so $\mu \equiv \pm 1 \pmod{p}$. Since $a^p = e$, we may take $\mu = \pm 1$. When $\mu = 1$ we have $bab^{-1} = a$, *i.e.*, $ba = ab$ and G is Abelian by Remark 1. When $\mu = -1$ we have $G \approx D_p$.

Before concluding this line of reasoning, let us describe an even further generalization of Example 2, which however requires a result we have not yet proved.

Example 8. Suppose $|G| = qp$ with $q < p$ prime numbers. G contains an element a of order p and an element b of order q, and $\langle a \rangle \triangleleft G$ by Corollary 5.7. Thus $G = \{a^i b^j : 0 \leq i < p,\ 0 \leq j < q\}$, and the generators a, b satisfy the relations

$$a^p = e, \tag{7}$$
$$b^q = e, \tag{8}$$
$$bab^{-1} = a^\mu \quad \text{for suitable} \quad 1 \leq \mu \leq p - 1. \tag{9}$$

As in Example 7, we can limit the possibilities for μ by noting

$$a = b^q a b^{-q} = b^{q-1}(bab^{-1})b^{-(q-1)} = \\ = b^{q-1} a^\mu b^{-(q-1)} = b^{q-2} a^{\mu^2} b^{-(q-2)} = \cdots = a^{\mu^q},$$

implying $\mu^q = 1$ in Euler(p). Since Euler(p) is cyclic (to be proved in Theorem 18.7), the subgroup $H = \{\mu : \mu^q = 1\}$ has order $d = (q, p-1)$, by Corollary 6.4 and Example 2.6′, so d is the number of solutions to $\mu^q \equiv 1 \pmod p$. Since q is prime, we see

$$d = 1 \quad \text{if} \quad q \nmid (p-1), \text{ in which case } \mu = 1 \text{ is the only solution;}$$
$$d = q \quad \text{if} \quad q | (p-1).$$

Of course, $\mu = 1$ corresponds to the Abelian group $\mathbb{Z}_q \times \mathbb{Z}_p \approx \mathbb{Z}_{qp}$. Thus, there exists another solution precisely when $q|(p-1)$, in which case the relations (7),(8),(9) define a new group (if it exists), called the *semidirect product* of $\mathbb{Z}_p$ and $\mathbb{Z}_q$. Actually we have $(q-1)$ possibilities, corresponding to the $q-1$ choices for $\mu \neq 1$, but in fact these all correspond to isomorphic groups, as indicated by the following argument:

Suppose G is a group corresponding to a particular relation $bab^{-1} = a^\mu$. Replacing b by $b' = b^j$ for any $1 \leq j < q$, we note that $\{a, b'\}$ generates G, and o$(b')= q$, but now $b'a(b')^{-1} = a^{\mu^j}$. Clearly, $\mu, \mu^2, \ldots, \mu^{q-1}$ are distinct mod q since o$(\mu) = q$ (in Euler (p)), so each of the different solutions $\mu \neq 1$ of (9) provides a different presentation of the same group G. (To formulate this result precisely, see the addendum and Exercise 23.)

In Exercise 6 we obtain a concrete realization of the semidirect product, and thereby see that it exists. The semidirect product will be put into a much wider context in Exercise 12.26.

Example 9. Every group of order 15 is isomorphic to $\mathbb{Z}_{15} \approx \mathbb{Z}_3 \times \mathbb{Z}_5$, since 3 does not divide $5 - 1 = 4$.

Our final example is the classification of groups of order 8. This will display the techniques we have developed until now.

Example 10. Suppose $|G| = 8$. If G is Abelian, then G is isomorphic to $\mathbb{Z}_8$, $\mathbb{Z}_4 \times \mathbb{Z}_2$, or $\mathbb{Z}_2 \times \mathbb{Z}_2 \times \mathbb{Z}_2$, by the fundamental theorem of finite Abelian groups. Thus, we shall assume G is not Abelian; in particular, $\exp(G) \neq 2$ (by Proposition 3.5) and $\exp(G) \neq 8$ (since otherwise G is cyclic). Hence $\exp(G) = 4$. Take $a \in G$ of order 4; then $\langle a \rangle$ has index 2 in G, and thus is normal. Take any $b \in G \setminus \langle a \rangle$. Then $bab^{-1} \in \langle a \rangle$. But $o(bab^{-1}) = o(a) = 4$, so $bab^{-1} \in \{a, a^3\}$. Furthermore, $|\langle a \rangle \langle b \rangle|$ is a power of 2 greater than 4, and so equals $8 = |G|$, *i.e.*, $G = \langle a \rangle \langle b \rangle$. Since G is non-Abelian, $bab^{-1} \neq a$, so we conclude $bab^{-1} = a^3$. Thus far we know the generators a and b satisfy the relations $a^4 = e$ and $bab^{-1} = a^3$, and next we want to determine the order of b. Clearly, there are two possibilities: 2 and 4. If $o(b) = 2$ then $G \approx D_8$, so we consider the remaining choice, $o(b) = 4$. By Remark 4 (iii), $\{e, a^2\} = \langle a \rangle \cap \langle b \rangle = \{e, b^2\}$ and thus $a^2 = b^2$.

Let us give this potential new group a name.

Definition 11. The *quaternion group* Q is the group of order 8, having generators a, b satisfying the relations $a^4 = b^4 = e$, $a^2 = b^2$, and $bab^{-1} = a^3$.

It is easy to build the multiplication table from these relations; Q is indeed a group (Exercise 9), and $Q = \{a^i b^j : 0 \leq i < 4, \ \ 0 \leq j \leq \ 1\}$ by Remark 4(iv), but we must wait until appendix B to see how Q derives its name. D_4 and Q both have exponent 4 and share many other properties in common. However, the number of elements of order 2 differs (*cf.* Exercises 1 and 8). We summarize our findings as

Remark 12. Any group of order 8 is isomorphic to one of the following (nonisomorphic) groups: $\mathbb{Z}_8$, $\mathbb{Z}_4 \times \mathbb{Z}_2$, $\mathbb{Z}_2 \times \mathbb{Z}_2 \times \mathbb{Z}_2$, D_4, or Q.

Combining Example 2 and Remark 12 with Corollary 6.16, we have classified all groups of order ≤ 8. Furthermore, we have classified all groups of order pq for $p \neq q$ prime, in Example 8. Thus the only remaining case for $|G|$ less than 12 is 9, to be dealt with in Example 9.9.

Addendum: Erasing Relations

Our point of view until now has been to start with a given group and determine its relations. This treatment bypassed the difficult question of exactly what a relation is, and we should like to consider this question, even though it removes us from the realm of finite groups. To gain intuition, let us see what happens when we start erasing relations. Perhaps the most natural thing to erase is the n from D_n.

Example 13. The *infinite dihedral* group D is defined as having generators a, b satisfying the two relations $b^2 = e$ and $bab^{-1} = a^{-1}$. Thus $D = \{a^i b^j : i \in \mathbb{Z},\ 0 \le j \le 1\}$. (Note that D is infinite.) As in Example 2 we see

$$(a^i b^j)(a^u b^v) = a^{i+(-1)^j u} b^{j+v},$$

implying for any n that the map $\varphi: D \to D_n$ sending $a^i b^j$ to its value in D_n is indeed a homomorphism, for

$$\varphi((a^i b^j)(a^u b^v)) = \varphi(a^{i+(-1)^j u} b^{j+v}) = \varphi(a^i b^j)\varphi(a^u b^v).$$

φ is onto, by inspection; $\ker \varphi = \{a^i b^j : i \equiv 0 \pmod n \text{ and } j = 0\} = \langle a^n \rangle$. Actually, we have not yet proved D is a group. This can be seen directly, *cf.* Exercises 12 and 13, but we can avoid many difficulties by erasing all the relations, thereby obtaining a group free of all relations. Ironically, the easiest method of constructing this group is by returning to monoids.

Example 14. The *free monoid* on a set $S = \{s_1, s_2, \dots\}$ is the set of "words" (*i.e.*, strings of elements of S) including the "blank word" (which has no letters in it); the operation is juxtaposition of words, *e.g.*, $(s_1 s_5)(s_2 s_1 s_4) = s_1 s_5 s_2 s_1 s_4$, and the neutral element is the blank word. Now take a disjoint copy $S' = \{s_1', s_2', \dots\}$ of S, and build the free monoid M on $S \cup S'$. We say two words in M are equivalent if we can obtain one from the other by successive insertions and/or deletions of various $s_i s_i'$ or $s_i' s_i$. For Example, $s_1 s_3 s_3' s_2$ is equivalent to $s_1 s_2 s_5' s_5$, since first we delete $s_3 s_3'$ and then insert $s_5' s_5$ at the end. Write $[h]$ for the equivalence class of the word h. Clearly, the equivalence classes of M form a monoid $\mathcal{G}(S)$ (also under juxtaposition), which is a group since we can obtain the inverse by switching s_i and s_i' and then reversing the order of the word. For example, the inverse of $[s_1 s_2' s_4]$ is $[s_4' s_2 s_1']$ since

$$[s_1 s_2' s_4 s_4' s_2 s_1'] = [s_1 s_2' s_2 s_1'] = [s_1 s_1'] = 1$$

(writing 1 for the equivalence class of the blank word) and, likewise,

$$[s_4' s_2 s_1' s_1 s_2' s_4] = [s_4' s_2 s_2' s_4] = [s_4' s_4] = 1.$$

In particular, $[s_i'] = [s_i]^{-1}$.

Remark 15. $\mathcal{G}(S)$ is called the *free group* on $|S|$ generators and satisfies the important property that for any group G and any set $\{a_1, a_2, \dots\}$ in G of the same cardinality as S, there is a group homomorphism $\psi: \mathcal{G}(S) \to G$ given by $[s_i] \mapsto a_i$. (See Exercise 15 for the proof.)

Now suppose $\{a_1, a_2, \dots\}$ is a generating set for G. The relations of this set are precisely the elements of $\ker \psi$, and so our quest for a minimal set of relations is simply a search for a minimal set A of elements in the free group, for which $\ker \psi$ is the smallest normal subgroup containing A.

Now we look at the coin from the other side. For any set A of words in $\mathcal{G}(S)$, let N be the intersection of all normal subgroups of $\mathcal{G}(S)$ that contain A. Then $N \triangleleft \mathcal{G}(S)$, and $\mathcal{G}(S)/N$ is the group defined by generators S and relations A, *cf.* Exercises 21 and 22. Thus any set of generators determines a group.

As important as this construction is, there are several shortcomings. Given a set of words A there is no comprehensive method of determining whether $\mathcal{G}(S)/N$ is a finite group (or even whether $\mathcal{G}(S)/N$ is trivial!). Indeed, it is impossible to find an algorithm that will always determine (for arbitrary A) when two given elements of $\mathcal{G}(S)$ have the same image in $\mathcal{G}(S)/N$; in technical language, the word problem is undecidable in general.

Also the structure of $\mathcal{G}(S)$ is considerably more complicated than the structure of the finite groups that led us to $\mathcal{G}(S)$. Nevertheless $\mathcal{G}(S)$ possesses certain interesting properties of its own. For example, one important theorem outside the scope of these notes is that any subgroup of a free group is free (but not on the same number of generators, *cf.* Exercise 17!).

Exercises

1. $o(a^i b) = 2$ in D_n, for every i. Conclude that $\exp(D_n) = 2n$ for n odd, and n for n even.
2. D_n is a group; and $|D_n| = 2n$. (*Hint*: Divide the circle into n equal parts, starting at the right. Label these points $0, 1, \dots, n-1$, and define the transformation a that rotates the circle from 0 to 1 (and thus from 1 to 2, and so on). Define the transformation b that flips the circle along the horizontal axis, *i.e.*, $1 \mapsto n-1$, $2 \mapsto n-2$, and so on). Show $bab^{-1} = a^{-1}$, *i.e.*, "flip rotate flip" corresponds to rotating the circle in the opposite direction. Thus the group generated by these rigid transformations corresponds to D_n.)

   One can show in fact that any finite group of symmetries of the circle has a cyclic subgroup of index 2 and thus is isomorphic to some D_n or $\mathbb{Z}_n$. The same general approach can be used in studying the finite subgroups of the symmetries of the sphere. What is the group of rigid transformations of the triangular pyramid? of the cube? See Weyl's book, *Symmetries*, Princeton University Press, for a beautiful treatment of symmetry in nature and art.

3. Identify D_n with the subgroup of S_n generated by the two permutations $(1\ n-1)(2\ n-2)(3\ n-3)\dots$ and $(1\ 2\ 3\dots n)$.
4. Any subgroup of D_n is either of the form $\mathbb{Z}_m$ or D_m (where $m|n$). (*Hint*: Exercise 1.)
5. Define the semidirect product of $\mathbb{Z}_p$ and $\mathbb{Z}_q$, in general, for any two natural numbers p, q such that q divides $\varphi(p)$.
6. Suppose p is a prime number, and $q|(p-1)$. The semidirect product of $\mathbb{Z}_p$ and $\mathbb{Z}_q$ can be identified with the subgroup of S_p generated by $a = (1\ 2\ \dots p)$ and $b = C_1 \dots C_{(p-1)/q}$, where the C_i are disjoint cycles of length q. (*Hint*: Take μ satisfying $\mu^q \equiv 1 \pmod{p}$, and $C_i = (j\ \mu j\ \mu^2 j\ \dots\ \mu^{q-1} j)$ for $1 \le j \le p-1$, throwing away repetitions.) Generalize this to the case p is not prime (*cf.* Exercise 5).
7. Suppose q is a prime number that does not divide n. There is a non-Abelian group of order nq having a normal cyclic subgroup of order n, iff q divides $\varphi(n)$.
8. Every element of Q other than e and $a^2 (= b^2)$ has order 4. Conclude that D_4 and Q are not isomorphic.
9. Q is a group. (*Hint*: Identify Q with the multiplicative subgroup of $M_2(\mathbb{C})$ generated by $a = \begin{pmatrix} i & 0 \\ 0 & -i \end{pmatrix}$ and $b = \begin{pmatrix} 0 & 1 \\ -1 & 0 \end{pmatrix}$.)
10. Generalize the quaternion group to define a non-Abelian group of order p^3 and exponent p^2, for any prime p.
11. Show that one cannot find $\sigma, \tau \in S_4$ such that $\langle\sigma\rangle\langle\tau\rangle = S_4$. (*Hint*: $|\langle\sigma\rangle||\langle\tau\rangle| \le 16$.)
12. Verify associativity in the infinite dihedral group D, namely that

$$((a^i b^j)(a^{i'} b^{j'}))(a^{i''} b^{j''}) = (a^i b^j)((a^{i'} b^{j'})(a^{i''} b^{j''})).$$

13. Avoid computation in Exercise 12 by noting that one need merely check it in D_n for n large enough, but D_n is already known to be a group. (This argument illustrates a general principle in mathematical logic.)
14. Show that the free monoid $\mathcal{M}(S)$ is free, insofar as for any monoid and any subset $\{a_1, a_2, \dots\}$ of M there is a monoid homomorphism $\mathcal{M}(S) \to M$ given by $s_{i_1} s_{i_2} \cdots \mapsto a_{i_1} a_{i_2} \cdots$.
15. Prove Remark 15. (*Hint*: In Example 14 take S to be a group; given $\{a_1, a_2, \dots\} \subseteq S$, define $\phi : \mathcal{M}(S) \to M$ by $s_i \mapsto a_i$ and $s_i' \mapsto a_i^{-1}$. Now $s_i s_i', s_i' s_i \in \ker \phi$, so ϕ induces a monoid homomorphism $\phi : \mathcal{M}(S) \to S$, and any monoid homomorphism of groups is a group homomorphism.)
16. The free group on 1 generator is isomorphic to $\mathbb{Z}$.

17. The free group on 2 generators contains a subgroup isomorphic to the free group on an arbitrary countable number of generators. (*Hint*: Consider $\{s_1 s_2^i s_1 : i \in \mathbb{Z}\}$.)

Explicit Generation of Groups by Arbitrary Subsets

18. Given any subset S of G, define the subgroup $\langle S \rangle$ *generated by* S, by the following inductive procedure:

$$(i) \quad S(1) = S;$$

$$(ii) \quad S(i+1) = S(i) \cup \{gh^{-1} : g, h \in S(i)\};$$

$$(iii) \quad \langle S \rangle = \cup_{i \geq 1} S(i) \quad (\text{in } G).$$

$\langle S \rangle < G$, for if $g, h \in \langle S \rangle$, then $g, h \in S(i)$ for some i, implying $gh^{-1} \in S(i+1) \subseteq \langle S \rangle$. On the other hand every subgroup of G containing S contains $\langle S \rangle$.

19. $\langle S \rangle$ is the subgroup generated by S, in the sense of Definition 6.5.
20. If all the generators of a group G commute with each other then G is Abelian. (*Hint*: induction using Exercise 18.)
21. If $N \triangleleft G$, then G/N satisfies all the relations of G. Conclude that $\mathcal{G}(S)$ satisfies the fewest relations of any group generated by S.
22. Suppose G is a group with generators S and relations A, and N is the intersection of all normal subgroups of $\mathcal{G}(S)$ containing A. Show there is a surjection $\mathcal{G}(S)/N \to G$. If this map is not an isomorphism, then G satisfies an extra relation not satisfied by $\mathcal{G}(S)/N$ (and thus not "implied" by A).
23. Use Exercise 22 to describe the semidirect product of $\mathbb{Z}_p$ and $\mathbb{Z}_q$ in the form $\mathcal{G}(S)/N$, and thereby verify formally the assertion made in the text that any two groups obtained in this way (for different values of μ) are isomorphic.

CHAPTER 9. WHEN IS A GROUP A GROUP? (CAYLEY'S THEOREM)

We have just constructed several new candidates for groups, *e.g.*, certain semidirect products and the quaternion group, and have shown by various *ad hoc* arguments in the exercises that they are indeed groups. Several of these proofs involved displaying the proposed group as a subgroup of S_n (*cf.* Exercises 8.3 and 8.6). We would like to see why these proofs work, and in doing so shall develop a general procedure that is guaranteed to work for any group.

First we must understand what is meant by "displaying as a subgroup." Philosophically, by "subset of S" we really mean a 1:1 map into S. Indeed, "$2\mathbb{Z} \subseteq \mathbb{Z}$" is a lazy way of saying there is an injection $2\mathbb{Z} \to \mathbb{Z}$ given by $2n \mapsto 2n$. Along the same lines, suppose H is a set with a given operation $\cdot$. Then, by "displaying H as a subgroup of G," we mean finding a 1:1 map $\varphi: H \to G$ such that

$$\varphi(h_1 \cdot h_2) = \varphi(h_1)\varphi(h_2)$$

with $\varphi(H) < G$; indeed, we shall see that H is a group because we can translate all the group axioms from $\varphi(H)$ to H. Our main result will display arbitrary groups of order n as subgroups of S_n.

Before proceeding, it is worth remarking that many good ideas in algebra are inspired by the terminology at hand. For example, many functions arising naturally between groups are homomorphisms.

In Remark 1.11 we saw that the left multiplication map $\ell_a: G \to G$ given by

$$\ell_a(g) = ag$$

is 1:1 and onto, for any fixed element a of G. Although ℓ_a itself is not a homomorphism, we do have the group $A(G)$ of 1:1 correspondences from G onto itself, *cf.* Example 1.5(5), and we have a function $\varphi: G \to A(G)$ given by $\varphi(a) = \ell_a$. The key idea is that φ is a homomorphism.

THEOREM 1. *(Cayley's Theorem.) Every group G is isomorphic to a subgroup of $A(G)$. In particular, every finite group of order n is isomorphic to a subgroup of S_n.*

Proof. Since $A(G)$ is identified with S_n when $|G| = n$, the second assertion follows from the first; thus, it is enough $\varphi: G \to A(G)$ given by $a \mapsto \ell_a$ is a group injection. To check $\ell_{ab} = \ell_a\ell_b$ we note

$$\ell_{ab}g = abg = a(bg) = a\ell_b g = \ell_a\ell_b g$$

for all g in G. Thus, φ is a homomorphism. Finally, $a \in \ker\varphi$ iff ℓ_a is the identity, implying $e = \ell_a e = a$, thereby proving $\ker\varphi = \{e\}$. □

Let us refine this argument, to check when a finite set G is a group under a given binary operation. We need to verify associativity and the existence of the neutral element and inverses. Once we have candidates for the neutral element and inverses, these are very easy to check by direct computation. On the other hand, associativity is difficult to verify directly, since it involves n^3 verifications, where $n = |G|$.

At this point, we find ourselves in the rather interesting situation that, although Theorem 1 itself is inapplicable (since G is not yet known to be a group), its statement and proof provide the guidance we need. If we knew G is a group we would have the natural injection $G \to S_n$ given by $a \mapsto \ell_a$; so why not exploit this map directly?

Suppose, more generally, we can succeed in finding some 1:1 function φ from G to a group $\mathcal{G}$, which preserves the operation, *i.e.*, $\varphi(ab) = \varphi(a)\varphi(b)$. Then associativity in G is instantaneous, since

$$\varphi((ab)c) = \varphi(ab)\varphi(c) = (\varphi(a)\varphi(b))\varphi(c) = \varphi(a)(\varphi(b)\varphi(c)) = \varphi(a(bc))$$

Likewise if $m = \mathrm{o}(\varphi(a))$, then $\varphi(a^m g) = \varphi(g) = \varphi(ga^m)$ for all g in G, implying a^m is the neutral element of G, and thus $a^{-1} = a^{m-1}$, so we have proved G is a group. Explicitly, we state

THEOREM 2. *Suppose G is a set with a binary operation, and the map $\varphi: G \to A(G)$ (given by $a \mapsto \ell_a$) is 1:1 and preserves the operation. Then G is a group, and φ is a group injection.*

Remark 2′. If $a \in G$ and $m = \mathrm{o}(a)$, then for any g in G the elements $g, ag, \ldots, a^{m-1}g$ are distinct; consequently the permutation corresponding to ℓ_a is a product of $\frac{|G|}{m}$ disjoint cycles, each of length m.

Example 3. The quaternion group Q. Recall Q has generators a and b such that $a^4 = b^4 = e$, $bab^{-1} = a^3$, and $a^2 = b^2$. Label the elements $e, a, a^2, a^3, b, ab, a^2b, a^3b$ of Q respectively by $1, 2, 3, 4, 5, 6, 7, 8$. To determine which permutations correspond to ℓ_a and ℓ_b we construct part of the multiplication table

	1	2	3	4	5	6	7	8
	e	a	a^2	a^3	b	ab	a^2b	a^3b
a	a	a^2	a^3	e	ab	a^2b	a^3b	b
b	b	$ba = a^3b$	$ba^2 = a^2b$	$ba^3 = ab$	$b^2 = a^2$	$bab = a$	e	a^3

The map ℓ_a sends $e \mapsto a$, $a \mapsto a^2$, $a^2 \mapsto a^3$, $a^3 \mapsto e$, $b \mapsto ab$, and so on. Translating this to the labels of the elements we see $\ell_a \leftrightarrow (1234)(5678)$

and likewise $\ell_b \leftrightarrow (1537)(2846)$ in S_n. Thus, letting $\sigma = (1234)(5678)$ and $\tau = (1537)(2846)$ in S_n, we define $\varphi: Q \to S_n$ by $\varphi(a^i b^j) = \sigma^i \tau^j$. To apply Theorem 2 we must verify that σ, τ satisfy the same relations as a, b, so that the multiplication tables are the same (and thus φ preserves multiplication).

$o(\sigma) = o(\tau) = 4$ since σ, τ are each products of disjoint cycles of length 4. $\sigma^2 = (13)(24)(57)(68) = (13)(57)(24)(68) = \tau^2$, and Remark 5.20 yields

$$\tau\sigma\tau^{-1} = (5876)(3214) = \sigma^{-1}.$$

Finally, we need to show that φ is 1:1. So suppose $\varphi(a^i b^j) = \varphi(a^{i'} b^{j'})$ for $0 \le i, i' \le 3$ and $0 \le j, j' \le 1$. Then $\sigma^{i-i'} = \tau^{j'-j} \in \langle\sigma\rangle \cap \langle\tau\rangle = \langle\tau^2\rangle$, so $2|(j'-j)$ implying $j = j'$, and thus $i = i'$.

Note. Of course, one would want a quicker, more intuitive way of determining the permutations σ and τ corresponding to a and b. The idea is to use Remark 5.20 judiciously. In view of Remark 2′, σ is a product of two disjoint cycles of length 4. Since we may replace σ by any conjugate (if we replace τ by its corresponding conjugate), we may assume that $\sigma = (1234)(5678)$. Then $\tau^2 = \sigma^2 = (13)(57)(24)(68)$, so $\tau = (1\ a_1\ 3\ a_2)(b_1\ c_1\ b_2\ c_2)$, where $\{a_1, a_2\}, \{b_1, b_2\}$, and $\{c_1, c_2\}$ are $\{2,4\}$, $\{5,7\}$, and $\{6,8\}$ (but not necessarily in that order). But also $\tau\sigma\tau^{-1} = \sigma^{-1} = (8765)(4321)$, which is impossible if $\{a_1, a_2\} = \{2,4\}$. Thus, we may assume $\tau = (1\ ?\ 3\ ?)(2\ ?\ 4\ ?)$. Noting $(8765)(4321) = (5876)(3214)$, we can apply Remark 5.20 and take $\tau = \begin{pmatrix} 1&2&3&4&5&6&7&8 \\ 5&8&7&6&3&2&1&4 \end{pmatrix} = (1537)(2846)$.

Let us try a similar approach for D_4. Now we want $\tau^2 = (1)$ and $\tau\sigma\tau^{-1} = \sigma^{-1} = (8765)(4321)$, so by inspection we take

$$\tau = \begin{pmatrix} 1&2&3&4&5&6&7&8 \\ 8&7&6&5&4&3&2&1 \end{pmatrix} = (18)(27)(36)(45).$$

Generalized Cayley's Theorem

We could apply our method at once to other groups, but would like first to make a slight adjustment to make the method more powerful, and inject a group G into S_n for suitable $n < |G|$. Take any $H < G$. Then ℓ_a acts on the left cosets $\{gH : g \in G\}$, by $\ell_a(gH) = agH$. Let $S = \{gH : g \in G\}$, whose order is $[G : H]$. Identifying $A(S)$ with S_k where $k = [G : H]$, we have

THEOREM 4. *Suppose* $[G : H] = k$. *Then there is a homomorphism* $\varphi: G \to S_k$ *given by* $a \mapsto \ell_a$, *and* $\ker\varphi \subseteq H$. *More precisely,* $a \in \ker\varphi$ *iff every conjugate of* a *is in* H.

Proof. (As in Theorem 1.) Each ℓ_a permutes the left cosets of H, and

$$\ell_{ab}gH = (ab)gH = a(b(gH)) = \ell_a\ell_b gH,$$

proving φ is a homomorphism. Obviously, $\ker\varphi \triangleleft G$. Also, $a \in \ker\varphi$ iff $agH = gH$ for all g in G, *i.e.*, $g^{-1}ag \in H$. □

Thus, $\ker\varphi$ is a normal subgroup of G contained in H. One way of making sure φ is an injection is by choosing H not to contain nontrivial normal subgroups of G. We cannot improve our previous argument for Q, since every nontrivial subgroup contains $\langle a^2 \rangle \triangleleft Q$. However, the improved method is applicable for D_n for each $n > 2$; we can take $H = \langle b \rangle$, since $aba^{-1} = a^2b \neq b$.

Example 5. D_n. Recall D_n has generators a, b satisfying $a^n = b^2 = e$ and $bab^{-1} = a^{n-1}$; taking $H = \langle b \rangle$ (of index $\frac{2n}{2} = n$ in D_n) we label the cosets $H, aH, \ldots, a^{n-1}H$ as $1, 2, \ldots, n$, and construct the multiplication table:

	1	2	3	...	$n-1$	n
coset	H	aH	a^2H	...	$a^{n-2}H$	$a^{n-1}H$
a	aH	a^2H	a^3H	...	$a^{n-1}H$	H
b	$bH = H$	$baH = a^{n-1}H$	$ba^2H = a^{n-2}H$	...	a^2H	aH

So $\ell_a \leftrightarrow (1\ 2 \ldots\ n)$ and $\ell_b \leftrightarrow (2\ n)(3\ n-1)(4\ n-2)\ldots$. Obviously, $o(\ell_a) = n$ and $o(\ell_b) = 2$, and $\ell_b\ell_a\ell_b^{-1} \leftrightarrow (1\ n\ n-1\ \ldots\ 2)$, so $\ell_b\ell_a\ell_b^{-1} = \ell_a^{-1}$. We have displayed D_n as a subgroup of S_n.

Although Theorem 4 inspires us to search for nonnormal subgroups of G in order to find injections into S_k, we can also apply the contrapositive and obtain normal subgroups of G.

COROLLARY 6. *Suppose $H < G$ with $|G| = n$ and $[G : H] = k$, and let $d = \gcd(k!, n)$. Then H contains a normal subgroup N of G, with $[G : N]$ dividing d. In particular, N is nontrivial if $n \nmid k!$*

Proof. In Theorem 4 take $N = \ker\varphi \leq H$, and let $t = n/|\ker\varphi|$. By Lagrange's theorem $t = [G : N] = |\varphi(G)|$ divides $|S_k| = k!$; but also $t|n$, so t divides d. □

COROLLARY 7. *Suppose p is the smallest prime number dividing $|G|$. If G has a subgroup H of index p, then $H \triangleleft G$.*

Proof. As in Corollary 6, take $d = \gcd(p!, |G|) = p$ (since $|G|$ has no prime factor dividing $(p-1)!$). Thus H contains a normal subgroup N of G of index dividing p, so $[H : N]$ is a proper divisor of $[G : N] = p$, implying $[H : N] = 1$, *i.e.*, $H = N$. □

However, G need not have any subgroup of index p; for example, we have seen that A_4 has no subgroup of index 2. Here is a cute application (to be generalized in Chapter 10).

COROLLARY 8. *Suppose p is prime. Any group G of order p^2 is isomorphic to $\mathbb{Z}_{p^2}$ or $\mathbb{Z}_p \times \mathbb{Z}_p$, and thus is Abelian.*

Proof. If $\exp(G) = p^2$, then G is cyclic, so we may assume $\exp(G) = p$. Take elements $a, b \neq e$, neither of which generates G, such that $a \notin \langle b \rangle$; then $|\langle a \rangle| = |\langle b \rangle| = p$, so $[G : \langle a \rangle] = [G : \langle b \rangle] = p^2/p = p$, implying $\langle a \rangle, \langle b \rangle \triangleleft G$. Hence $G \approx \langle a \rangle \times \langle b \rangle$ by proposition 6.13. □

Example 9. Coupled with our results of Chapter 8, this result completes the classification of all groups of order < 16 (except 12), as follows (listing only nonprime orders, since any group of prime order is cyclic):

Order	Group
4	$\mathbb{Z}_4,\ \mathbb{Z}_2 \times \mathbb{Z}_2$
6	$\mathbb{Z}_3 \times \mathbb{Z}_2,\ S_3$
8	$\mathbb{Z}_8,\ \mathbb{Z}_4 \times \mathbb{Z}_2,\ \mathbb{Z}_2 \times \mathbb{Z}_2 \times \mathbb{Z}_2,\ D_4,\ Q$
9	$\mathbb{Z}_9,\ \mathbb{Z}_3 \times \mathbb{Z}_3$
10	$\mathbb{Z}_5 \times \mathbb{Z}_2,\ D_5$
14	$\mathbb{Z}_7 \times \mathbb{Z}_2,\ D_7$
15	$\mathbb{Z}_5 \times \mathbb{Z}_3$

(Note $\mathbb{Z}_3 \times \mathbb{Z}_2 \approx \mathbb{Z}_6$; $\mathbb{Z}_5 \times \mathbb{Z}_2 \approx \mathbb{Z}_{10}$; $\mathbb{Z}_7 \times \mathbb{Z}_2 \approx \mathbb{Z}_{14}$; $\mathbb{Z}_5 \times \mathbb{Z}_3 \approx \mathbb{Z}_{15}$.)

Group Representations

Cayley's theorem and its generalization motivate us to define all finite groups directly as sets of permutations. (In fact, this "concrete" description historically preceded the abstract theory.) In principle, then, we could use information about S_n to tell us about arbitrary groups. Let us continue this process by injecting S_n into a well-known group.

Example 10. There is an injection $S_n \to \mathrm{GL}(n, F)$ (for any field F), which sends π to the "permutation matrix" having 1 in the $\pi i, i$ position for each i, and 0 everywhere else. Thus, any group G of order n can be viewed as a subgroup of $\mathrm{GL}(n, F)$ via the composition $G \to S_n \to \mathrm{GL}(n, F)$; this is called the *regular representation.* For example, there is a representation $Q \to M_2(\mathbb{C})$ given in Exercise 8.9.

In general, a group homomorphism $\varphi: G \to \mathrm{GL}(n, F)$ is called a *group representation* of *degree n (over F)*, also *cf.* Exercise 10.19. The advantage of studying G through group representations is that we can bring in the trace of a matrix as a powerful tool; *i.e.*, we obtain information about g in G by studying the trace of the matrix $\varphi(g)$. This powerful method, called *character theory,* lies outside the scope of this book.

Another useful application of Example 10 is obtained by taking $F = \mathbb{Z}_p$, since then $\mathrm{GL}(n, F)$ is a finite group that can be studied in terms of various known finite subgroups (such as the subgroups of diagonal matrices, of upper triangular matrices, and of upper triangular matrices having 1 on each diagonal entry, *cf.* Exercises 11.2, 11.3, 11.4.

Exercises

1. Given a monoid M, define the *opposite monoid* M^{op} to be the same set as M, but with multiplication in the opposite order. Show that M^{op} indeed is a monoid, and is a group if M is a group.
2. If G is a group, then G^{op} (as well as G) is isomorphic to a subgroup of $A(G)$. (*Hint*: Use the right multiplication map.)
3. If G is a group and $a, b \in G$ then each left multiplication map ℓ_a commutes with each right multiplication map r_b (in $A(G)$).
4. Show that there is an injection from $(\mathbb{Z}_6, +)$ to S_5.
5. $D_6 \approx S_3 \times \mathbb{Z}_2$. (*Hint*: Take $a^6 = e = b^2$ with $bab^{-1} = a^{-1}$. Then $\langle a^2, b\rangle \approx S_3$ and $\langle a^3 \rangle \approx \mathbb{Z}_2$.)
6. Show how Exercise 8.6 can be proved "intuitively," without knowing the permutations in advance.
7. If G is displayed as a subgroup of S_n, find a natural injection of $G \times G$ into S_{2n} (by applying Exercise 6.10) and, more generally, of $G \times \cdots \times G$ (taken m times) into S_{mn}.
8. Show that for any field F there is a group representation of D_4 given by $a \mapsto \begin{pmatrix} 0 & 1 \\ -1 & 0 \end{pmatrix}$ and $b \mapsto \begin{pmatrix} 1 & 0 \\ 0 & -1 \end{pmatrix}$.
9. Generalize Cayley's theorem to monoids: For any monoid S, left multiplication by any element s in S yields a map $\ell_s \colon S \to S$, thereby injecting S into $\mathrm{Map}(S, S)$, the set of functions from S to itself.

CHAPTER 10. RECOUNTING: CONJUGACY CLASSES AND THE CLASS FORMULA

Recall that b is a *conjugate* to a in a group G if $b = gag^{-1}$ for suitable g in G. Having seen already in Chapter 5 how conjugacy arises in determining when a subgroup is normal, we want to study conjugacy now in its own right. Sometimes it is useful to note that $g^{-1}ag = g^{-1}a(g^{-1})^{-1}$ is also conjugate to a.

Remark 1. Conjugacy is an equivalence relation. (Indeed we check reflexivity, symmetry, and transitivity: $g = ege^{-1}$; if $b = gag^{-1}$ then $a = g^{-1}bg$; if $b = gag^{-1}$ and $c = hbh^{-1}$, then $c = (hg)a(hg)^{-1}$.)

The equivalence class of a under this relation is called the *conjugacy class* of a. We usually designate a conjugacy class by one of its representatives.

Example 2. The conjugacy classes in S_n. By Remark 5.20, each conjugacy class of a permutation σ is determined by its placement of parentheses, writing σ as a product of disjoint cycles. The reader should try independently to list the conjugacy classes in S_4 before reading on. The big worry is to make sure that one has not forgotten some conjugacy class. But the number of possible ways of choosing a cycle of length t from m possible letters is $n(n-1)\dots(n-t+1)/t$; we divided by t since the same cycle can be started at any of its entries, *i.e.* $(1234) = (2341) = (3412) = (4123)$. Thus, the conjugacy classes of S_4 are

Conjugacy class	Number of elements
{(1234), (1243), (1324), (1342), (1423), (1432)}	6
{(123), (124), (132), (134), (142), (143), (234), (243)}	8
{(12)(34), (13)(24), (14)(23)}	3
{(12), (13), (14), (23), (24), (34)}	6
{(1)}	1
TOTAL	24

The total number of elements in all conjugacy classes is $24 = |S_4|$. It would be awkward to list all the elements of S_5 according to conjugacy class, so instead we make the following table:

Representative	Number of elements in class (determined combinatorically)
(12345)	$\frac{5!}{5} = 24$
(1234)	$\frac{5!}{4} = 30$
(123)(45)	$\frac{5\cdot 4\cdot 3}{3}\frac{2}{2} = 20$
(123)	$\frac{5\cdot 4\cdot 3}{3} = 20$
(12)(34)	$(\frac{5\cdot 4}{2}\frac{3\cdot 2}{2})/2 = 15$
(1)	1
	TOTAL 120

(In calculating the size of the class of (12)(34) we had to divide again by 2, because (12)(34) = (34)(12).)

In both of the examples above, the size of each conjugacy class divides the order of the group. Let us now consider the size of a conjugacy class in an arbitrary group. We call a conjugacy class *trivial* if it consists of exactly one element.

The Center of a Group

Definition 3. The *center* of a group G (denoted $Z(G)$) is $\{z \in G : gz = zg$ for all g in $G\}$.

For example, $Z(G) = G$ iff G is Abelian. On the other hand, $e \in Z(G)$ for every group G.

Remark 4. $z \in Z(G)$ iff $gzg^{-1} = z$ for all g in G, iff the conjugacy class of z is trivial.

Partitioning G into its disjoint conjugacy classes $C_1 \cup \cdots \cup C_t$, we thus have

$$|G| = \sum_{1 \le i \le t} |C_i| = |Z(G)| + \sum_{\text{nontrivial conjugacy classes}} |C_i| \qquad (1)$$

(1) is called the *Class Formula*, and, although obvious, can be made quite useful by means of a few observations. We need some method of computing the conjugacy class of a given element a in G. To this end we define the *centralizer* $C(a) = \{g \in G : ag = ga\}$.

Proposition 5. *The following facts hold for for any $a \in G$:*

(i) $C(a) \leq G$;
(ii) $a \in C(a)$;
(iii) $C(a) = G$ *iff* $a \in Z(G)$.

Proof. (i) Suppose $ag = ga$ and $ah = ha$. Then $a(gh) = gah = (gh)a$; furthermore

$$gg^{-1}a = a = agg^{-1} = gag^{-1},$$

implying $g^{-1}a = ag^{-1}$, *cf.*, $g^{-1} \in C(a)$.

(ii) and (iii) are immediate. □

Thus, we see $|C(a)| = |G|$ iff the conjugacy class of a is trivial, and we want to generalize this (to Proposition 7).

Lemma 6. *There is a 1:1 correspondence ϕ between the conjugates of a and the right cosets of $C(a)$, given by $\phi(g^{-1}ag) = C(a)g$.*

Proof. Clearly ϕ is onto, so we want to check ϕ is well-defined and 1:1. But $g^{-1}ag = h^{-1}ah$ iff $hg^{-1}agh^{-1} = a$, iff $gh^{-1} \in C(a)$, iff $C(a)g = C(a)h$. □

Proposition 7. *The number of conjugates of a is $[G : C(a)]$ (and thus divides $|G|$).*

Proof. Count each side in Lemma 6. □

Corollary 8. $|G| = |Z(G)| + \sum[G : C(a)]$, *where the right-hand sum is taken over representatives of the nontrivial conjugacy classes (and, in particular, each $[G : C(a)] > 1$).*

Corollary 9. *If G is a p-group, then $Z(G) \neq \{e\}$.*

Proof. Each term $[G : C(a)]$ of the right-hand summation in Corollary 8 is a power of p other than 1, and thus is a multiple of p. Hence p divides $\sum[G : C(a)]$ as well as $|G|$, so Corollary 8 implies p divides $|Z(G)|$. □

$Z(G)$ plays a special role, largely because of the following property (also see Exercise 1).

Proposition 10. $Z(G) \triangleleft G$.

Proof. Let $Z = Z(G)$. $Z < G$, as in the proof of Proposition 5(i) (or just by noting that $Z = \cap_{a \in G} C(a)$). Moreover, $Z \triangleleft G$, for if $z \in Z$ and $g \in G$ then $gzg^{-1} = z \in Z$. □

One application of the center is a strong converse of Lagrange's theorem for p-subgroups, *cf.* Exercise 5; related results are given in Sylow's Theorem (11.8(i)) and Frobenius' Theorem (Exercise 11.12). Let us determine more information concerning how $Z(G)$ sits inside G.

LEMMA 11. *$G/Z(G)$ cannot be a nontrivial cyclic group.*

Proof. Otherwise, letting $\bar{g}$ denote the canonical image in $G/Z(G)$ of any element g of G, we write $G/Z(G) = \langle \bar{a} \rangle$, for suitable a in G. Let $H = Z(G)\langle a \rangle$, an Abelian subgroup of G. For any g in G we have $\bar{g} = \overline{a^i}$ in $G/Z(G)$, for suitable i, implying $g = a^i z$ for suitable z in $Z(G)$; hence $g \in H$. Therefore $G = H$ is Abelian, so $Z(G) = G$, contradiction. □

Corollary 12. $[G : Z(G)]$ cannot be a prime number.

Example 13. We apply these techniques to p-groups G, where p is prime. Let $Z = Z(G)$.

(i) Reproof of Corollary 9.8. Suppose $|G| = p^2$. $|Z| \neq 1$ by Corollary 9, and $|Z| \neq p$ by Corollary 12. Thus, $|Z| = p^2$, so $Z = G$.

(ii) Suppose G is non-Abelian of order p^3. Then $|Z| = p$, and $G/Z \approx \mathbb{Z}_p \times \mathbb{Z}_p$, since G/Z is noncyclic of order p^2.

Groups Acting on Sets: A Recapitulation

The following recurrent theme has pervaded several of the more complicated proofs:

Definition 14. An *action* of a group G on a set S is a map $G \times S \to S$, sending (a, s) to $a \cdot s$, satisfying the following properties:

(i) $$e \cdot s = s \quad \text{for all } s \text{ in } S;$$

(ii) $$(ab) \cdot s = a \cdot (b \cdot s) \quad \text{for all } a, b \text{ in } G, \ s \text{ in } S.$$

Remark 15. Suppose the group G acts on a set S. Any a in G yields the left multiplication map $\ell_a : S \to S$, given by $\ell_a(s) = a \cdot s$. Then $a \mapsto \ell_a$ defines a group homomorphism $\phi : G \to A(S)$. Indeed ℓ_e is the identity, by (i), and $\ell_a^{-1} = \ell_{a^{-1}}$ by (ii).

As usual, the ideas of group theory guide us along the correct path. For any s in S we let $G_s = \{g \in G : g \cdot s = s\}$, the *stabilizer* of s, and we call $G \cdot s$ the *orbit* of s. Then, as in Lemma 6, we have

$$|G \cdot s| = [G : G_s] \tag{2}$$

Example 16. Let us see which group actions have occurred so far.

(1) The usual group multiplication in G defines an action of G on itself, which was the basis for the proof of Cayley's theorem (9.1).

(2) Theorem 9.4 provides an action of G on the left cosets of a given subgroup H, given by $a \cdot bH = (ab)H$.

(3) G acts on itself by conjugation, *cf.* $g \cdot a = gag^{-1}$. Proposition 7 then can be viewed as a consequence of (2).
(4) The *trivial* action is given by $g \cdot s = s$ for every s in S, all g in G.

Actually, Example 16(2) is quite general, *cf.* Exercise 14. Rather than recast our proofs in terms of group actions, we refer the interested reader to exercises 14ff.

Exercises

1. $\varphi(Z(G)) \subseteq Z(H)$, for any homomorphism $\varphi: G \to H$.
2. Suppose G is generated by elements $a_1, \dots, a_t$. Then $z \in Z(G)$ iff $za_i = a_i z$ for $1 \le i \le t$.
3. Using the usual presentations of Q and D_n in generators and relations, show $Z(Q) = \langle a^2 \rangle$; and $Z(D_n) = \{e\}$ for n odd, $\{e, a^{n/2}\}$ for n even. (*Hint:* $a^i b^j$ commutes with b iff $n|2i$.)
4. $Z(S_n) = \{e\}$ for $n > 2$; $Z(A_n) = \{e\}$ for $n > 3$. What happens for $n = 2, 3$?
5. If G is a p-group and m divides $|G|$, then G has a normal subgroup of order m. (*Hint:* $Z(G)$ has an element z of order p. By induction on m, $G/\langle z \rangle$ has a normal subgroup $H/\langle z \rangle$ of order $\frac{m}{p}$.)
6. If $G = S_n$ and $a = (1\ 2 \dots n)$ or $a = (1\ 2 \dots n-1)$, then $C(a) = \langle a \rangle$. (*Hint:* Example 2.)
7. What are the conjugacy classes in D_n? What are the conjugacy classes in A_n?
8. Determination of all non-Abelian groups of order p^3, p prime. Let $Z = Z(G)$. Then there are a, b in $G \setminus Z$ such that $a^p \in Z$, $b^p \in Z$, and $e \ne aba^{-1}b^{-1} \in Z$. Letting $z = aba^{-1}b^{-1}$ show G is generated by a, b and z. (Actually z is redundant, but is retained for convenience.)

   Case I. $\exp(G) = p$. The relations $a^p = b^p = z^p = e$, $zaz^{-1} = a$, $zbz^{-1} = b$, and $ab = zba$ permit a presentation of the group, for $p > 2$. (What goes wrong for $p = 2$?)

   Case II. $\exp(G) = p^2$. Then one may assume $o(a) = p^2$, so $\langle a \rangle \triangleleft G$ and $bab^{-1} = a^i$ where $i^p \equiv 1 \pmod{p^2}$; since $\varphi(p^2) = p(p-1)$ this has a solution $i \ne 1$, by Cauchy's theorem.

   Use Cayley's theorem to display these groups concretely.
9. There are two nonisomorphic finite groups G and H such that, for each n, G and H have the same number of elements of order n. (*Hint:* Use the non-Abelian group of exponent p, found in Exercise 8.) Compare with Exercise 7.2.
10. If $Z(G) = \{e\}$ and $|G| = n$, then S_n contains a subgroup isomorphic

to $G \times G^{\text{op}}$, cf., exercise 9.1 (*Hint*: The canonical copies of G and G^{op} in S_n intersect in $Z(G)$.)

Double Cosets

11. Given subgroups H, K of G and $g \in G$ define the *double coset* HgK to be $\{hgk : h \in H, k \in K\}$. Show that the double cosets (with respect to H, K) comprise a partition of G.
12. Notation as in Exercise 11, show for k_1, k_2 in K that $Hgk_1 = Hgk_2$ iff $k_1k_2^{-1} \in K \cap g^{-1}Hg$. Conclude that the number of ordinary cosets of H in HgK is $[K : K \cap g^{-1}Hg]$.
13. Prove the "double coset" version of the class formula, for any subgroups H, K of G:

$$[G : H] = \sum_{g \in G} [K : K \cap g^{-1}Hg].$$

Group Actions on Sets

(In these exercises, we suppress the $\cdot$ in the notation of the group action.) An action of G on a set S is said to be *transitive* if for every s_1, s_2 in S there is g in G such that $gs_1 = s_2$; more generally, the action is *t-fold transitive* if for any $s_1, \ldots, s_t, s'_1, \ldots, s'_t$ in S there is g in G such that $gs_1 = s'_1, \ \ldots, \ gs_t = s'_t$.

14. We say actions of G on sets S and T are *equivalent* if there is a 1:1 correspondence $\varphi: S \to T$ which "preserves" the action, *i.e.*, $\varphi(gs) = g\varphi(s)$ for all s in S. Any transitive group action of G on S is equivalent to the one of Example 16(2), where $H = G_{s_0}$ for any given s_0 in S. (*Hint*: Taking $T = \{\text{left cosets of } H\}$, define $\varphi: S \to T$ by $\varphi(s) = gH$, where g is taken such that $s = gs_0$.)

A *G-partition* of S is a disjoint union $S = S_1 \cup \cdots \cup S_t$ of nonempty subsets stable under the given action, *i.e.*, such that for each g in G, $gS_i = S_j$ for suitable j. Clearly, S has two "trivial" G-partitions:

(i) $t = 1$, and $S_1 = S$;

(ii) $t = |S|$, and writing $S = \{s_1, \ldots, s_t\}$, take each $S_i = \{s_i\}$.

We say the action is *primitive* if these are the only G-partitions. The next exercise shows that the definition depends only on S_1:

15. An action is *not* primitive iff S has a proper subset S' of order > 1, such that, for each g in G, either $gS' = S'$ or $gS' \cap S' = \emptyset$. (*Hint*: $\{gS' : g \in G\}$ is a G-partition.)
16. A transitive action is primitive iff G_s is a maximal subgroup of G, for any s in S. (*Hint*: By Exercise 14, one may assume the action is as in Example 16(2); apply Exercise 15)

17. Any two-fold transitive action is primitive. Conversely, if $H < G$ is not contained in the kernel of the action and if the action is primitive, then the action of H on S is either transitive or trivial.
18. If $H < G$ acts transitively on S, then for any s in S, one has $G = HG_s$.
19. Suppose $\varphi: G \to \mathrm{GL}(n, F)$ is a group representation. Viewing $M_n(F)$ as the set of linear transformations of an n-dimensional vector space V over F, one defines an action of G on V, by $g \cdot v = \varphi(g)(v)$, for g in G and v in V. Verify the following properties:

    (i) $g \cdot (v + w) = g \cdot v + g \cdot w$ for all g in G, v, w in V;

    (ii) $g \cdot (\alpha v) = \alpha g \cdot v$ for all g in G, α in F, v in V.

    A vector space V over F is called a *G-module* if there is an action of G on V satisfying (i) and (ii). Thus, every representation gives rise naturally to a G-module. On the other hand, given a G-module V that is n-dimensional over F, one can define $\varphi: G \to \mathrm{GL}(n, F)$ by taking $\varphi(g)$ to be the linear transformation $v \mapsto g \cdot v$. In this way conclude that there is a 1:1 correspondence between the group representations of G (over F) and the G-modules that are finite dimensional as vector spaces over F.
20. Two G-modules are isomorphic iff their G-actions are equivalent.

CHAPTER 11. SYLOW SUBGROUPS: A NEW INVARIANT

In the discussion of finite Abelian groups we saw that every Abelian group is isomorphic to the direct product of its Sylow subgroups. Of course this fails terribly for most non-Abelian groups, but amazingly enough, one can salvage a good part of this theorem. In the process we shall come up with a powerful new invariant — the number of p-Sylow subgroups. Our basic tool is a generalization of conjugacy classes. First we generalize centralizers of elements.

Definition 1. Given any set $A \subset G$, define its *normalizer* $N(A) = \{g \in G : gAg^{-1} = A\}$.

Remark 2. $N(A) \leq G$, for if $gAg^{-1} = A$ and $hAh^{-1} = A$, then $A = h^{-1}Ah$ and, thus, $gh^{-1}A(gh^{-1})^{-1} = gh^{-1}Ahg^{-1} = gAg^{-1} = A$.

Of course, this can be viewed in terms of group actions: G acts on {subgroups of A} by conjugation; the normalizer of A is just the stabilizer of A under this action.

Note that $N(\{a\}) = C(a)$. Thus, Remark 2 includes Proposition 10.5(i). However, we shall be interested mainly when $A < G$, in which case $A \triangleleft N(A)$, thereby justifying the name "normalizer." (In particular, if G is Abelian, then $N(A) = G$.) We want to generalize this definition a bit further. If $H < G$ we define a *conjugate* of A *with respect to* H to be a set of the form $h^{-1}Ah$ for h in H.

LEMMA 3. *There is a 1:1 correspondence between the conjugates of A with respect to H, and the cosets of $H \cap N(A)$ in H, given by*

$$h^{-1}Ah \to (H \cap N(A))h.$$

Proof. In parallel to the proof of Lemma 10.6 we note that the correspondence is onto, so we need merely note

$$\begin{aligned} h_1^{-1}Ah_1 = h_2^{-1}Ah_2 \quad &\text{iff} \quad h_2h_1^{-1}A(h_2h_1^{-1})^{-1} = A, \\ \text{iff} \quad h_2h_1^{-1} \in H \cap N(A), \quad &\text{iff} \quad (H \cap N(A))h_2 = (H \cap N(A))h_1. \quad \square \end{aligned}$$

PROPOSITION 4. *The number of conjugates of A with respect to H is* $[H : H \cap N(A)]$.

Proof. Just count in Lemma 3. □

COROLLARY 5. *The number of conjugates of A (in G) is* $[G : N(A)]$.

Now let us try to take as much of the Sylow theory from Abelian groups as possible.

PROPOSITION 6. *Suppose K is a p-Sylow subgroup of G, and H is a p-subgroup of G. If $H \subseteq N(K)$, then $H \subseteq K$.*

Proof. $K \triangleleft N(K)$, so $HK \leq N(K)$, and $|HK| = \frac{|H||K|}{|H \cap K|}$ is a power of p. Thus, HK is a p-subgroup containing K, so $HK = K$, implying $H \subseteq K$. □

Note that when G is Abelian $H \subseteq N(K) = G$, so we have a generalization (albeit modest) of Theorem 7.11(ii), which suffices for the following key computation.

COROLLARY 7. *Suppose K is a p-Sylow subgroup of G, and H is any p-subgroup of G. Let m be the number of conjugates of K with respect to H. Then m is a power of p, which is 1 iff $H \subseteq K$.*

Proof. $m = [H : H \cap N(K)]$, a power of p, and is 1 iff $H \subseteq N(K)$, iff $H \subseteq K$ by Proposition 6. □

We are ready for possibly the most important basic theorem in group theory.

THEOREM 8. *(Sylow's Theorems.) Suppose p is a prime dividing $|G|$. Write m_p for the number of p-Sylow subgroups of G.*

(i) *If p^t divides $|G|$, then G has a subgroup of order p^t; in particular, G has a p-Sylow subgroup, so $m_p \geq 1$.*
(ii) *Every p-subgroup is contained in a p-Sylow subgroup.*
(iii) *Any two p-Sylow subgroups of G are conjugate.*
(iv) *$m_p \equiv 1 \pmod{p}$.*
(v) *m_p divides $\frac{|G|}{p^t}$, where p^t is the largest power of p that divides $|G|$. More precisely, m_p is the index of the normalizer of any p-Sylow subgroup (in G).*

Proof. The proof is comprised of two main parts. First we use the Class Formula to prove (i). Then, knowing that a p-Sylow subgroup exists, we use it to set up a machine that grinds out the rest of the theorem by means of a clever counting argument based on Corollary 7. We shall use repeatedly the following easy observation : If $u = \frac{n}{m}$ and $p^t \nmid m$, then $p \mid u$. Let us turn to the proof itself.

(i) By induction on $|G|$: if p^t divides $|H|$ for some $H < G$, then by induction H has a subgroup of order p^t, which is then a p-Sylow subgroup of G. Thus, we may assume that p^t does not divide $|H|$, for every $H < G$. In particular, $p^t \nmid |C(a)|$ for every $a \in G \setminus Z(G)$; hence $p|[G : C(a)]$ whenever

$a \notin Z(G)$. But p divides $|G|$, so Corollary 10.8 implies p divides $|Z(G)|$. Cauchy's theorem implies $Z(G)$ contains an element z of order p.

Clearly, $\langle z \rangle \triangleleft G$, and $|G/\langle z \rangle| = \frac{|G|}{p}$. By induction, $G/\langle z \rangle$ contains a subgroup of order p^{t-1}, whose preimage in G is then a subgroup of order $p \cdot p^{t-1} = p^t$, yielding (i) as desired.

Now that we have proved G has a p-Sylow subgroup, which we denote as K, we continue with the following set-up: Let $\mathcal{C} = \{\text{conjugates of } K\}$, which clearly are p-Sylow subgroups of G. Let $m = |\mathcal{C}|$, the number of subgroups of G conjugate to K. Hence $m = [G : N(K)]$, by Corollary 5. But $p \nmid [G : K]$, and $|K|$ divides $|N(K)|$, by Lagrange, so $p \nmid m$.

Now take any p-subgroup H of G. Conjugation with respect to H defines an equivalence relation on $\mathcal{C}$, thereby partitioning $\mathcal{C}$ into $\mathcal{C}_1 \cup \cdots \cup \mathcal{C}_k$ for some k. Then

$$|\mathcal{C}| = |\mathcal{C}_1| + \cdots + |\mathcal{C}_k|,$$

so $p \nmid |\mathcal{C}_j|$ for some j.

For each i let $u_i = |\mathcal{C}_i|$, and pick K_i arbitrarily in $\mathcal{C}_i$. By Corollary 7, each u_i is a power of p, which is 1 iff $H \subseteq K_i$. But $p \nmid u_j$ so $u_j = 1$, implying $H \subseteq K_j$. This proves (ii).

Next, we take H to be a p-Sylow subgroup, and continue from where we just left off. $|H| = |K_j|$ implies $H = K_j$, proving H is conjugate to K, yielding (iii). In particular $m_p = m$. On the other hand, $H \not\subseteq K_i$ for all $i \neq j$ (since otherwise $K_j = H = K_i$, which is false). Thus, for each $i \neq j$ we see $u_i \neq 1$, implying $p | u_i$. Hence

$$\begin{aligned} m_p &= u_1 + \ldots + u_{j-1} + u_j + u_{j+1} + \ldots + u_k \\ &\equiv 0 + \ldots + 0 \quad + 1 + 0 \quad + \cdots + 0 \quad \equiv 1 \pmod{p}, \end{aligned}$$

proving (iv).

Finally, $m_p = m = [G : N(K)]$, which we saw divides $[G : K] = \frac{n}{p^t}$, yielding (v). □

Theorem 8(iv) should remind us of an analogous result obtained earlier, in Exercise 3.2, and lead us to wonder whether there is a general theorem concerning the number of subgroups of order p^t. In fact, such a result (due to Frobenius) exists, *cf.* Exercise 12, which has a very slick proof using group actions.

Remark 9. One point of Theorem 8 is that $m_p = 1$ iff the Sylow p-group is normal in G, in view of Proposition 5.7. This gives us a powerful tool for obtaining nontrivial normal subgroups.

Groups of Order Less Than 60

What can we say so far about classifying finite groups? Let p, q denote distinct prime numbers.

1. If $|G| = p$ then $G \approx \mathbb{Z}_p$ is cyclic, by Theorem 6.2.
2. If G is Abelian, then G is a direct product of cyclic groups, by Theorem 7.2.
3. If $|G| = p^2$, then G is Abelian, so $G \approx \mathbb{Z}_p^2$ or $\mathbb{Z}_p \times \mathbb{Z}_p$ by Corollary 9.8.
4. If $|G| = pq$ for $p > q$, then either $G \approx \mathbb{Z}_p \times \mathbb{Z}_q \approx \mathbb{Z}_{pq}$ or G is the semidirect product of $\mathbb{Z}_p$ by $\mathbb{Z}_q$; this latter group only exists if q divides $p-1$, and then is non-Abelian and uniquely determined (*cf.* Example 8.8).

Assume in the remainder that $|G|$ is not prime. We shall prove that G has a nontrivial normal subgroup if $|G| < 60$. By what we have just seen, this is clear if G is Abelian or has order pq, so we exclude these cases from the subsequent discussion.

5. If $|G| = p^t$ for $t \geq 3$, then the normal subgroup $Z(G)$ has order $p, p^2, \ldots, p^{t-1}$, or p^{t-2}. (This covers the cases $|G| = 8, 16, 27, 32$.)
6. If $|G| = p^t v$ for $p > v > 1$, then $m_p = 1$, by Theorem 8(iv),(v), so the p-Sylow subgroup is normal, by Remark 9. (This covers the cases $|G| = 18, 20, 28, 42, 44, 50, 52, 54$.) Note that v was taken not prime when $t = 1$, in view of (4) above.
7. If $|G| = 40$ or 45, then $m_5 = 1$, so the 5-Sylow subgroup is normal, by Remark 9. (Also *cf.* Exercise 6.)
8. If $|G| = 12$, 24, or 48, then any 2-Sylow subgroup of G either is normal or contains a nontrivial normal subgroup of G, by Corollary 9.6. (Also *cf.* Exercise 8.)
9. If $|G| = 36$, then any 3-Sylow subgroup of G either is normal or contains a nontrivial normal subgroup of G, by Corollary 9.6.

   This leaves us with 30 and 56.
10. If $|G| = 56$, then $m_7 = 1$ or 8. If there are eight subgroups $H_1, \ldots, H_8$ of order 7, then noting $|H_i \cap H_j| = 1$ for each $i \neq j$ (since 7 is prime) we see each H_i has six elements of order 7, and these are all distinct, so G has $6 \cdot 8 = 48$ elements of order 7; hence only eight elements have order $\neq 7$, and these are needed for a unique subgroup of order $8 = 2^3$. We have proved $m_7 = 1$ or $m_2 = 1$, so G has a nontrivial normal subgroup.
11. If $|G| = 30$ then $m_3 = 1$ or $m_5 = 1$, as argued above. (Otherwise, $m_3 = 10$ and $m_5 = 6$, so G would require twenty elements of order 3 and twenty-four elements of order 5.) We can push this much further, *cf.* Exercise 7.

In summary, we have proved the following useful fact:

PROPOSITION 10. *Every group G of nonprime order < 60 has a nontrivial normal subgroup.*

Simple Groups

There are groups failing to have any nontrivial normal subgroups; such groups are called *simple*. Obviously the only simple Abelian groups are the cyclic groups of prime order. The easiest non-Abelian example of a simple group is A_5, which has order 60 (thereby showing that Proposition 10 is sharp); in Exercise 5.6 we saw that A_n is simple for all $n \geq 5$.

Two famous theorems that can be formulated about simple groups are:

BURNSIDE'S THEOREM. *If G is nonabelian and simple then $|G|$ has at least three distinct prime factors.*

FEIT-THOMPSON THEOREM. *Every non-Abelian simple group has even order.*

(The original proof of the Feit-Thompson Theorem required 250 pages in a research journal!) Actually these two theorems are usually formulated in terms of "solvable" groups, as will be explained in Chapter 12.

Recently the complete classification of simple groups has been finished, and consists of 18 infinite families (one of which is the family of cyclic groups of prime order, and another of which is $\{A_n : n \geq 5\}$), as well as 26 "sporadic" groups, some of which crop up for rather mysterious reasons. This theorem is a major triumph for mathematics in the second half of the twentieth century, having required the combined efforts of dozens of research mathematicians aided by computers; however, it also raises questions as to what a "proof" is, since a complete proof of this result has not yet been written down from start to finish.

In the exercises of Chapter 3 we were introduced to the classical groups, which are certain subgroups of $\mathrm{GL}(n, F)$ and their homomorphic images. These provide several important classes of simple groups, which are finite when F is a finite field. Some of this theory will be outlined in the exercises of Chapter 12; details may be found in Chapter 6 of Jacobson's *Basic Algebra I* (1985), Freeman, San Francisco. The reader interested in The Classification could pursue the matter in Gorenstein's tomes on group theory.

Exercises

1. Show how Lemma 3 is really a special case of Remark 10.15.

2. Let H be the subgroup of G =GL(n, F) consisting of all "upper unipotent" matrices, *i.e.*, upper triangular matrices whose diagonal entries are all 1. If $|F| = q$ is a power of p, show H is a p-Sylow subgroup of G. (*Hint*: $|H| = q^{n(n-1)/2}$; compare with Exercise 1.6.)
3. If H is a p-Sylow subgroup of G and $K < G$, then $K \cap gHg^{-1}$ is a Sylow p-subgroup of K, for suitable g in G. (*Hint*: By the class formula for double cosets, *cf.* Exercise 10.13, some $[K : K \cap gHg^{-1}]$ must be prime to p.)
4. Reprove Theorem 8(i) by applying Exercise 3 to Exercise 2, noting that any finite group of order n can be embedded into GL$(n, \mathbb{Z}_p)$.
5. If S is a p-Sylow subgroup of G, then $N(N(S)) = N(S)$. (*Hint*: S is the only p-Sylow subgroup of $N(S)$, implying $gSg^{-1} = S$ for all g in $N(N(S))$.)

Classification of Groups of Various Orders

6. Every group of order 45 is isomorphic to $\mathbb{Z}_3 \times \mathbb{Z}_3 \times \mathbb{Z}_5$ or $\mathbb{Z}_9 \times \mathbb{Z}_5$. (*Hint*: $m_3 = m_5 = 1$.)
7. The groups of order 30 (up to isomorphism) are $\mathbb{Z}_{30}$, D_{15}, $\mathbb{Z}_3 \times D_5$, and $\mathbb{Z}_5 \times D_3$. (*Hint*: Let $|H| = 3$ and $|K| = 5$. Then H or K is normal in G, implying $N = HK < G$. But $|N| = 15$ so $N \approx \mathbb{Z}_{15}$. Write $N = \langle a \rangle \triangleleft G$. Take b of order 2. $bab^{-1} = a^i$ where $i^2 \equiv 1 \pmod{15}$; there are four solutions.)
8. Every group of order 12 is isomorphic to one of the following five groups: $\mathbb{Z}_4 \times \mathbb{Z}_3$, $\mathbb{Z}_2 \times \mathbb{Z}_2 \times \mathbb{Z}_3$, $D_6 \approx S_3 \times \mathbb{Z}_2$, A_4, and a fifth group to be described in the hint. (*Hint*: Case I. $m_2 = 1$ and $m_3 = 1$. Then G is Abelian.

   Case II. $m_2 = 1$ and $m_3 \neq 1$. Let N be the 2-Sylow subgroup of G. Suppose $o(a) = 3$. If N were cyclic, then take $o(b) = 4$ and note $aba^{-1} = b^{-1}$ contrary to $o(a) = 3$. Thus, $N = \{e, b_1, b_2, b_3\}$ with each $o(b_i) = 2$. One may assume $ab_ia^{-1} = b_{i+1}$ (subscripts mod 3), so G is uniquely determined, and thus $G \approx A_4$.

   Case III. $m_2 \neq 1$. Then $m_3 = 1$, by counting the number of elements in G. By Corollary 9.6 any Sylow 2-subgroup contains a normal subgroup $\langle z \rangle$ of G of order 2, so $z \in Z(G)$. Suppose $o(a) = 3$, and let $c = az$, which has order 6. Take $b \in G$ with $bcb^{-1} \neq c$. Replacing b by a suitable odd power of b, one may assume $o(b)$ is a power of 2. If $o(b) = 2$ then $G \approx D_6$; if $o(b) = 4$, then $b^2 = c^3 = z$, and one has an analog of Q.)
9. Determine every group G of order 18. (*Hint*: Let N be the Sylow 3-subgroup of G. If N is cyclic, then $G \approx \mathbb{Z}_{18}$ or D_9. If N is noncyclic, then $N \approx \mathbb{Z}_3 \times \mathbb{Z}_3$, which has generators a_1, a_2 of order 3.

Take b of order 2.

Case I. b commutes with both of the a_i. Then $G \approx \mathbb{Z}_{18}$.

Case II. b commutes with one of the a_i, but not the other. Then $G \approx D_3 \times \mathbb{Z}_2$.

Case III. $ba_1b = a_2$. Then $ba_1a_2b = a_1a_2$, so this case degenerates to case II.

Case IV. $ba_1b = a_1^{-1}$ and $ba_2b = a_2^{-1}$. Describe this group as a subgroup of S_6.

10. Classify all groups of order < 32, except for those of order 16. (*Hint*: The previous results cover all orders except 24 and 28. $|G| = 28$ is rather like Case III of Exercise 8 (since $m_7 = 1$); 24 is trickier, since there are five possibilites for the 2-Sylow subgroup.)
11. Q cannot be isomorphic to a subgroup of S_n for $n \leq 5$. (*Hint*: D_4 is a 2-Sylow subgroup of S_5.)
12. (Frobenius' theorem) If p^k divides $n = |G|$, then the number m of subgroups of G having order p^k is congruent to 1 (mod p). (*Hint*: Let $|G| = p^k q$. Let $\mathcal{S} = \{$subsets of G of order $p^k\}$, and let G act on $\mathcal{S}$ by left translation. If $S \in \mathcal{S}$, then,writing S as a disjoint union of G_S-orbits, note that $|G_S|$ divides $|S| = p^k$.

    Let $\mathcal{S}_0 = \{S \in \mathcal{S} : |G_S| = p^k\}$. If $S \in \mathcal{S} \setminus \mathcal{S}_0$, then pq divides $|G \cdot S|$; so $|\mathcal{S}| \equiv |\mathcal{S}_0|$ (mod pq).

    On the other hand, if $S \in \mathcal{S}_0$, then for any s in S, note that $G_S s$ is a subset of S of order p^k, so $S = G_S s$ is a coset of G_S; conversely, for any coset Hg of any subgroup H having order p^k, show that $Hg \in \mathcal{S}_0$. Thus, $|\mathcal{S}_0| = m\frac{n}{p^k} = mq$.

    Conclude $m \equiv \frac{|\mathcal{S}|}{q}$ (mod p). But $|\mathcal{S}|$ depends only on n and p^k, *i.e.*, is independent of the particular group structure of G. Thus, it suffices to prove the result for any group G of order n; conclude by taking G cyclic.)

CHAPTER 12. SOLVABLE GROUPS: WHAT COULD BE SIMPLER?

Certain groups are considerably more general than Abelian groups, but possess some of their nice properties. We shall discuss these groups now, in preparation for an application in Chapter 26. The underyling question is, "How far is a given group from being Abelian?" The answer lies in the rather trivial fact that $ab = ba$ iff $aba^{-1}b^{-1} = e$, which we needed in Proposition 6.13. This leads us to view the subset of elements of the form $aba^{-1}b^{-1}$ as the obstuction to a group being Abelian.

Commutators

Definition 1. The *group commutator* $[a, b]$ is defined as $aba^{-1}b^{-1}$. The *commutator subgroup* G' is the subgroup of G generated by all group commutators in G.

Remark 2. (i) If $H < G$ then $H' < G'$.

(ii) $[a, b]^{-1} = bab^{-1}a^{-1} = [b, a]$. Thus, every element of G' is a product of group commutators. (See, however, Exercise 19.)

PROPOSITION 3. (i) *$f(G') \subseteq H'$, for any group homomorphism $f: G \to H$.*

(ii) *$G' \triangleleft G$.*

Proof. (i) $f([a_1, b_1] \dots [a_n, b_n]) = [f(a_1), f(b_1)] \dots [f(a_n), f(b_n)] \in H'$.

(ii) Taking $f_g: G \to G$ to be the homomorphism given by $a \mapsto gag^{-1}$, we see by (i) that $gG'g^{-1} \subseteq G'$, for each g in G. Hence $G' \triangleleft G$. □

This thread is carried further in the addendum, but meanwhile let us get to the main point of commutators.

PROPOSITION 4. (i) *G/G' is an Abelian group.*

(ii) *Conversely, if $N \triangleleft G$ and G/N is an Abelian group, then $G' \subseteq N$.*

Proof. (i) For any a, b in G we see $[a, b] \in G'$ by definition, so the images of a and b commute in G/G'.

(ii) For any a, b in G we have $[a, b] \in N$, so $G' \subseteq N$. □

In particular, G is Abelian iff $G' = \{e\}$. On the other hand, here is an example with $G' = G$.

Example 5. (i) If G is simple non-Abelian, then $G' = G$ (since $G' \neq \{e\}$). In particular, $A'_n = A_n$ for all $n \geq 5$.

(ii) $S'_n = A_n$ for all $n \geq 5$. Indeed, S_n/A_n is Abelian, implying $S'_n \subseteq A_n$, but $S'_n \supseteq A'_n = A_n$.

Solvable Groups

Definition 6. A *subnormal series* from G to a subgroup N is a chain of subgroups,

$$G = G_0 > G_1 > G_2 > \cdots > G_t = N, \tag{1}$$

where each $G_{i+1} \triangleleft G_i$. We call the G_i/G_{i+1}, $1 \le i < t$, the *factors* of the series. G is *solvable* if there is some subnormal series from G to $\{e\}$, with each factor G_i/G_{i+1} Abelian.

In particular, every Abelian group is solvable. The key to solvability lies in the *derived subgroups* $G^{(i)}$ of G, defined by $G^{(0)} = G$ and inductively $G^{(i+1)} = G^{(i)\prime}$.

PROPOSITION 7. *G is solvable iff some $G^{(t)} = \{e\}$.*

Proof. ($\Leftarrow$) Take $G_i = G^{(i)}$. Then $G_i/G_{i+1} = G_i/G_i'$ is Abelian, by Proposition 4(i).

($\Rightarrow$) Given a subnormal series $G = G_0 > G_1 > G_2 > \cdots > G_t = \{e\}$, we shall prove by induction on i that $G^{(i)} \le G_i$ for $1 \le i \le t$. (Then $G^{(t)} \le G_t = \{e\}$). Indeed, $G^{(0)} = G = G_0$. Now assume $G^{(i-1)} \le G_{i-1}$. Then $G^{(i)} = G^{(i-1)\prime} \le G_{i-1}{}' \le G_i$, since G_{i-1}/G_i is Abelian. □

Remark 8. Assume $N \triangleleft G$. We can understand subnormal series better by recalling that $H \to H/N$ defines a 1:1, onto, order-preserving correspondence between {subgroups of G containing N} and {subgroups of G/N}; furthermore, by Noether II (Theorem 5.17), $K/H \approx (K/N)/(H/N)$ if $N \le H \le K$ with $H \triangleleft K$. Thus, any subnormal series from G to N,

$$G = G_0 > G_1 > G_2 > \cdots > G_t = N,$$

corresponds to the following subnormal series from G/N to $N/N = \{e\}$:

$$G/N = G_0/N > G_1/N > G_2/N > \cdots > G_t/N = \{e\},$$

where the corresponding factors G_i/G_{i+1} and $(G_i/N)/(G_{i+1}/N)$ are isomorphic.

THEOREM 9. *Suppose $N \triangleleft G$.*

(i) *If G is solvable, then N and G/N are solvable.*

(ii) *If N and G/N are solvable, then G is solvable.*

Proof. (i) We use Proposition 7. By induction, $N^{(i)} \le G^{(i)}$ for each i; if $G^{(t)} = \{e\}$, then $N^{(t)} = \{e\}$, proving N is solvable. Similarly $(G/N)^{(i)}$

is the image of $G^{(i)}$ in G/N (seen by taking cosets at every stage), so if $G^{(t)} = \{e\}$, then $(G/N)^{(t)} = \{e\}$.

(ii) We are given a subnormal series for G/N having Abelian factors, which we translate to a subnormal series from G to N, and then continue with a given subnormal series from N to $\{e\}$ having Abelian factors, thereby obtaining a subnormal series from G to $\{e\}$ having Abelian factors, implying G is solvable. □

COROLLARY 10. *Every p-group is solvable.*

Proof. By induction on $|G|$. $Z(G) \neq \{e\}$ and is Abelian and, thus, solvable. $G/Z(G)$ also is a p-group, so by induction is solvable; hence G is solvable. □

(See Exercise 4 for an improved version of this result.)

COROLLARY 11. *Every group of order* < 60 *is solvable.*

Proof. We may assume G is non-Abelian. But by Proposition 11.10, G has a nontrivial normal subgroup N. N and G/N have smaller order than $|G|$, so by induction are solvable; hence G is solvable. □

A similar induction argument enables one to extend the two deep theorems quoted for simple groups in Chapter 11, to the formulation in which they are usually remembered:

BURNSIDE'S THEOREM. *If* $|G| = p^i q^j$ *for* p, q *prime, then* G *is solvable.*

FEIT-THOMPSON THEOREM. *Every group of odd order is solvable.*

Let us close with one final application of Remark 8, which will be needed in Chapter 27.

PROPOSITION 12. *A finite group* G *is solvable iff it has a subnormal series each of whose factors is cyclic of prime degree.*

Proof. We start with a subnormal series

$$G = G_0 > G_1 > G_2 > \cdots > G_t = \{e\},$$

having Abelian factors. We want to insert enough entries between each G_i and G_{i+1} for the factors to be cyclic of prime degree; *i.e.*, we want a subnormal series from G_i to G_{i+1}, having cyclic factors of prime degree. In view of Remark 8, it is equivalent to find a subnormal series from the Abelian group $A = G_i/G_{i+1}$ to $\{e\}$ having cyclic factors of prime degree. But this is done easily by induction on $n = |A|$. If n is prime, then A already is cyclic; if n is not prime, take a prime factor p of n and an element a in A of order p, and apply induction to $A/\langle a\rangle$, to obtain the desired subnormal series from A to $\langle a\rangle$. □

Addendum: Automorphisms of Groups

An *automorphism* of a group G is an isomorphism from G to G. Group automorphisms play an important role in the theory of groups, and so we would like to discuss them briefly here. We have already encountered one type of automorphism — for any a in G one has the *inner automorphism* given by $g \mapsto aga^{-1}$, which we denote here as τ_a.

The set of automorphisms of G is easily seen to form a group called $\text{Aut}(G)$, whose group operation is the composition of maps; the inner automorphisms, denoted Inn $\text{Aut}(G)$, comprise a normal subgroup (for if $\sigma \in \text{Aut}(G)$ and $\tau_a \in \text{Inn Aut}(G)$, then $\sigma\tau_a\sigma^{-1} = \tau_{\sigma(a)}$). Of course, if G is Abelian, then Inn $\text{Aut}(G) = (1)$.

Example 13. Suppose $G = \langle g \rangle$ is cyclic of order n. Then any homomorphism $\sigma: G \to G$ satisfies $\sigma(g) = g^m$ for suitable m; conversely, given m we can define a homomorphism $\sigma: G \to G$ given by $g^i \mapsto g^{im}$ for each i. Then σ is an automorphism iff $o(g^m) = o(g) = n$, which is true iff m and n are relatively prime. It follows easily that $\text{Aut}(G) \approx \text{Euler}(n)$.

Definition 14. A subgroup H of G is *characteristic*, written H char G, if $\sigma(H) \subseteq H$ for every σ in $\text{Aut}(G)$.

Noting that $H \triangleleft G$ iff $\sigma(H) \subseteq H$ for every σ in Inn $\text{Aut}(G)$, we see that every characteristic subgroup is normal, and we draw a few parallels between characteristic subgroups and normal subgroups, omitting the straightforward proofs.

Remark 15. (i) H char G iff $\sigma(H) = H$ for every σ in $\text{Aut}(G)$.

(ii) If H is the unique subgroup of G having order m, then H char G. In particular, every subgroup of a cyclic group is characteristic.

On the other hand, Exercise 23 shows a significant way in which characteristic subgroups behave "better" than normal subgroups.

Exercises

1. Show $A_4' = \{(1), (12)(34), (13)(24), (14)(23)\}$. (*Hint*: This is the only nontrivial normal subgroup of A_4.)
2. $D_n' = \langle a^2 \rangle$, under the notation of definition 8.5. (*Hint*: $\langle a^2 \rangle$ is a normal subgroup, and the residue group has order 2 or 4.) Of course, $\langle a^2 \rangle = \langle a \rangle$ iff n is odd.

Nilpotent Groups

3. Define $G^1 = G'$, and G^{i+1} to be the subgroup of G generated by $\{[a, g] : a \in G^i,\ g \in G\}$. G is called *nilpotent* of *class* t if some $G^t = \{e\}$ (t minimal such). Show that $G^{(t)} \subseteq G^t$; hence every

nilpotent group is solvable. The center of every nilpotent group is nontrivial. Thus A_4 is an example of a solvable group that is not nilpotent, since its center is trivial.

4. The *central series* is defined via $Z_1 = Z(G)$, and inductively $Z_{i+1}(G)$ is that subgroup $H \supseteq Z_i(G)$ of G for which $H/Z_i(G) = Z(G/Z_i(G))$. Show by induction on t that G is nilpotent of class t, iff $Z_t(G) = G$. Conclude that every p-group is nilpotent.
5. If G is a nilpotent group, then $N(H) > H$ for every $H < G$. (*Hint*: If $Z_i(G) \leq H$, then $Z_{i+1}(G) \leq N(H)$.) Conclude from Exercise 11.5 that G has a unique Sylow p-subgroup, for each prime p dividing $|G|$.
6. A finite group is nilpotent iff it is a direct product of p-groups. (*Hint*: Exercise 5.)
7. Let $Z = Z(G)$. G is nilpotent of class 2 iff $G' \subseteq Z$. Conclude that the following identities hold for all a, b, c in G:

$$[ab, c] = [a, c][b, c]; \qquad [a, bc] = [a, b][a, c];$$

$$(ab)^n = a^n b^n [a, b]^{-n(n-1)/2}.$$

(*Hint*: $bc = cbz$ for suitable z in Z.)
8. If G is nilpotent of class 2 and is generated by $a_1, \ldots, a_n$, then every element of G has the form $a_1^{k_1} \ldots a_n^{k_n} z$ where $z \in G' \subseteq Z$, and consequently G' is generated by the $\binom{n}{2}$ commutators $[a_i, a_j]$, for $1 \leq i < j \leq n$.

The Special Linear Group SL(n, F), for a Field F

9. $\mathrm{SL}(n, F) \triangleleft \mathrm{GL}(n, F)$, and $\mathrm{GL}(n, F)/\mathrm{SL}(n, F) \approx F \setminus \{0\}$. (*Hint*: By Exercise 4.3.) Using Exercise 1.6, compute $|\mathrm{SL}(n, F)|$ when F is finite.
10. For $i \neq j$, define the *elementary matrices* $T_{ij}(\alpha) = I + \alpha e_{ij}$ in $\mathrm{SL}(n, F)$. (I denotes the identity matrix.) $T_{ij}(\alpha)^n = T_{ij}(n\alpha)$, for all n in $\mathbb{Z}$. In particular, $T_{ij}(\alpha)^{-1} = T_{ij}(-\alpha)$; if $F = \mathbb{Z}_p$, then $T_{ij}(\alpha)^p = I$. Conclude that $[T_{ij}(\alpha), T_{jk}(\beta)] = T_{ik}(\alpha\beta)$ if $i \neq k$; $[T_{ij}(\alpha), T_{k\ell}(\beta)] = I$ if $i \neq \ell$ and $j \neq k$.
11. Multiplying a matrix A on the left (resp. right) by $T_{ij}(\alpha)$ takes A and adds on α times the jth row (resp. ith column) of A to the ith row (resp. jth column). Prove $\mathrm{SL}(n, F)$ is generated by elementary matrices. (*Hint*: There are elementary row and column transformations taking $A \in \mathrm{SL}(n, F)$ to a diagonal matrix; equivalently there are suitable products P and Q of elementary matrices such that PAQ is diagonal, so assume A is diagonal. But for $0 \neq d \in F$ there are elementary transformations taking $\begin{pmatrix} c & 0 \\ 0 & d \end{pmatrix}$ to $\begin{pmatrix} cd & 0 \\ 0 & 1 \end{pmatrix}$ (in five steps);

writing $d_1, \ldots, d_n$ for the diagonal entries of A, one can thus transform $\operatorname{diag}\{d_1, \ldots, d_n\}$ to $\operatorname{diag}\{d_1, \ldots, d_{n-1}d_n, 1\}$, and continue by induction to arrive at the identity matrix.)

12. Suppose $n \geq 2$. Then $\mathrm{SL}(n,F)' = \mathrm{SL}(n,F)$ unless $n = 2$ with $|F| \leq 3$. (*Hint:* By Exercise 11 it suffices to display $T_{ij}(\alpha)$ as a commutator. $T_{12}(\alpha) = [T_{13}(\alpha), T_{32}(1)]$, if $n \geq 3$. If $n = 2$, then assuming $|F| > 3$, take $\gamma \neq 0$ with $\gamma^2 \neq 1$, and solve for $T_{12}(\alpha) = [\begin{pmatrix} \gamma^{-1} & 0 \\ 0 & \gamma \end{pmatrix}, T_{12}(\beta)]$.)
13. $\mathrm{GL}(n,F)' = \mathrm{SL}(n,F)$, under the hypotheses of Exercise 12.
14. $Z(\mathrm{SL}(n,F)) = \{\alpha I : \alpha^n = 1 \text{ in } F\}$.

The Projective Special Linear Group PSL(n, F)

15. Define $\mathrm{PSL}(n,F) = \mathrm{SL}(n,F)/\mathrm{Z}(\mathrm{SL}(n,F))$. $\mathrm{PSL}(n,F)$ is trivial when $n = 1$, so we assmue $n > 1$. If $|F| = q$ is finite, $|\mathrm{PSL}(n,F)| = (q^n - 1)(q^n - q)\ldots(q^n - q^{n-2})q^{n-1}/d$, where $d = \gcd(n, q-1)$. In particular, $\mathrm{PSL}(2,F)$ has order $(q^2-1)q$ for q even, and $\frac{(q^2-1)q}{2}$ for q odd. (*Hint:* Exercises 9 and 14.)
16. Let $F^{(n)}$ denote the standard n-dimensional vector space over the field F, and let $S = \{1\text{-dimensional subspaces of } F^{(n)}\}$. $\mathrm{SL}(n,F)$ acts on S by the rule $A(Fv) = Fw$ where $w = Av$ (for $A \in \mathrm{SL}(n,F)$, $0 \neq v \in F^{(n)}$); show that $(AB)(Fv) = A(B(Fv))$ and

$$Z(\mathrm{SL}(n,F)) = \{A \in \mathrm{SL}(n,F) : A(Fv) = Fv \text{ for all } v \neq 0 \in F^{(n)}\}.$$

(*Hint:* If v, w are F-independent and $Av = \alpha v$, $Aw = \beta w$, and $A(v+w) = \gamma(v+w)$ with $Fv \neq Fw$, then $(\alpha - \gamma)v = (\gamma - \beta)w$ so $\alpha = \gamma = \beta$.)

17. $\mathrm{PSL}(n,F)$ is a simple group, unless $n = 2$ with $|F| \leq 3$. (*Hint:* Let $G = \mathrm{SL}(n,F)$. Suppose $Z(G) \subset N \triangleleft G$. Show $N = G$, in the following series of steps, notation as in Exercise 16.

1. If $Fv_1 \neq Fv_2$ and $Fw_1 \neq Fw_2$, then there is A in G such that $A(Fv_i) = Fw_i$ for $i = 1, 2$. (This is because any two independent rows can be extended to an invertible matrix.)

2. For any v, w in $F^{(n)}$ there is $B \in N$ for which $B(Fv) = Fw$. (*Hint:* Since $Z(G) \subset N$, there is x in $F^{(n)}$ and $B_0 \in N$ for which $B_0(Fx) \neq Fx$. Take A in G such that $A(Fx) = Fv$ and $A(B_0(Fx)) = Fw$. Then $AB_0A^{-1} \in N$, and $AB_0A^{-1}(Fv) = AB_0(Fx) = Fw$.)

3. Let $v_1 = (1, 0, \ldots, 0) \in F^{(n)}$ and $H = \{A \in G : Av_1 \in Fv_1\}$. Then $G = NH$. (*Hint:* Given any A in G take B in N such that $AFv_1 = BFv_1$; then $B^{-1}A \in H$.

4. Identify H with the set of matrices having $(i, 1)$-position 0 for every $i > 1$. Let K be the subgroup of G generated by all elementary matrices of the form $T_{1j}(\alpha)$, $\alpha \in F$. Then $K \triangleleft H$, so $NK \triangleleft NH = G$. But every $T_{ij}(\alpha)$ is conjugate to $T_{1j}(\alpha)$, and thus is in NK, implying $NK = G$; thus, $G/N \approx K/(N \cap K)$.

Since K is Abelian, conclude that $N \supseteq G' = G$.)

18. Using Exercises 10.17 and 10.18, show that the proof of Exercise 17 can be applied to obtain the following more general result: Suppose we are given a group $G = G'$ and a primitive action of G on a set S. Suppose, moreover, for some s in S that G_s contains some normal Abelian subgroup A of G, for which $\{\cup gAg^{-1} : g \in G\}$ generates G. Then G/K is a simple group, where K is the set of elements of G acting trivially on S. (There are several important instances where this holds, in groups arising from bilinear forms.)
19. If $n = [G : Z(G)]$, then the number of distinct commutators in G is at most $n(n-1)$. (*Hint*: $[z_1 a, z_2 b] = [a, b]$.) Conclude that some element of G' is *not* a commutator, if $|G'| > n(n-1)$. To construct such an example, find a subgroup G of $\mathrm{SL}(m, \mathbb{Z}_p)$ (for suitable $m > n$), generated by suitable matrices $A_1, \dots, A_n$, such that G is nilpotent of class 2, $|G/Z(G)| = p^n$, and $|G'| = p^{n(n-1)/2}$ (*Hint*: Refer to exercises 7 and 8. In the notation of Exercise 10 write E_{ij} for $T_{ij}(1)$. For $n = 3$ take $A_1 = E_{12}$, $A_2 = E_{34}E_{27}$, and $A_3 = E_{56}E_{28}E_{49}$. Generalize to arbitrary n.)

Exercises for the Addendum

20. $\mathrm{Inn\,Aut}(G) \approx G/Z(G)$.
21. $G = \mathbb{Z}_2 \times \mathbb{Z}_2$ has subgroups that are not characteristic (but are normal since G is Abelian).
22. Prove that if $\mathrm{Aut}\, G = \{1\}$, then $|G| \le 2$. (*Hint*: $\mathrm{Inn\,Aut}\, G = \{1\}$ implies G is Abelian, and thus has the automorphism $g \mapsto g^{-1}$; hence $\exp(G) \le 2$.)
23. If H char K and K char G, then H char G. If H char K and $K \triangleleft G$, then $H \triangleleft G$. However, $H \triangleleft K$ and $K \triangleleft G$ do not necessarily imply $H \triangleleft G$.
24. The derived subgroups $G^{(i)}$ are characteristic. (*Hint*: Apply induction to Exercise 23.)
25. A Sylow p-subgroup of G is characteristic iff it is normal.

Semidirect Products, Also *Cf.* Example 8.8

26. Suppose H, G are groups, with a homomorphism $H \to \mathrm{Aut}(G)$ given by $\sigma \mapsto \overline{\sigma}$. Define a binary operation on the cartesian product

$K = G \times H$ by

$$(g_1, \sigma)(g_2, \tau) = (g_1\overline{\sigma}(g_2), \sigma\tau).$$

Show this defines a group structure on K, called the *semidirect product of G by H*. Note that K has subgroups $\overline{G} = G \times \{e\}$ and $\overline{H} = \{e\} \times H$ satisfying the following properties:
$K = \overline{G}\overline{H} = \overline{H}\overline{G}$; $\overline{G} \triangleleft K$; $\overline{G} \cap \overline{H} = \{e\}$.

27. Conversely to Exercise 26, suppose K is any group having subgroups G and H satisfying $K = GH$, $G \triangleleft K$, and $G \cap H = \{e\}$. Then K is isomorphic to the semidirect product of G by H. (In this situation we say K is the *internal semidirect product* of G by H.)
28. Suppose a group K has a subgroup G of index p, where p is the smallest prime number dividing $|K|$, and $K \setminus G$ has an element h of order p. Then K is the internal semidirect product of G by $\langle h \rangle$.
29. For p an odd prime, display the non-Abelian group of order p^3 and exponent p as a semidirect product of $\mathbb{Z}_p \times \mathbb{Z}_p$ by $\mathbb{Z}_p$.

The Wreath Product

30. Supppose A, G are groups. The *wreath product* W of G by A is defined as follows: Let $B = \{$functions from A to $G\}$, viewed as a group by means of "pointwise multiplication;" *i.e.*, if we write (g_a) for the element of B given by $a \mapsto g_a$ for each a in A; then for each $(g_a), (h_a)$ in B we define $(g_a)(h_a) = (g_a h_a) \in B$. G acts as a group of automorphisms of B, by "translation," *i.e.*, each h in G induces the homomorphism σ_h of B given by $\sigma_h((g_a)) = ((hg)_a)$; W is defined as the semidirect product of B by G. Show this is a group and compute its order. The wreath product is an important tool for constructing weird groups, but in its easiest instance (see the next exercise) produces quite manageable examples.
31. Suppose $|G| = n$. The wreath product of G by $\mathbb{Z}_m$ can be defined explicitly as follows: Viewing G as a subgroup of S_n, let W be the subgroup of S_{mn} generated by $G \times \cdots \times G$ (viewed naturally in S_{mn}, *cf.* Exercise 9.7) and the cycle
$(1\; n{+}1\; \ldots\; (m{-}1)n{+}1)(2\; n{+}2\; \ldots\; (m{-}1)n{+}2) \ldots (n\; 2n\; \ldots\; mn)$.
Show that W is a group of order mn^m. What group do we get for $m = n = 2$?
32. In general, one can describe the wreath product of G by A as follows, where $m = |A|$: By means of Cayley's theorem, view A as a subgroup of S_m and thus as a group of permutation matrices, which acts by conjugation (via the usual matrix multiplication) on

the set of "diagonal matrices with entries in G." Of course, this description is rather imprecise, since in general one cannot define multiplication of matrices whose entries are in groups, but there is no difficulty in this instance because the matrix multiplication does not involve addition.

REVIEW EXERCISES FOR PART I

1. True or false? (Prove or give a counterexample) The set of left invertible elements in a monoid forms a group.
2. (i) What are the last two digits of 2^{42}?

   (ii) What are the last two digits of 3^{42}?

   (iii) What are the last two digits of 6^{42}?
3. Write down all the subgroups of Euler(n) for $n = 5, 6, 7, 8, 12$.
4. True or false? A group can be the union of two proper subgroups.
5. Give an example of a group of even order, that does not have a subgroup of index 2.
6. Give an example of a function $f: M \to N$ of monoids, such that $f(ab) = f(a)f(b)$ for all a, b in M, but such that $f(e) \neq e$. Give an example where M, N are finite.
7. Prove from scratch: In any group of even order, the number of elements of order 2 must be odd.
8. Compute the exponent of the following groups:

   D_n, S_n, and A_n, for $n = 4, 5, 6, 7, 8$.
9. True or false? (Prove or give counterexample) The product of two subgroups of D_5 must be a subgroup.
10. Prove that if $|G| = 2n$ for n odd then any two elements of order 2 in G cannot commute.
11. The direct product of two cyclic groups of respective orders m and n is cyclic, iff m and n are relatively prime.
12. Prove $(G_1 \times G_2)/(N_1 \times N_2) \approx (G_1/N_1) \times (G_2/N_2)$, for $N_1 \triangleleft G_1$ and $N_2 \triangleleft G_2$.
13. Give an example of two nonisomorphic, nonabelian groups having the same order and the same exponent.
14. True or false: If $a = (2, 1, 4, 4)$ in $A = \mathbb{Z}_4 \times \mathbb{Z}_8 \times \mathbb{Z}_8 \times \mathbb{Z}_8$ then $\langle a \rangle$ is a direct summand of A.
15. Prove in $\mathrm{SL}(n, \mathbb{Z}_p)$ that the subset $A = \left\{ \begin{pmatrix} 1 & m \\ 0 & 1 \end{pmatrix} : m \in \mathbb{Z}_p \right\}$ is a cyclic subgroup. What is its generator?
16. Give an example of a subgroup of an Abelian group that is *not* a direct summand.
17. Prove: The Sylow p-subgroup of an Abelian group must be unique.
18. Find a maximal Abelian subgroup of S_6, and write it as a direct product of cyclic subgroups.
19. An Abelian p-group is cyclic iff it has precisely $p - 1$ elements of order p.
20. In a group of odd order, can an element be conjugate to its inverse?

21. Prove that every element in S_n is conjugate to its inverse.
22. List the conjugacy classes of S_6.
23. List the conjugacy classes of A_5.
24. How many subgroups does S_5 have of order 4?
25. Give an example of a finite group that is generated by 3 elements but is not generated by two elements.
26. Construct a nonabelian group of order 27 and exponent 3.
27. Prove that there is a group injection from any group of order n into $\mathrm{GL}(n, \mathbb{Q})$.
28. Prove: If G has a subgroup H of index 2, with $|H|$ prime, then G is either cyclic or dihedral.
29. What happens to exercise 28 when $|H|$ is not prime?
30. What is the centralizer of $(1\ 2\ \ldots\ n)$ in S_n?
31. True or false: If G is a solvable group and $\exp(G) = |G|$, then G is cyclic.
32. Prove: If $|G| = p^t$ for p prime, then G has a subgroup of order p^k for each $1 \leq k < t$.
33. What are the Abelian groups of order 2000, up to isomorphism?
34. Prove that all groups of order 1994 are solvable.
35. Prove that any group of order 1995 has a normal cyclic group of index 3.
36. List all the groups of order 175, up to isomorphism.

PART II — RINGS AND POLYNOMIALS

Having examined groups (and monoids) so thoroughly, we are ready to study sets with two operations — addition and multiplication. Such sets satisfying certain basic axioms are called "rings," and turn out to be a beautiful illustration of the principle of abstraction. We start with $\mathbb{Z}$, one of the most familiar objects in mathematics, and write down its basic algebraic properties. A few extra properties define a certain kind of ring (PID) in which one can prove virtually all the arithmetic properties of $\mathbb{Z}$, including "unique" factorization into prime numbers. But there is another important example of PID — the ring $F[x]$ of polynomials with entries in a given field F. The theory of fields is closely connected to roots of polynomials (for example, $\sqrt{2}$ is a root of the polynomial $x^2 - 2$); furthermore, considering the ring $F[x]$ as a whole (rather than limiting ourselves to one particular polynomial) yields surprising applications to field theory, to be considered in Part III. In other words, the process of abstraction enables us to transfer basic arithmetical properties of $\mathbb{Z}$ to $F[x]$, which then yield applications to field extensions of F. We shall pause along the way to note other instances of PIDs, and see some fascinating applications in number theory, which historically inspired the discovery of ideals (which play such a crucial role in ring theory). Also we shall take note of related rings that share some of the important properties of PIDs.

In this Chapter we lay out the foundations of rings, focusing on domains and (skew) fields; in the next, we complement the discussion with the theory of homomorphisms and ideals.

Definition 1. A *ring* is a set R together with binary operations $+$ and $\cdot$ and elements 0 and 1, such that

(i) $(R, +, 0)$ is a group.

(ii) $(R, \cdot, 1)$ is a monoid.

(iii) Distributivity of multiplication over addition holds on both sides, *i.e.*, $a(b+c) = ab + ac$ and $(b+c)a = ba + ca$.

1 is called the *unit element* of R. The ring R is *commutative* if $(R, \cdot, 1)$ is a commutative monoid.

Usually the operations $+$ and $\cdot$ are understood. The set $\{0\}$ is a ring (taking $1 = 0$), called the *trivial* ring. We shall only consider nontrivial rings in the sequel. Examples include:

$\mathbb{Z}$;

$\mathbb{Z}_m$, for any positive number m, where $+$ and $\cdot$ are taken modulo m;

$\mathbb{Q}, \mathbb{R}, \mathbb{C}$, and so on; clearly, any field is a ring.

(Note that all these examples are commutative. For a noncommutative example, see Exercise 1.)

Other than $\mathbb{Z}$, the most significant example for us is the ring of polynomials $F[x]$, to be defined in Chapter 16, which will be seen to share many properties with $\mathbb{Z}$. We turn to $\mathbb{Z}$ for intuition and start off with some easy facts for all a, b in an arbitrary ring R.

LEMMA 2. (i) $0 = 0a = a0$;

(ii) $(-a)b = a(-b) = -(ab)$.

Proof. (i) $0a = (0+0)a = 0a + 0a$, so $0 = 0a$; likewise $0 = a0$.

(ii) $0 = a0 = a(-b+b) = a(-b) + ab$, implying $a(-b) = -(ab)$; likewise $(-a)b = -(ab)$. □

Lemma 2 implies that $1 \neq 0$ in any (nontrivial) ring. Indeed if $1 = 0$ then any r in R satisfies $r = r1 = r0 = 0$.

Domains and Skew Fields

A *skew field* (or *division ring*) is a ring D for which $(D \setminus \{0\}, \cdot)$ is a group, *i.e.*, every nonzero element d in D has a left and right inverse. In other words, a skew field satisfies all the axioms of a field except perhaps commutativity of multiplication. Although our principal interest lies in fields, we consider this more general situation as a very brief introduction to noncommutative techniques, to be continued in Appendix B.

A *domain* is a ring R for which $(R \setminus \{0\}, \cdot)$ is a monoid, *i.e.*, if $a, b \neq 0$, then $ab \neq 0$. A commutative domain is called an *integral domain*, in analogy to $\mathbb{Z}$, the ring of integers.

Remark 3. If R is a domain, then the monoid $(R \setminus \{0\}, \cdot)$ is cancellative. (For if $ab = ac$ then $a(b - c) = 0$ in R, implying $b - c = 0$, so $b = c$.)

Remark 4. Using Theorem 1.7 we see that any finite domain is a skew field and any finite integral domain is a field; in fact, a deeper theorem of Wedderburn (Theorem 15 in Appendix B) says any finite skew field is commutative, so we see that any finite domain is a field. Since our interest in this part lies in integral domains that are not fields, further discussion of finite rings is postponed until Chapter 24.

A *subring* of a ring R is a subset containing 0,1, that is a ring under the given $+$ and $\cdot$ of R. To verify that T is a subring of R we need only check that $(T, +)$ is subgroup of $(R, +)$ and that $(T, \cdot, 1)$ is a submonoid of $(R, \cdot, 1)$, for distributivity in T is a direct consequence of distributivity in R. This observation is enhanced in exercise 4.

Clearly, any subring of a commutative ring is commutative, and any subring of a domain is a domain, *cf.* Exercise 5. In particular, any subring of a field is an integral domain.

On the other hand, the integral domain $\mathbb{Z}$ certainly is not a field, so the question arises as to how to check in general whether an (infinite) domain is a skew field. To this end, for any a in R define $Ra = \{ra : r \in R\}$. Then we have

PROPOSITION 5. *R is a skew field iff $Ra = R$ for all $0 \neq a \in R$.*

Proof. ($\Rightarrow$) If $a \neq 0$, then $r = (ra^{-1})a \in Ra$ for any r in R.

($\Leftarrow$) If $a \neq 0$, then $1 \in Ra$ implying a is left invertible; hence $R \setminus \{0\}$ is a group, by Lemma 1.8. □

Let us dwell on this result a bit.

Left Ideals

Definition 6. A *left ideal* of a ring R is a subgroup L of $(R, +)$ satisfying the extra property

$$ra \in L \text{ for all } r \text{ in } R \text{ and } a \text{ in } L.$$

The left ideal L is called *proper* if $L \neq R$. *Right ideals* are defined symmetrically, using ar instead of ra.

Remark 6′. The left ideal L is proper iff $1 \notin L$. (Indeed, if $1 \in L$ then $r = r1 \in L$ for all r in R.)

Example 7. For any a in R, Ra is a left ideal of R, called the *principal left ideal* generated by a. (Indeed $r_1a \pm r_2a = (r_1 \pm r_2)a \in Ra$, proving Ra is a subgroup of $(R, +)$; also $r(r_1a) = (rr_1)a \in Ra$.) Note $a = 1a \in Ra$. Also, Ra is the smallest left ideal containing a; for if L is any left ideal containing a, then $ra \in L$ for each r in R, proving $Ra \subseteq L$.

PROPOSITION 8. *R is a skew field iff R has no proper nonzero left ideals.*

Proof. ($\Rightarrow$) If $L \neq 0$ is a left ideal take $0 \neq a \in L$. Then $L \supseteq Ra = R$ by Proposition 5, so L is improper.

($\Leftarrow$) For any nonzero a in R we have $Ra \neq 0$, so $Ra = R$ by hypothesis; hence R is a skew field by Proposition 5. □

Here are some facts about combining left ideals that we shall need repeatedly.

PROPOSITION 9. *Suppose L_1, L_2 are left ideals of R.*

(i) *$L_1 \cap L_2$ is the largest left ideal contained in both L_1 and L_2;*

(ii) *$L_1 + L_2 = \{a_1 + a_2 : a_i \in L_i\}$ is the smallest left ideal containing both L_1 and L_2.*

Proof. (i) $L_1 \cap L_2$ is an additive subgroup; if $r \in R$ and $a \in L_1 \cap L_2$, then $ra \in L_1$ and $ra \in L_2$ so $ra \in L_1 \cap L_2$, proving $L_1 \cap L_2$ is a left ideal. Obviously any left ideal contained in L_1 and L_2 is contained in $L_1 \cap L_2$.

(ii) $L_1 + L_2$ is an additive subgroup; if $r \in R$ and $a_i \in L_i$, then $r(a_1 + a_2) = ra_1 + ra_2 \in L_1 + L_2$, proving $L_1 + L_2$ is a left ideal. If L is a left ideal containing both L_1 and L_2, then for any $a_i \in L_i$ we see $a_1 + a_2 \in L$ so $L_1 + L_2 \subseteq L$. □

Remark 10. To illustrate Proposition 9, suppose $L_1 = Ra$ and $L_2 = Rb$ are principal left ideals. Then $c \in L_1 + L_2$ iff $c = ra + sb$ for suitable r, s in R. For example, $7\mathbb{Z} + 5\mathbb{Z} = \mathbb{Z}$, since $1 = -2 \cdot 7 + 3 \cdot 5$. In general, one should see without difficulty that $m\mathbb{Z} + n\mathbb{Z} = d\mathbb{Z}$, where $d = \gcd(m, n)$.

The union of two left ideals need not be a left ideal; *e.g.*, $3\mathbb{Z} \cup 2\mathbb{Z}$ contains 3 and 2 but not $3 - 2 = 1$. Nevertheless, there is a positive result along these lines.

Remark 11. If $L_1 \subseteq L_2 \subseteq L_3 \subseteq \dots$ are left ideals of R, then so is $\cup_{i \geq 1} L_i$. (For if $a \in \cup L_i$ then $a \in L_i$ for some i implying $ra \in L_i \subseteq \cup L_i$ for each r in R; likewise if $a, b \in \cup L_i$ then $a, b \in L_i$ for some i, implying $a \pm b \in L_i \subseteq \cup L_i$.)

Exercises

Rings of Matrices

1. For any field F and any $n \geq 1$, the set of $n \times n$ matrices $M_n(F)$ is a ring, under usual matrix multiplication and addition. More generally, let R be any ring and define $M_n(R)$ to be the set of $n \times n$ matrices with entries in R, endowed with the usual matrix addition and multiplication. Show that this is a ring. (*Hint*: The easiest way to manage the computations is by using matric units, *cf.* Chapter 0.) Show that $M_n(R)$ is neither commutative nor a domain, for $n \geq 2$.
2. An element e of a ring R is called *idempotent* if $e^2 = e$. (For example, 0,1 are idempotents, and these are the only idempotents if R is a domain.) Show that if $e \in R$ is idempotent, then eRe is a ring whose unit element is e.
3. The set of continuous functions from $\mathbb{R}$ to $\mathbb{R}$ is a ring. Likewise for differentiable functions, and so on.
4. Recall that a sentence in elementary logic is called *universal* if its normal form involves only the quantifier $\forall$. Prove that any universal sentence holding in a ring also holds in all of its subsets. In particular, this holds for associativity and distributivity. Conclude that an additive subgroup R of a ring T containing 1 of T and closed under the multiplication of T is necessarily a subring. On the other hand, give an example of a subset of $\mathbb{Z}$, closed under addition and multiplication and containing 0 and 1, which is not a subring.
5. Show that the following properties for a ring pass to subrings: being commutative; being a domain.

Direct Product of Rings

6. Define the *direct product* R of rings $R_1, \ldots, R_t$ to be the cartesian product $R_1 \times \cdots \times R_t$, operations defined componentwise, and prove it is a ring. If $e_i \in R_i$ are idempotents, then $(e_1, \ldots, e_t)$ is idempotent in R; in particular, any element $(0, \ldots, 0, 1, 0, \ldots, 0)$ is idempotent. Can R be an integral domain?
7. The direct product of commutative rings is a commutative ring.
8. Any left ideal of $R_1 \times R_2$ has the form $A_1 \times A_2$ where A_i is a left ideal of R_i for $i = 1, 2$. Similarly for right ideals. (*Hint*: Use the idempotents $(1, 0)$ and $(0, 1)$.)
9. $\mathbb{Z}_{mn} \approx \mathbb{Z}_m \times \mathbb{Z}_n$ as rings, if m, n are relatively prime.
10. $\text{Unit}(R_1 \times \cdots \times R_t) \approx \text{Unit}(R_1) \times \cdots \times \text{Unit}(R_t)$ as groups, for any rings $R_1, \ldots, R_t$. (*Hint*: View R_i as monoids.)

11. Use Exercises 9 and 10 to rederive Exercise 6.7.
12. Generalizing Proposition 9, define the sum of an arbitrary number of left ideals $L_i : i \in I$ to be the set of *finite* sums of elements taken from various L_i; show that this is the smallest left ideal containing each L_i. What is the largest left ideal contained in each L_i?

CHAPTER 14. THE STRUCTURE THEORY OF RINGS

Although our emphasis in group theory was on finite groups, our attention here is mainly on infinite rings, because many of the rings of greatest interest to us are infinite, such as $\mathbb{Z}$, $\mathbb{Q}$, $\mathbb{R}$, and $\mathbb{C}$. (Finite fields also are of interest, and will be classified in Chapter 24.) However, the classification problem for infinite rings becomes so formidable as to be nigh impossible. In providing manageable partial results, the structure theory assumes an even more important role than before and pervades ring theory. After a while the structure theory takes on a form of its own, guiding us to the most profitable avenues of inquiry. Availing ourselves of the structure theory of groups, we can make use of the fact that every ring is an Abelian group (under $+$), and the transition to ring theory is easy. Thus, at the outset, we pass through the following stages:

(1) homomorphism,
(2) kernel of homomorphism $\leftrightarrow$ ideals (analogous to normal subgroups),
(3) residue ring = factor ring R/A where A is an ideal of R,
(4) structure of R/A in terms of R, and
(5) Noether isomorphism theorems.

We shall also give the ring-theoretic analog of Cayley's theorem.

Definition 1. A *ring homomorphism* $\varphi: R \to T$ is a group homomorphism (under $+$) that is also a monoid homomorphism (under $\cdot$). A *ring isomorphism* is a ring homomorphism that is 1:1 and onto.

Remark 2. By Remark 4.2, it suffices to check the following properties for all a, b in R, in order to verify that $\varphi: R \to T$ is a ring homomorphism:

(i) $\varphi(a+b) = \varphi(a) + \varphi(b)$;
(ii) $\varphi(ab) = \varphi(a)\varphi(b)$;
(iii) $\varphi(1) = 1$.

Condition (iii) is not redundant, by the example in Digression 4.3.

Much of the structure theory can be obtained readily from the theory of Abelian groups, since ring homomorphisms are also group homomorphisms (of the additive structure). Thus, we have

Remark 2′. If $\varphi: R \to T$ is a ring isomorphism, then $\varphi^{-1}: T \to R$ is also a ring isomorphism. Indeed, by group theory φ^{-1} is an isomorphism from $(T,+)$ to $(R,+)$, so we need only check conditions (ii) and (iii) for φ^{-1}. (iii) is obvious; to check (ii) for a, b in T we let $u = \varphi^{-1}(a)$ and $v = \varphi^{-1}(b)$ and note that

$$\varphi^{-1}(ab) = \varphi^{-1}(\varphi(u)\varphi(v)) = \varphi^{-1}\varphi(uv) = uv = \varphi^{-1}(a)\varphi^{-1}(b).$$

Another instance of this philosophy:

Remark 3. If $\varphi: R \to T$ is a ring homomorphism, then $\varphi(R)$ is a subring of T. (Indeed, we already know $\varphi(R)$ is an additive subgroup; furthermore, $1_T = \varphi(1_R)$ and $\varphi(r_1)\varphi(r_2) = \varphi(r_1 r_2)$ for r_i in R.)

Now we turn to ring homomorphisms that need not be isomorphisms. Again drawing from group theory, we define the *kernel of* φ, denoted $\ker\varphi$, to be $\varphi^{-1}(0)$, and recall that φ is 1:1 iff $\ker\varphi = 0$. Furthermore, $\ker\varphi$ is a subgroup of $(R, +)$, and is a left and right ideal since for any $a \in \ker\varphi$ and r in R we have

$$\varphi(ra) = \varphi(r)\varphi(a) = \varphi(r)0 = 0$$
$$\text{and} \quad \varphi(ar) = \varphi(a)\varphi(r) = 0\varphi(r) = 0.$$

As with groups, a 1:1 ring homomorphism $\varphi: R \to T$ is called an *injection* (of rings), and enables us to view R as a subring of T.

Ideals

Let us now describe the structural properties of the kernel of a ring homomorphism and define the ring-theoretic analog of a normal subgroup.

Definition 4. An *ideal* A of R is a left and right ideal.

Clearly, 0 and R are ideals of R, called the *trivial* ideals. Thus, "nontrivial" means "proper nonzero." Any ideal of R containing 1 is all of R, by Remark 13.6′. Hence a proper ideal *cannot* be a subring, thereby breaking the analogy with groups. This failure has caused several authors (most notably Herstein) to discard the unit element 1 from the definition of a ring (since then every ideal would be a subring). However, this complicates the theory in several key places such as Remark 13.6′ itself, and the standard practice has come to require $1 \in R$ in the definition.

Note 4′. Proposition 13.8 says a commutative ring R is a field iff R has nontrivial ideals.

Let us consider this result more carefully, especially in view of Exercise 2, which shows that $M_n(F)$ is not a skew field (and even has zero-divisors) whenever $n > 1$, but has no nontrivial ideals, for any field F. If we scratch beneath the surface, we uncover one of the great dilemmas of noncommutative ring theory: Although the structure theory of rings involves ideals, an important role is also played by left ideals, and left ideals often are more manageable than ideals. (Exercise 1 indicates the difficulty of computing with ideals.) The way this dilemma was resolved in Note 4′ was that we worked with commutative rings, in which left ideals and ideals are the same. In general the theory of commutative rings is much more accessible than the theory of noncommutative rings, so after this Chapter we shall concentrate on commutative rings, postponing the treatment of noncommutative rings until Appendix B.

Since our study of commutative rings will hinge on their ideals, the case of fields is in some sense "trivial;" we shall be interested here in rings which are *not* fields. In analogy to normal subgroups, we write $A \triangleleft R$ to denote that A is a proper ideal of R.

Remark 4″. Many assertions for left ideals also hold analogously for right ideals and thus for ideals. For example, the sum $A+B$ of two ideals A, B is an ideal, and is the smallest ideal containing A and B (*cf.* Proposition 13.9). Likewise the intersection of ideals is the largest ideal contained in each of them. In general the union of ideals is not an ideal, but if $A_1 \subseteq A_2 \subseteq \dots$ are ideals of R, then $\cup_{i\geq 1} A_i \triangleleft R$, by Remark 13.11.

Our next task is to show that every ideal is indeed the kernel of a suitable ring homomorphism.

Definition 5. If $A \triangleleft R$, define $R/A = \{r + A : r \in R\}$, which is given the additive group structure of the residue group and multiplication

$$(r_1 + A)(r_2 + A) = r_1 r_2 + A.$$

THEOREM 6. *Let $A \triangleleft R$. Then R/A as defined above is a ring, with unit element $1 + A$. There is an onto homomorphism $\varphi: R \to R/A$ given by $\varphi(r) = r + A$, and $\ker \varphi = A$.*

Proof. Multiplication in R/A is well-defined, since if $r_1 + A = r_1' + A$ and $r_2 + A = r_2' + A$, then writing $r_i' = r_i + a_i$ for a_i in A we see

$$\begin{aligned} r_1' r_2' + A &= (r_1 + a_1)(r_2 + a_2) + A \\ &= r_1 r_2 + (r_1 a_2 + a_1 r_2 + a_1 a_2) + A = r_1 r_2 + A. \end{aligned}$$

Associativity and distributivity are easy to verify in R/A, as a consequence of the respective axioms in R. Moreover,

$$(1 + A)(r + A) = r + A = (r + A)(1 + A),$$

so $1 + A$ is the unit element of R/A. We already know φ is a group homomorphism with respect to $+$, and $\varphi(1) = 1 + A$, and

$$\varphi(r_1)\varphi(r_2) = (r_1 + A)(r_2 + A) = r_1 r_2 + A = \varphi(r_1 r_2),$$

proving φ is a ring homomorphism. From group theory, φ is onto and $\ker \varphi = \varphi^{-1}(0) = A$. □

Here is a useful result, which transfers the ideal structure from R to R/A.

PROPOSITION 7. *Let $A \triangleleft R$. If $A \subseteq B \triangleleft R$ then $B/A \triangleleft R/A$. Conversely, every ideal of R/A can be written uniquely in the form B/A, where $A \subseteq B \triangleleft R$.*

Proof. We know B/A is an additive subgroup of R/A, and we check that $(r+A)(b+A) = rb+A \in B/A$ and $(b+A)(r+A) \in B/A$ for all b in B and r in R. Conversely, if $I \triangleleft R/A$ we know from group theory that $I = B/A$, where B is uniquely determined as the preimage of I in R, and B is an additive subgroup of R. But if $b+A \in I$ and $r \in R$, then $rb+A = (r+A)(b+A) \in I$, and likewise $br + A \in I$, implying $rb \in B$ and $br \in B$, so $B \triangleleft R$. □

This has an immediate consequence. Let us say a proper ideal A is a *maximal* ideal (in R) if no proper ideal strictly contains A. For example, $3\mathbb{Z}$ is a maximal ideal of $\mathbb{Z}$.

COROLLARY 8. *Suppose R is a commutative ring and $A \triangleleft R$. A is a maximal ideal iff R/A is a field.*

Proof. Apply Note 4′ to Proposition 7. □

Noether's Isomorphism Theorems

Let us turn now to ring-theoretic analogs of the Noether isomorphism theorems for groups.

LEMMA 9. *Suppose $\varphi: R \to T$ is any ring homomorphism, and $A \triangleleft R$ with $A \subseteq \ker \varphi$. Then there is a ring homomorphism $\bar{\varphi}: R/A \to T$ given by $\bar{\varphi}(r + A) = \varphi(r)$, and $\ker \bar{\varphi} = (\ker \varphi)/A$.*

Proof. $\bar{\varphi}$ is an additive group homomorphism with the correct kernel, by Lemma 5.14; in fact, φ is a ring homomorphism, since

$$\begin{aligned}\bar{\varphi}((r_1 + A)(r_2 + A)) = \bar{\varphi}(r_1 r_2 + A) &= \varphi(r_1 r_2) = \varphi(r_1)\varphi(r_2) \\ &= \bar{\varphi}(r_1 + A)\bar{\varphi}(r_2 + A). \quad \square\end{aligned}$$

THEOREM 10. NOETHER I. *If $\varphi: R \to T$ is an onto ring homomorphism then $T \approx R/\ker \varphi$.*

Proof. As in Theorem 5.16. □

Example 11. $\mathbb{Z}_n \approx \mathbb{Z}/n\mathbb{Z}$ as rings (compare with Example 5.16′).

THEOREM 12. NOETHER II. *$R/B \approx (R/A)/(B/A)$, for any ideals $A \subseteq B$ of R.*

Proof. As in Theorem 5.17. □

Noether III does not carry over nicely to rings, since an ideal of a ring is not a subring.

Exercises

1. Recall that the smallest left ideal containing a is Ra. Show that the smallest ideal containing a is $\{\sum_{i=1}^{k} r_i a s_i : k \in \mathbb{N},\ r_i, s_i \in R\}$, a much more complicated object.
2. Any proper ideal of $M_n(R)$ has the form $M_n(A)$ where $A \triangleleft R$. In particular, if D is a skew field, then $M_n(D)$ has no nontrivial ideals. On the other hand, $M_n(D)$ has nontrivial left ideals for $n > 1$.

The Regular Representation

3. Suppose $(S, +)$ is an Abelian group. Then, replacing $\mathrm{Map}(S, S)$ (of Exercise 9.9) by the set of group homomorphisms from S to S, denoted as $\mathrm{Hom}(S, S)$, show that $\mathrm{Hom}(S, S)$ is a submonoid of $\mathrm{Map}(S, S)$. In fact, $\mathrm{Hom}(S, S)$ is a ring, where addition is given by $(f + g)(s) = f(s) + g(s)$.
4. If S is a ring then there is a ring injection $S \to \mathrm{Hom}(S, S)$ given by $a \mapsto \ell_a$, where $\ell_a: S \to S$ is "left multiplication by a," given by $\ell_a(s) = as$. This injection is called the *left regular representation* (cf. also Exercise 21.15). Note the parallel to Cayley's theorem from group theory. If we took the right multiplication map r_a instead of the left multiplication map ℓ_a, we would have an anti-injection, in the sense that multiplication would be reversed, *i.e.*, $r_{ab} = r_b r_a$.
5. For any group $(S, +)$, any additive subgroup of $\mathrm{Hom}(S, S)$ closed under composition of maps and containing the identity map $S \to S$ is a subring, by Exercise 13.4.

   As with Cayley's theorem, one uses Exercise 13.5 to prove that a given additive group S is indeed a ring, by displaying it as a subring of $\mathrm{Hom}(S, S)$, *cf.* Exercise 16.3.

General Structure Theory

6. If $\varphi: R \to T$ is a ring homomorphism, with R commutative, then $\varphi(R)$ is a commutative ring.
7. State and prove that the analogs of Lemma 13.9 and Proposition 13.11, for ideals instead of left ideals.
8. (Abstract Chinese Remainder Theorem) Ideals $A_1, \dots, A_t$ are called *comaximal* in R if $A_i + A_j = R$ for all $i \neq j$. In this case prove

$$R/\cap_{i=1}^{t} A_i \ \approx\ R/A_1 \times \cdots \times R/A_t.$$

   (*Hint*: Define $\varphi: R \to R/A_1 \times \cdots \times R/A_t$ by $r \mapsto (r + A_1, \dots, r + A_t)$. One needs φ onto, or, equivalently, each $(0, \dots, 0, 1, 0, \dots, 0) \in \varphi(R)$. Fixing i, write $1 = a_{ij} + a_j$ for $a_{ij} \in A_i$, $a_j \in A_j$. For convenience take $i = 1$. Then $1 = \prod_{j \neq 1}(a_{1j} + a_j) = a + a_2 \dots a_t$ for suitable a in A; consequently $\varphi(a) = (1, 0, \dots, 0)$.)

9. (Chinese Remainder Theorem) Using Exercise 8, show that for any relatively prime numbers $n_1, \dots, n_t$ and any integers $a_1, \dots, a_t$, there is $a \in \mathbb{Z}$ such that $a \equiv a_i \pmod{n_i}$, $1 \leq i \leq t$.

Chapter 15. The Field of Fractions — A Study in Generalization

We have seen that any subring of a field is an integral domain, and every finite integral domain is a field. On the other hand, although $\mathbb{Z}$ is an integral domain that is not a field, $\mathbb{Z}$ is a subring of the field $\mathbb{Q}$. This leads us to ask whether every integral domain need be a subring of a suitable field. Actually, the standard construction of $\mathbb{Q}$ from $\mathbb{Z}$ can be modified to yield a positive response to this query.

Construction 1. The field of fractions. Let R be any integral domain, and $S = R \setminus \{0\}$. Define an equivalence relation $\sim$ on $R \times S$ by "cross-multiplication," *i.e.*,

$$(r_1, s_1) \sim (r_2, s_2) \qquad \text{if} \qquad r_1 s_2 = r_2 s_1.$$

Reflexivity and symmetry of this relation are apparent; to verify transitivity suppose that $r_1 s_2 = r_2 s_1$ and $r_2 s_3 = r_3 s_2$; we need to check $r_1 s_3 = r_3 s_1$. But

$$(r_1 s_3) s_2 = (r_1 s_2) s_3 = r_2 s_1 s_3 = r_2 s_3 s_1 = r_3 s_2 s_1 = r_3 s_1 s_2,$$

so $r_1 s_3 = r_3 s_1$, seen by canceling s_2 on the right. Now write $\frac{r}{s}$ for the equivalence class of (r, s) in $R \times S$, and define $R[S^{-1}]$ (also denoted $S^{-1}R$ and R_S in the literature) to be the set of equivalence classes $\frac{r}{s}$, where $r \in R$ and $s \in S$.

Remark 2. $\frac{r's}{s's} = \frac{r'}{s'}$ for all s in S. In particular, $\frac{0}{s} = \frac{0}{1}$ (taking $r' = 0$, $s' = 1$), and $\frac{s}{s} = \frac{1}{1}$ (taking $r' = s' = 1$).

Next, we define operations

$$\frac{r_1}{s_1} + \frac{r_2}{s_2} = \frac{r_1 s_2 + r_2 s_1}{s_1 s_2}$$
$$\text{and} \qquad \frac{r_1}{s_1}\frac{r_2}{s_2} = \frac{r_1 r_2}{s_1 s_2}.$$

(These are both in $R[S^{-1}]$ since $(S, \cdot)$ is a monoid.) Let us show that these operations are well-defined. Suppose

$$\frac{r_1'}{s_1'} = \frac{r_1}{s_1} \quad \text{and} \quad \frac{r_2'}{s_2'} = \frac{r_2}{s_2};$$

then $r_1' s_1 = r_1 s_1'$ and $r_2' s_2 = r_2 s_2'$, so

$$\begin{aligned}(r_1' s_2' + r_2' s_1') s_1 s_2 &= r_1' s_1 s_2' s_2 + r_2' s_2 s_1' s_1 = r_1 s_1' s_2' s_2 + r_2 s_2' s_1' s_1 \\ &= (r_1 s_2 + r_2 s_1) s_1' s_2',\end{aligned}$$

implying $\frac{r_1'}{s_1'} + \frac{r_2'}{s_2'} = \frac{r_1}{s_1} + \frac{r_2}{s_2}$. Also

$$\frac{r_1'}{s_1'}\frac{r_2'}{s_2'} = \frac{r_1}{s_1}\frac{r_2}{s_2}$$

since $(r_1'r_2')s_1s_2 = r_1's_1r_2's_2 = r_1s_1'r_2s_2' = (r_1r_2)s_1's_2'$.

PROPOSITION 3. *$R[S^{-1}]$ is a ring, and there is a natural ring injection $\psi: R \to R[S^{-1}]$ given by $\psi(r) = \frac{r}{1}$.*

Proof. Associativity (both of addition and of multiplication) and distributivity of multiplication over addition are checked easily, and we see at once that $\frac{0}{1}$ is the 0 element and $\frac{1}{1}$ is the unit element. Finally $\frac{-r}{s} = -\frac{r}{s}$, since $\frac{r}{s} + \frac{-r}{s} = \frac{0}{s^2} = \frac{0}{1}$, by Remark 2. ψ is a homomorphism since

$$\psi(r + r') = \frac{r + r'}{1} = \frac{r}{1} + \frac{r'}{1} = \psi(r) + \psi(r'),$$
$$\psi(rr') = \frac{rr'}{1} = \frac{r}{1}\frac{r'}{1} = \psi(r)\psi(r'),$$
$$\psi(1) = \frac{1}{1}.$$

Finally, $r \in \ker\psi$ iff $\frac{r}{1} = \frac{0}{1}$, iff $r \cdot 1 = 0 \cdot 1 = 0$, implying $\ker\psi = 0$. □

PROPOSITION 4. *With notation as above, $R[S^{-1}]$ is actually a field.*

Proof. Suppose $\frac{r}{s} \neq \frac{0}{1}$. Then $r \neq 0$ (by Remark 2), so $r \in S$, implying $\frac{s}{r} \in R[S^{-1}]$, and, clearly, $(\frac{r}{s})(\frac{s}{r}) = \frac{rs}{rs} = \frac{1}{1}$. □

We separated Propositions 3 and 4, since Proposition 3 soon will be generalized.

Intermediate Rings

Thus far, we have succeeded in finding an injection from an arbitrary integral domain R to the field $R[S^{-1}]$, where $S = R \setminus \{0\}$. The reader may have noticed however that in the proof of Proposition 3 we have used only the following properties of S:

1. The elements of S commute with all elements of R;
2. S is a multiplicative submonoid of R;
3. Cancellation works in R for elements of S (needed for transitivity).

Thus, we can reformulate Proposition 3 as

PROPOSITION 6. *Let R be any commutative ring, and let S be a submonoid such that $rs \neq 0$ for every $0 \neq r \in R$ and $s \in S$. Then $R[S^{-1}]$ as constructed above is a ring, and there is a natural injection $\psi: R \to R[S^{-1}]$ given by $r \mapsto \frac{r}{1}$.*

See exercise 3 for a further generalization.

Remark 6$'$. Given $A \subset R$, let us write AS^{-1} for $\{\frac{a}{s} : a \in A,\ s \in S\}$. If $A \triangleleft R$, then clearly $AS^{-1} \triangleleft R[S^{-1}]$. The converse to this assertion is given in Exercise 10.

Now we come to a rather delicate issue. Suppose that we already have a ring containing R, in which every element of S is invertible. How does this compare with the abstract construction $R[S^{-1}]$? For example, there are various ways of viewing $\mathbb{Q}$. How do we compare these with the abstract construction? The answer to this question takes us to the key property of $R[S^{-1}]$.

PROPOSITION 7. *Suppose R, T are commutative rings, and $S \subset R$ is as in Proposition 6; suppose $\varphi: R \to T$ is a homomorphism of commutative rings, such that $\varphi(s)$ is invertible in T for every s in S. Then φ extends to a homomorphism $\hat{\varphi}: R[S^{-1}] \to T$ given by*

$$\hat{\varphi}(\frac{r}{s}) = \varphi(r)\varphi(s)^{-1},$$

and $\ker \hat{\varphi} = (\ker \varphi)S^{-1}$.

Proof.

$$\begin{aligned}
\hat{\varphi}(\frac{r_1}{s_1}) + \hat{\varphi}(\frac{r_2}{s_2}) &= \varphi(r_1)\varphi(s_1)^{-1} + \varphi(r_2)\varphi(s_2)^{-1} \\
&= (\varphi(r_1)\varphi(s_2) + \varphi(r_2)\varphi(s_1))(\varphi(s_1)^{-1}\varphi(s_2)^{-1}) \\
&= \varphi(r_1s_2 + r_2s_1)\varphi(s_1s_2)^{-1} = \hat{\varphi}(\frac{r_1}{s_1} + \frac{r_2}{s_2}); \\
\hat{\varphi}(\frac{1}{1}) &= \varphi(1)\varphi(1)^{-1} = 1; \\
\hat{\varphi}(\frac{r_1}{s_1})\hat{\varphi}(\frac{r_2}{s_2}) &= \varphi(r_1)\varphi(s_1)^{-1}\varphi(r_2)\varphi(s_2)^{-1} \\
&= \varphi(r_1r_2)\varphi(s_1s_2)^{-1} = \hat{\varphi}(\frac{r_1}{s_1}\frac{r_2}{s_2});
\end{aligned}$$

proving $\hat{\varphi}$ is a homomorphism. Now $\frac{r}{s} \in \ker \hat{\varphi}$ iff $\varphi(r)\varphi(s)^{-1} = 0$, iff $\varphi(r) = 0$, so $\ker \hat{\varphi} = (\ker \varphi)S^{-1}$. □

COROLLARY 8. *Notation as in Proposition 6, if R is a subring of T, then $R[S^{-1}]$ is canonically isomorphic to $\{rs^{-1} : r \in R,\ s \in S\}$ taken in the ring T.*

Proof. Let $\varphi: R \to T$ be the natural injection. Then, with notation as in Proposition 7, $\hat{\varphi}: R[S^{-1}] \to T$ is also an injection, and thus $R[S^{-1}]$ is isomorphic to its image in T, which is $\{rs^{-1} : r \in R, s \in S\}$. □

Example 9. Any subring of $\mathbb{Q}$ contains $\mathbb{Z}$, since $n = 1 + \cdots + 1$, taken n times. Thus, any *subfield* of $\mathbb{Q}$ contains the field of fractions of $\mathbb{Z}$, which can be identified naturally with $\mathbb{Q}$, so the only subfield of $\mathbb{Q}$ is $\mathbb{Q}$ itself.

Next one could ask, "Is there a ring W such that $\mathbb{Z} \subset W \subset \mathbb{Q}$?" To answer this, we can try $W = \mathbb{Z}[S^{-1}]$ when S is any submonoid of $\mathbb{Z}$, for we can view $T \subseteq \mathbb{Q}$ by Corollary 8. Here is one instance when T lies properly between $\mathbb{Z}$ and $\mathbb{Q}$.

Example 10. Take $s > 1$ in $\mathbb{N}$ and let $S = \{s^i : i \in \mathbb{N}\}$. Write $\mathbb{Z}[\frac{1}{s}]$ for $\mathbb{Z}[S^{-1}]$. Then $\mathbb{Z} \subset \mathbb{Z}[\frac{1}{s}]$ since $s^{-1} \notin \mathbb{Z}$. Also if $s|t$, then $\mathbb{Z}[\frac{1}{s}] \subseteq \mathbb{Z}[\frac{1}{t}]$, and the inclusion is certainly proper if t does not divide any power of s (for if $\frac{1}{t} = \frac{n}{s^i}$ then $s^i = tn$). In particular, we have an infinite chain

$$\mathbb{Z} \subset \mathbb{Z}[\frac{1}{2}] \subset \mathbb{Z}[\frac{1}{6}] \subset \mathbb{Z}[\frac{1}{30}] \subset \mathbb{Z}[\frac{1}{210}] \subset \ldots$$

These results are examined further in Exercises 4*ff*.

Exercises

1. Write the details of the proofs of associativity and distributivity for Proposition 3.
2. $\mathbb{Z}[\mathbb{N} \setminus \{0\}]^{-1} \approx \mathbb{Z}[\mathbb{Z} \setminus \{0\}]^{-1} \approx \mathbb{Q}$.
3. Suppose S is an arbitrary submonoid of a commutative ring R. Define $\sim$ on $R \times S$ by $(r_1, s_1) \sim (r_2, s_2)$ if $r_1 s_2 s = r_2 s_1 s$ for some s in S. This is an equivalence relation, and generalizing the construction in the text one can build $R[S^{-1}]$ together with a homomorphism $R \to R[S^{-1}]$ whose kernel is $\{r \in R : rs = 0 \text{ for some } s \text{ in } S\}$. (In particular, if $0 \in S$ then $R[S^{-1}] = 0$.) Prove the analog of Corollary 8.

Subrings of $\mathbb{Q}$

4. If p is a prime number then there is no ring W properly contained between $\mathbb{Z}$ and $\mathbb{Z}[\frac{1}{p}]$.
5. For any prime number p define $P = p\mathbb{Z}$ and $\mathbb{Z}_P = \{\frac{m}{n} : p \text{ does not divide } n\}$. Write this in the form $\mathbb{Z}[S^{-1}]$, for a suitable submonoid S of $\mathbb{N}$.
6. With notation as in Exercise 5, show that $\mathbb{Z}_P$ has the unique maximal ideal $p\mathbb{Z}_P = \{\frac{pm}{n} : p \text{ does not divide } n\}$.
7. Every subring of $\mathbb{Q}$ has the form $\mathbb{Z}[S^{-1}]$ for a suitable submonoid S of $\mathbb{Z}$. (Hint: Arguing as in Exercise 4, show that S is the set of denominators of fractions expressed in lowest terms.)
8. Given any finite set S_0 of distinct prime numbers $\{p_1, \ldots p_t\}$, let S be the submonoid of $\mathbb{N}$ "generated" by S_0; *i.e.*, S is comprised of

those natural numbers of the form $\{p_1^{m_1} \dots p_t^{m_t} : p_i \in S_0,\ m_i \in \mathbb{N}\}$. Prove $\mathbb{Z}[S^{-1}] \approx \mathbb{Z}[\frac{1}{n}]$ where $n = p_1 \dots p_t$.

9. Every subring of $\mathbb{Q}$ can be obtained by taking a suitable set S_0 of distinct prime numbers (not necessarily finite), letting S be the submonoid of $\mathbb{N}$ "generated" by S_0, and forming $\mathbb{Z}[S^{-1}]$. (Hint: Start with Exercise 7.)
10. Assumptions as in Proposition 6. Every ideal of $R[S^{-1}]$ has the form AS^{-1} for suitable $A \triangleleft R$.
11. Weaken the hypothesis in Proposition 7 to R commutative and S a submonoid of R, *cf.* Exercise 14.6
12. Propositions 6 and 7 and Corollary 8 remain valid, if we replace the condition R commutative by "$sr = rs$ for all r in R, s in S." Furthermore, under this condition, one need not assume that T is commutative. (There is a further weakening of this condition for noncommutative rings, which is outside the scope of these notes.)

CHAPTER 16. POLYNOMIALS AND EUCLIDEAN DOMAINS

As indicated in the introduction to Part II, some of the most beautiful results of algebra involve polynomials. Upon first acquaintance, a polynomial seems rather ethereal – a formal expression involving an indeterminate x and coefficients from a given ring R – and one might expect that we must substitute some value for x in order to obtain meaningful results. However, it turns out that the collection of all these polynomials can be given the structure of a ring having many nice properties which are inherited from R, and in this sublime transition from chaos to algebraic structure, x becomes a very meaningful element of the new ring.

The polynomial ring over a field shares several key properties with $\mathbb{Z}$, one of which is the "Euclidean algorithm," which enables us to divide one element into another and check whether the remainder is 0. This property is so important that we use it to define a class of rings, called *Euclidean domains*, our first common generalization of polynomial rings and of $\mathbb{Z}$; we conclude this section by developing the basic properties of number theory in these rings.

To start with, we must settle on the "correct" definition of polynomial with coefficients in a ring R. In all innocence one might write down a typical polynomial, such as $2x^3 + 3x + 7$, but we must be careful, since then $0x^4 + 2x^3 + 3x + 7$ certainly should be considered to be the "same" polynomial. To avoid complications involving equivalence relations, we define instead a polynomial (over R) to be a formal infinite sum $\sum_{i\in \mathbb{N}} a_i x^i$, where each $a_i \in R$ and almost all $a_i = 0$. The nonzero a_i are called the *coefficients* of the polynomial. Thus, $2x^3 + 3x + 7$ is considered shorthand notation for $\sum_{i\in \mathbb{N}} a_i x^i$ where $a_0 = 7$, $a_1 = 3$, $a_2 = 0$, $a_3 = 2$, and $a_i = 0$ for all $i \geq 4$.

The Ring of Polynomials

Write $R[x]$ for the set of polynomials over R. Defining addition componentwise, *i.e.*, $\sum a_i x^i + \sum b_i x^i = \sum (a_i + b_i) x^i$, we have

Remark 1. $(R[x], +, 0)$ is an Abelian group, where

$$0 = \sum 0x^i \quad \text{and} \quad -(\sum a_i x^i) = \sum (-a_i) x^i.$$

It remains to define multiplication of polynomials. Clearly, we want $x^i x^j = x^{i+j}$, and thus

$$(\sum a_i x^i)(\sum b_j x^j) = \sum_{i,j} a_i b_j x^{i+j}.$$

However, the expression on the right is not written as a polynomial, since each power of x does not have a unique coefficient; instead we note

$$\sum_{i,j} a_i b_j x^{i+j} = \sum_u (\sum_{i=0}^{u} a_i b_{u-i}) x^u,$$

and thus define formally

$$(\sum a_i x^i)(\sum b_j x^j) = \sum c_u x^u, \tag{1}$$

where $c_u = \sum_{i=0}^{u} a_i b_{u-i}$ is calculated in R.

PROPOSITION 2. *Suppose R is a ring. Then $R[x]$ is a ring under the given addition and multiplication, and is commutative if R is commutative.*

Proof. The ring axioms are routine to check, where the unit element is $1 + 0x + 0x^2 + \dots$; associativity of multiplication is the most cumbersome and is left to the reader, *cf.* Exercise 1. More streamlined approaches are given in Exercises 2 and 3. □

Digression. One might wonder what would happen if we dropped the condition for $\sum_{i \in \mathbb{N}} a_i x^i$ that almost all the a_i are 0, *i.e.*, if we dealt with "power series" instead of polynomials. Surprisingly, one still winds up with a ring, which is very useful and is described in Exercises 7ff.

Now that we have defined $R[x]$, let us study some of its properties as a ring. The key is the *degree* of a nonzero polynomial $\sum a_i x^i$, defined to be the largest number u for which $a_u \neq 0$; a_u is called the *leading coefficient.* We say f is *monic* if its leading coefficient is 1. The degree of a polynomial f is denoted as $\deg f$. For example, if $f = 2x^3 + 3x + 8$, then $\deg f = 3$. A polynomial of degree 0 is called a *constant polynomial.*

Remark 3. There is an injection $R \to R[x]$ given by $r \mapsto r + 0x + 0x^2 + \dots$; in this way we can identify R with the constant polynomials.

Remark 4. Suppose R is a domain, and $0 \neq f, g \in R[x]$. A glance at (1) shows $fg \neq 0$, and

$$\deg(fg) = \deg f + \deg g. \tag{2}$$

COROLLARY 5. *If R is a domain, then $R[x]$ is also a domain.*

Example 6. $\mathbb{Z}[x]$ is an integral domain. $F[x]$ is an integral domain, for any field F.

Now recall the Euclidean algorithm for $\mathbb{Z}$ (which says that for any numbers $a, b \neq 0$ there are integers q, r (for "quotient" and "remainder" respectively) with $0 \leq r < b$, such that $a = bq + r$.

PROPOSITION 7 (EUCLIDEAN ALGORITHM FOR POLYNOMIALS). *Suppose R is any ring, and $f, g \in R[x]$, with g monic. Then there are polynomials q, r in $R[x]$ such that $f = qg + r$, where either $r = 0$ or $\deg r < \deg g$.*

The proof is obtained by examining division for polynomials, learned (without proof) in high school, as exemplified by

$$\begin{array}{r l}
 & \quad 2x^2 + 6x + 21 \\
x - 3 & \overline{)2x^3 + 0x^2 + 3x + 7} \\
 & \underline{2x^3 - 6x^2} \\
 & \qquad 6x^2 + 3x \\
 & \qquad \underline{6x^2 - 18x} \\
 & \qquad\qquad 21x + 7 \\
 & \qquad\qquad \underline{21x - 63} \\
 & \qquad\qquad\qquad 70
\end{array}$$

where here $f = 2x^3 + 3x + 7$ and $g = x - 3$.

Proof of Proposition 7. Induction on n, where $n = \deg f$ and $m = \deg g$. If $n < m$, then we are done, by taking $q = 0$ and $r = f$, so assume $n \geq m$. Let $h = f - ax^{n-m}g$, where a is the leading coefficient of f. The leading terms of f and $ax^{n-m}g$ are each ax^n, and thus cancel each other in h, so $\deg h < n$. By induction on n we can write $h = qg + r$ where $\deg r < \deg g$; hence

$$f = h + ax^{n-m}g = (q + ax^{n-m})g + r,$$

which has the desired form. □

Euclidean Domains

Having seen that the Euclidean algorithm is the key to studying polynomials, we are ready for a major breakthrough — namely, we use the Euclidean algorithm to define a common generalization of $\mathbb{Z}$ and $F[x]$. Then, proving the basic theorems of arithmetic (well-known for $\mathbb{Z}$) in this more general setting, we shall have them instantly for $F[x]$.

Definition 8. A *Euclidean domain* is an integral domain R together with a "degree function" $d: R \setminus \{0\} \to \mathbb{N}$ satisfying:

(Euc1) $d(a) \leq d(ba)$ for any a, b in $R \setminus \{0\}$;

(Euc2) for any a, b in $R \setminus \{0\}$ there are q, r in R such that

$$b = aq + r,$$

with either $r = 0$ or $d(r) < d(a)$.

Example 9. (i) $\mathbb{Z}$ is Euclidean, when we take $d(a)$ to be $|a|$; indeed, (2) is the familiar Euclidean algorithm for integers, and (1) follows from the fact that $|b| \geq 1$ for any integer $\neq 0$ (for then $|ab| = |a||b| \geq |a|$).

(ii) $F[x]$ is Euclidean for any field F, where $d(\ )$ is the degree of the polynomial. Indeed, we have seen that $F[x]$ is an integral domain. $d(fg) = d(f) + d(g) \geq d(f)$ for any polynomial $g \neq 0$, yielding (Euc1); (Euc2) is the Euclidean algorithm for polynomials. (Since F is a field, we may divide through by the leading coefficient of the polynomial g and assume g is monic.)

The case $r = 0$ leads us to the following basic notion.

Definition 10. a "divides" b, written $a|b$, if $b = qa$ for some q in R, or, equivalently, if $b \in Ra$. We say a is an *associate* of b if $Ra = Rb$, *i.e.*, if $a|b$ and $b|a$.

Remark 10′. In an integral domain R, if a and b are associates then $a = ub$ for some invertible u in R. (Indeed, write $a = ub$ and $b = va$. Then $a = u(va)$, implying $uv = 1$.)

Remark 11. If $A = Ra$ in a Euclidean domain R, then $d(b) \geq d(a)$ for every $0 \neq b \in A$. (Indeed, write $b = qa$ for $q \in R$, and apply (Euc1).)

Consequently, if $Ra = Rb$, then $d(a) = d(b)$. The converse also holds.

PROPOSITION 12. *Suppose R is a Euclidean domain with degree function d, and $A \triangleleft R$. If a has minimal degree of all nonzero elements in A, then $A = Ra$.*

Proof. For any $0 \neq b \in A$, write $b = qa + r$ for $r = 0$ or $d(r) < d(a)$. But $r = b - qa \in A$ since $b \in A$ and $qa \in A$, so we cannot have $d(r) < d(a)$. Hence $r = 0$. □

COROLLARY 13. *If $b \in Ra$ and $d(a) = d(b)$ then $Ra = Rb$.*

Our next corollary is the key to the theory, since it gives the connection between the degree function (which is rather far removed from the abstract notion of ring) and the ideals of the ring. In fact, one has

COROLLARY 14. *In any Euclidean domain R, if $A \triangleleft R$, then $A = Ra$ for suitable a in R.*

Proof. Take $0 \neq a \in A$ of minimal possible degree. □

Let us turn to the building blocks of arithmetic. The invertible elements are "too good," so we usually disregard them.

Unique Factorization

Definition 15. A noninvertible element $p \neq 0$ of a ring R is *irreducible* if whenever $p = ab$ we have a or b invertible.

Our first goal is to obtain the factorization of an arbitrary elements into irreducibles.

PROPOSITION 16. *In a Euclidean domain, any noninvertible element $r \neq 0$ can be factored into irreducibles.*

Proof. By induction on $d(r)$. The result is obvious unless r is reducible, i.e., $r = ab$ for a, b not associates of r. By Corollary 13, $d(a) < d(r)$ and $d(b) < d(r)$, so, by induction, each of a and b has a factorization into irreducibles; putting these together yields the desired factorization for r. □

Now we turn to uniqueness of the factorization. We say two factorizations into irreducibles, $r = p_1 \dots p_k$ and $r = q_1 \dots q_m$, are *equivalent* if $m = k$ and there is a permutation π of $\{1, \dots, k\}$ such that $q_{\pi i}$ and p_i are associates for $1 \leq i \leq k$. The element r has *unique factorization* if all factorizations of r into irreducibles are equivalent.

Since "uniqueness" is only up to associates, it would make more sense to deal with the monoid of equivalence classes of elements (under associates), which, in fact, is what we do in practice, although we rarely admit it in public. (For example, we normally deal with $\mathbb{Z}^+$ instead of $\mathbb{Z}$, taking the positive prime number p as the representative of the class $\{\pm p\}$.) Thus, when we talk of "the" irreducible p in a factorization we really are referring to its class of associates. To analyze the uniqueness of factorizations, we need a new concept.

Definition 17. A noninvertible element $p \in R$ is *prime* if it satisfies the following property: If $p|ab$ then $p|a$ or $p|b$.

Remark 18. In an integral domain, any nonzero prime p is irreducible. (Indeed, suppose $p = ab$; then $a|p$ and $b|p$. But also $p|ab$, implying $p|a$ or $p|b$, so a or b is an associate of p.)

The point of this definition is seen in the next lemma.

LEMMA 19. *If a prime p divides $r_1 \dots r_m$, then p divides some r_i; if furthermore r_i is prime, then p and r_i are associates.*

Proof. By definition $p|r_1 \dots r_{m-1}$ or $p|r_m$, so the first assertion follows by induction on m. The second assertion is clear, by Remark 18. □

PROPOSITION 20. *In an integral domain, all factorizations of a given element r into primes are equivalent.*

Proof. Induction on k, where we write $r = p_1 \dots p_k$ with each p_i prime. Suppose $r = q_1 \dots q_m$ with each q_j prime. Then p_k divides some q_j; since primes are irreducible, we see p_k is an associate of q_j. Writing $q_j = up_k$ we have

$$p_1 \cdots p_{k-1} = uq_1 \cdots q_{j-1}q_{j+1} \cdots q_m. \tag{3}$$

Noting uq_1 is prime, we apply induction on k to see that the factorizations in (3) are equivalent; thus $k-1 = m-1$, and $uq_1, q_2, \ldots, q_{j-1}, q_{j+1}, \ldots, q_m$ is a permutation of associates of $p_1, \ldots, p_{k-1}$. Then we are done, since q_1 and uq_1 are associates. □

To obtain unique factorization in Euclidean domains, it remains to show that every irreducible is prime. One way is to translate the definitions of "irreducible" and "prime" to the structure of rings. We say a divisor a of b is *proper* if a is not an associate of b.

Remark 21. (i) An element p is irreducible, iff every proper divisor of p is necessarily invertible. In other words, p is irreducible iff whenever $Rp \subset Ra$ one necessarily has $Ra = R$.

(ii) An element p is prime iff whenever $ab \in Rp$ we have $a \in Rp$ or $b \in Rp$ (*i.e.*, iff R/Rp is an integral domain).

PROPOSITION 22. *In a Euclidean domain R, any irreducible element p is prime.*

Proof. Otherwise R/Rp is not an integral domain, and thus not a field, so by Proposition 14.7 R/Rp has a proper ideal $A/Rp \neq 0$, *i.e.*, $Rp \subset A \triangleleft R$. But then A has the form Ra by Proposition 12, contrary to Remark 21(i). □

Let us put everything together.

Definition 23. A *unique factorization domain* (UFD) is an integral domain for which each noninvertible element has unique factorization.

THEOREM 24. *Every Euclidean domain is a UFD.*

Proof. Combine Propositions 16, 20, and 22. □

It is easy to see that unique factorization is the key to many results about number theory. For example, we can compute the *greatest common divisor* of a and b as the product of those irreducibles (counting multiplicity) that are common to the factorizations of a and b; a and b are *relatively prime* precisely when they have no common irreducible divisors. The following observation is critical:

Remark 25. In a UFD, every irreducible element is prime. (Indeed, we want to show that a given irreducible p is prime. Suppose p divides ab. Then p appears in the factorization of ab into irreducibles. But by uniqueness this is the product of the respective factorizations of a and b into irreducibles; hence p (or an associate) appears in one of these factorizations, and so divides a or b.)

This remark can be strengthened to characterize UFDs *cf.* Exercise 21. An application:

Example 26. Suppose $ab = c^n$ in a UFD, where a and b are relatively prime. Then $a = uc_1^n$ and $b = u^{-1}c_2^n$, where u is invertible and c_1, c_2 are relatively prime. (Proof: Any prime p dividing a must appear with multiplicity divisible by n, since p divides c^n and does not appear in the factorization of b. Thus a and b are associates of nth powers, *i.e.*, $a = uc_1^n$ and $b = vc_2^n$ with u, v invertible. Hence $c^n = ab = uv(c_1c_2)^n$, and comparing factorizations we see c and c_1c_2 are associates. Write $c = wc_1c_2$ where w is invertible; replacing c_1 by wc_1 we may assume $w = 1$. But then $uv = 1$, so $v = u^{-1}$.)

Exercises

1. Verify the ring axioms for $R[x]$. (*Hint*: The hardest verification is associativity of multiplication, which follows from

$$((\sum a_ix^i)(\sum b_jx^j))(\sum c_kx^k) = (\sum_u(\sum_{i=0}^{u} a_ib_{u-i})x^u)(\sum c_kx^k)$$
$$= (\sum_v(\sum_{u=0}^{v}\sum_{i=0}^{u} a_ib_{u-i}c_{v-u}x^k).)$$

2. (Another way of writing polynomials) Write the infinite vector (a_i) for the polynomial $\sum a_ix_i$. Then the ring operations in $R[x]$ become $(a_i) + (b_i) = (a_i + b_i)$ and $(a_i)(b_j) = (c_u)$, where $c_u = \sum_{i=0}^{u} a_ib_{u-i}$.
3. Prove slickly that $R[x]$ is a ring, by means of the regular representation (Exercise 14.4): Identify $R[x]$ as a subring of $\mathrm{Hom}(R[x], R[x])$.
4. Describe explicitly the field of fractions of $F[x]$, where F is a field.
5. Give an example in a Euclidean domain R where $d(a) = d(b)$ but $Ra \neq Rb$.
6. In a Euclidean domain, an element b is invertible iff $d(b) = d(1)$.

Formal Power Series

Exercises 7 through 20 introduce formal power series, and illustrate some of their varied applications.

7. The *ring of formal power series* $R[[x]]$ is defined as the set of all formal (infinite) sums $\sum_{i\in\mathbb{N}} a_ix^i$, where each $a \in R$, with addition defined componentwise, *i.e.*, $\sum a_ix^i + \sum b_ix^i = \sum(a_i + b_i)x^i$, and

multiplication given via the formula

$$(\sum_{i\in\mathbb{N}} a_i x^i)(\sum_{j\in\mathbb{N}} b_j x^j) = \sum_{u\in\mathbb{N}} c_u x^u,$$

where $c_u = \sum_{i=0}^{u} a_i b_{u-i}$ is calculated in R. Show that $R[[x]]$ is indeed a ring, and $R[x]$ is a subring of $R[[x]]$. Arguing by means of the lowest order term, show that $R[[x]]$ is an integral domain if R is an integral domain.

The point of working in $R[[x]]$ is that often one has a concrete way of describing the inverse of a polynomial.

8. $(1-x)^{-1} = 1 + x + x^2 + \dots$, in $R[[x]]$; more generally if $f \in R[[x]]$ has constant term 0 then $(1-f)^{-1} = 1 + f + f^2 + \dots$. (The point here is to explain why the right-hand side makes sense.)
9. If F is a field then any power series g in $F[[x]]$ with nonzero constant term is invertible. (*Hint*: Write $g = \alpha(1-f)$.)
10. Noting that $\frac{1}{(a+x)(b+x)} = \frac{1}{a-b}(\frac{1}{b+x} - \frac{1}{a+x})$, give an explicit formula (over $\mathbb{C}$) for the power series corresponding to the inverse of a quadratic polynomial having nonzero constant term.
11. As in exercise 2, any infinite sequence $(a_0, a_1, \dots)$ corresponds to the formal power series $f = a_0 + a_1 x + \dots$. In particular, take f corresponding to the Fibonacci series $(1, 1, 2, 3, 5, 8, \dots)$, and note $f = 1 + xf + x^2 f$. Hence $f = (1 - x - x^2)^{-1}$; use Exercise 10 to produce a closed formula for the Fibonacci coefficients.
12. Given a power series $f = \sum_{i\in\mathbb{N}} a_i x^i$, define $f(0)$ formally as a_0, and define the formal derivative $f' = \sum_{i\geq 1} i a_i x^{i-1}$; define inductively $f^{(0)} = f$ and $f^{(n)} = f^{(n-1)'}$. Prove $(f+g)' = f' + g'$ and $(fg)' = f'g + fg'$. Prove the formal version of Maclaurin's expansion:

$$f = \sum_{n\in\mathbb{N}} \frac{1}{n!} f^{(n)}(0) x^n.$$

13. If $f(0) = 1$, then $(f^n)' = nf^{n-1}f'$, for any integer n.
14. Prove the "binomial expansion" for formal power series, for any rational number n:

$$(1+ax)^n = 1 + nax + \frac{n(n-1)}{2!}a^2x^2 + \frac{n(n-1)(n-2)}{3!}a^3x^3 + \dots$$

(*Hint*: apply Exercise 12 to $f = 1 + ax$.)

15. If $f(0) = 0$, define $\exp(f) = 1 + f + f^2 + \dots$. Show $\exp(f)' = \exp(f)f'$. Likewise if $g(0) = 1$, define $\log(g) = g - \frac{1}{2}g^2 + \frac{1}{3}g^3 \pm \dots$; show $(\log g)' = g'g^{-1}$. Also show $\log\exp(f) = f$ and $\exp\log(g) = g$.

The Partition Number

In Exercises 16 through 20 we shall use formal power series, coupled with a clever combintorial argument, to compute the partition number P_n (*cf.* Exercise 7.13). We shall write each partition of n in descending order, *i.e.*, $n = n_1 + n_2 + \cdots + n_k$ with $n_1 \geq n_2 \geq \cdots \geq n_k$.

16. If $f_i(0) = 1$ for each i, then the infinite product $f_1 f_2 f_3 \ldots$ makes sense as a formal power series, since one can determine each coefficient as a finite sum.
17. If $f_i = (1 - x^i)^{-1} = 1 + x^i + x^{2i} + \ldots$, then $f_1 f_2 \cdots = \sum_{n=0}^{\infty} P_n x^n$, where formally we take $P_0 = 1$.
18. Let us call a partition $n = n_1 + n_2 + \cdots + n_k$ *strictly descending* if $n_1 > n_2 > \cdots > n_k$. Let Q_n denote the number of strictly descending partitions of n. The partition $n = n_1 + n_2 + \cdots + n_k$ is called *even* (resp. *odd*) if k is even (resp. odd). Define Q_n^+ (resp. Q_n^-) to be the number of even (resp. odd) strictly descending partitions of n. For example, $P_4 = 5$, and $Q_4^+ = Q_4^- = 1$. Show

$$(1-x)(1-x^2)(1-x^3)\cdots = \sum_{n=0}^{\infty}(Q_n^+ - Q_n^-)x^n$$
$$=1 - x - x^2 + x^5 + x^7 - x^{12} - x^{15} \pm \ldots .$$

Conclude that $(\sum_{n=0}^{\infty} P_n x^n)(\sum_{n=0}^{\infty}(Q_n^+ - Q_n^-)x^n) = 1$. Deduce the formula $P_n = \sum_{m=1}^{n}(Q_m^- - Q_m^+)P_{n-m}$.

19. Prove that $Q_m^+ = Q_m^-$ unless m has the form $\frac{3j^2 \pm j}{2}$, in which case $Q_m^+ - Q_m^- = (-1)^j$. Calculate $Q_m^- - Q_m^+$ for all values of $m \leq 20$. (Hint: Given any strictly descending partition $m_1 > m_2 > \cdots > m_k$ of m, take j maximal such that $m_j = m_1 - (j-1)$, *i.e.*, such that $m_1 = m_2 + 1 = m_3 + 2 = \cdots = m_j + (j-1)$. Let $i = m_k$. If $i < j$, then one can produce a partition of opposite parity, namely $m_1 + 1 > m_2 + 1 > \cdots > m_i + 1 > m_{i+1} > \cdots > m_{k-1}$; similarly if $i = j$ and $k > j$. On the other hand, if $j < i - 1$, one can reverse the procedure and get the partition $m_1 - 1 > m_2 - 1 > \cdots > m_{i-1} - 1 > m_i > \cdots > m_k > i - 1$; similarly if $j = i - 1$ and $k > j$. Thus all strictly descending partitions pair off in opposing parities, *except* in the case $j = i - 1$ or $j = i$, where also $k = j$. In each of these two cases, the unpaired partition must be $i + j - 1 > i + j - 2 > \cdots > i$, so $m = ij + \frac{j(j-1)}{2}$; substitute $i = j + 1$ and $i = j$.)
20. Using exercises 18 and 19, conclude *Euler's formula*:

$$P_n = P_{n-1} + P_{n-2} - P_{n-5} - P_{n-7} + P_{n-12} + P_{n-15} \pm \ldots .$$

(Precisely, $P_n = \sum_{m=(3j^2 \pm j)/2}(-1)^{j+1}P_{n-m}$.)

Unique Factorization Domains

21. The following two conditions are equivalent for an integral domain R to be a UFD:

    (i) Every irreducible is prime;

    (ii) There is no infinite sequence $r_1, r_2, \ldots,$ in R such that r_{i+1} is a proper divisor of r_i for each i. (*Hint*: ($\Rightarrow$) Write $\ell(r)$ for the number of primes in the factorization of r. Clearly, if $r = ab$, then $\ell(r) = \ell(a) + \ell(b)$; by hypothesis $\ell(r_1) > \ell(r_2) > \ldots$, so any such sequence has $\leq \ell(r_1)$ terms. ($\Leftarrow$) Mimic the proof of Proposition 16. Compare with Theorem 17.18).

Chapter 17. Principal Ideal Domains: Induction without Numbers

Although the applications given in the previous section are quite beautiful, the basic theorems on unique factorization might seem somewhat ad hoc, based on the fortuitous appearance of the Euclidean algorithm and the degree function, and are rather cumbersome to state. Actually one can rework the arithmetic theory more intrinsically in terms of ideals, so that many of the main results become easier to state, and in greater generality.

Definition 1. A PID (*principal ideal domain*) is an integral domain R in which every ideal is principal, *i.e.*, has the form Ra.

Example 2. Every Euclidean domain is a PID, by Corollary 16.14.

Actually, one can broaden Corollary 16.14 to get PID's which are not Euclidean, *cf.* Exercises 3,4, and 5.

The point of the PID is that it is defined concisely in terms of ideals and permits us to utilize ideal theory in its study. To this end, we shall compile a dictionary for integral domains that "translates" the relevant notions of number theory from elements to ideals. In an integral domain R, the element a corresponds to the ideal Ra, so it is appropriate to note at the outset that $RaRb = Rab$ (verification left to reader), which implies that the set of principal ideals of R inherits a monoid structure from the multiplication in R, *cf.* Exercise 1.

Elements	*Ideals*	*Comments*
$a \mid b$	$Rb \subseteq Ra$	$b \in Ra$ iff $Rb \subseteq Ra$
a associate to b	$Ra = Rb$	Immediate
p irreducible	$Rp \subset Ra \subset R$ is impossible.	Remark 16.21
p prime	If $RaRb \subseteq Rp$ then $Ra \subseteq Rp$ or $Rb \subseteq Rp$.	Remark 16.21
c divides a and b	$Ra + Rb \subseteq Rc$	Remarks 13.10, 14.4″
a and b both divide c	$Ra \cap Rb \supseteq Rc$	Remarks 13.10, 14.4″

Remark 2′. Let us continue this analysis. We say c is a *common divisor* of a and b if $c|a$ and $c|b$. The *greatest common divisor* $\gcd(a, b)$ of a and b is defined to be that common divisor d (of a and b) for which every common divisor c of a and b divides d. If $\gcd(a, b)$ exists then it must be unique up to associate.

Clearly, c is a common divisor of a and b iff $Ra \subseteq Rc$ and $Rb \subseteq Rc$, or equivalently $Ra + Rb \subseteq Rc$. It follows that if $Ra + Rb = Rd$, then $d = \gcd(a, b)$. Consequently, $d = \gcd(a, b)$ exists in any PID. Furthermore, we have the useful consequence that there are r, s in R for which $d = ra+sb$. (See Exercise 2 for a refinement of this observation.)

One could continue this reasoning and describe the lcm (*cf.* Exercise 12). However, we satisfy ourselves here by noting that two elements a and b are relatively prime if $Ra + Rb = R$; the converse is true when R is a PID. One method of checking the relative primeness of polynomials is given in Exercise 13.

Although "irreducible" and "prime" elements have been described in terms of principal ideals in arbitrary rings, they are especially significant in PIDs. In a PID, the criterion for p to be irreducible states precisely that Rp is a maximal ideal (*cf.*, however, Examples 10 and 11.) To obtain a similar neat condition for p to be prime, we need to introduce a new kind of ideal.

Prime Ideals

Definition 3. A (proper) ideal P of an arbitrary ring R is called *prime* if the following condition holds for arbitrary ideals A, B of R :

$$\text{If} \quad AB \subseteq P \quad \text{then} \quad A \subseteq P \quad \text{or} \quad B \subseteq P.$$

"Prime ideal" is one of the basic concepts of ring theory, involving the structure theory in several ways.

LEMMA 4. *The following conditions are equivalent for a commutative ring R:*

(i) *R is an integral domain;*

(ii) *The element 0 is a prime element of R;*

(iii) *The ideal 0 is a prime ideal of R;*

(iv) *The product of any two nonzero principal ideals of R is nonzero.*

Proof. (Note that an ideal $A \subseteq 0$ iff $A = 0$.)

$(i) \Leftrightarrow (ii)$ By Remark 16.21(ii).

$(ii) \Rightarrow (iii)$ Suppose $0 \neq A, B \triangleleft R$ with $AB = 0$. Taking $0 \neq a \in A$ and $0 \neq b \in B$, we see $RaRb \subseteq AB = 0$. Hence $Ra = 0$ or $Rb = 0$, contradiction.

$(iii) \Rightarrow (iv)$ Obvious.

$(iv) \Rightarrow (ii)$ By definition of prime element. □

LEMMA 5. *Suppose $I \subseteq P$ are ideals of R. Then P is a prime ideal of R, iff P/I is a prime ideal of R/I.*

Proof. We use Proposition 14.7 repeatedly. Write $\overline{R} = R/P$ and $\overline{P} = P/I$.

($\Rightarrow$) Suppose $\overline{A}, \overline{B} \triangleleft \overline{R}$ with $\overline{AB} \subseteq \overline{P}$. Then there are $A, B \triangleleft R$ with $\overline{A} = A/I$, $\overline{B} = B/I$. Clearly, $AB \subseteq P$ so $A \subseteq P$ or $B \subseteq P$; *i.e.*, $\overline{A} \subseteq \overline{P}$ or $\overline{B} \subseteq \overline{P}$.

($\Leftarrow$) Suppose $A, B \triangleleft R$ with $AB \subseteq P$. Then $(A/I)(B/I) \subseteq \overline{P}$ in $\bar{R}$, so $A/I \subseteq \overline{P}$ or $B/I \subseteq \overline{P}$. We conclude $A \subseteq P$ or $B \subseteq P$. □

PROPOSITION 6. *An ideal P of a commutative ring R is prime iff R/P is an integral domain.*

Proof. Combine Lemmas 4 and 5. □

Thus an element p of a PID R is prime, iff Rp is a prime ideal.

Remark 7. Recall (Corollary 14.8) that A is a maximal ideal of a commutative ring R iff R/A is a field. Since every field is an integral domain, we conclude that every maximal ideal is prime.

PROPOSITION 8. *A nontrivial ideal of a PID is prime iff it is maximal.*

Proof. ($\Rightarrow$) If $Rp \neq 0$ is a prime ideal then $p \neq 0$ is a prime element and thus is irreducible, by Remark 16.18; hence Rp is a maximal ideal.

($\Leftarrow$) by Remark 7. □

COROLLARY 9. *In a PID, a nonzero element is prime iff it is irreducible.*

Example 10. The polynomial ring $R = \mathbb{Z}[x]$ is *not* a PID. Indeed the substitution $x \mapsto 0$ yields an onto homomorphism $\psi: R \to \mathbb{Z}$. Then $R/\ker\psi \approx \mathbb{Z}$ by Noether I, so $\ker\psi$ is a prime ideal of R that is not maximal (since $\mathbb{Z}$ is not a field). But $f(x) \in \ker\psi$ iff $f(0) = 0$, iff $x|f$, so $\ker\psi = Rx \neq 0$. In view of Proposition 8 we conclude that R is not a PID. Note that here x is irreducible although Rx is not a maximal ideal.

Example 11. Define the polynomial ring $R = F[x_1, x_2]$ in two indeterminates over a field F to be $(F[x_1])[x_2]$. R is not a PID. Indeed define the homomorphism $\psi: R \to F[x_1]$ by $x_2 \mapsto 0$; as in Example 10, $\ker\psi = Rx_2$ is a nonzero prime, nonmaximal ideal.

Fact 12. In a PID, "unique factorization" means that every nonzero ideal A can be written as a product $P_1 \dots P_t$ of maximal ideals, which is uniquely determined in the sense that if $P_1 \dots P_t = Q_1 \dots Q_u$ for maximal ideals $P_1, \dots, P_t, Q_1, \dots, Q_u$ then $t = u$ and the $Q_1, \dots, Q_t$ are a reassortment of the $P_1, \dots, P_t$.

Noetherian Rings

We want to reprove Theorem 16.24, under the weaker hypothesis that R is a PID. Uniqueness of any factorization holds by Proposition 16.20 (since in any PID, irreducibles are prime). However, we need to prove the existence of a factorization into irreducibles, thereby requiring a substitute for the induction argument used in the proof of Proposition 16.16. This is a tall order, since in general a PID has no number with which to build an induction argument; nevertheless there is is a way that bears the name of E. Noether, its discoverer.

To motivate the argument, let us review the proof of Proposition 16.16 applied to some positive number, say 48. To prove 48 is factorizable into primes, we could first factor 48 into two proper divisors, say $48 = 3 \cdot 16$, and then note by induction that 3 and 16 are factorizable into primes. But we did not use the full induction hypothesis — indeed we only needed to apply induction to *divisors* of 48. Concerning an element a in general, our hypothesis need apply only to proper divisors of a, or, translating to ideals in a PID, we need consider only principal ideals properly containing Ra. Carrying this idea further, we want to be sure that a given process must terminate when we pass on to larger and larger principal ideals. This leads us to the following definition.

Definition 13. A ring R satisfies the *ascending chain condition* on ideals (denoted ACC(ideals)) if any chain $A_1 \subseteq A_2 \subseteq A_3 \subseteq \dots$ of ideals necessarily terminates, *i.e.*, there is n such that $A_n = A_{n+1} = \dots$. Rings satisfying ACC(ideals) are also called *Noetherian.*

Example 14. Any Euclidean domain R is Noetherian; indeed if $Ra_1 \subseteq Ra_2 \subseteq Ra_3 \subseteq \dots$ then $\deg a_1 \geq \deg a_2 \geq \deg a_3 \geq \dots$, so this chain must terminate. More generally we have

PROPOSITION 15. *Any PID is Noetherian.*

Proof. Suppose $A_1 \subseteq A_2 \subseteq \dots$ are ideals of R. Then Remark 14.4″ implies $\cup_{i \geq 1} A_i \triangleleft R$ and so has the form Ra, by hypothesis. Hence $a \in A_n$ for some n, implying $A \subseteq A_n \subseteq A_{n+1} \subseteq \dots \subseteq A$, yielding $A_n = A_{n+1} = \dots$. □

ACC(ideals) implies the following formally stronger property:

PROPOSITION 16. *R is Noetherian iff every set $\mathcal{S}$ of ideals of R has a maximal member (i.e., some ideal of $\mathcal{S}$ is maximal among those in $\mathcal{S}$).*

Proof. (⇒) Take any ideal A_1 in $\mathcal{S}$. If A_1 is maximal, then we are done; otherwise $\mathcal{S}$ has some $A_2 \supset A_1$. Continuing in this way we take

$$A_1 \subset A_2 \subset A_3 \subset \dots$$

and by definition are forced to stop at some A_n; thus A_n is maximal in $\mathcal{S}$.

($\Leftarrow$) Given $A_1 \subseteq A_2 \subseteq A_3 \subseteq \ldots$ let $\mathcal{S} = \{A_1, A_2, A_3, \ldots\}$; some A_n is maximal in $\mathcal{S}$, so $A_n = A_{n+1} = \ldots$. □

COROLLARY 17. *(But see Aside 20.) Any ideal of a Noetherian ring is contained in a maximal (proper) ideal.*

Proof. In Proposition 16, take $\mathcal{S}$ to be the set of proper ideals. □

ACC provides a suitable substitute to mathematical induction. One technique of verifying a given property $\mathcal{P}$ on a class of ideals of a Noetherian ring R, is to assume on the contrary that $\mathcal{P}$ does not hold for all such ideals; then take an ideal maximal with respect to *not* satisfying $\mathcal{P}$, and arrive at a contradiction. This method is called *Noetherian Induction.* Here is a good illustration.

THEOREM 18. *Every PID is a UFD.*

Proof. As observed above, we only need to prove that a factorization into irreducibles exists. If not, take Ra_0 maximal of all principal ideals Ra for which a fails to have a factorization into irreducibles. Clearly, a_0 itself is not irreducible, so Ra_0 is properly contained in a maximal ideal $P = Rp$. Write $a_0 = bp$. Then $Ra_0 \subset Rb$, so by hypothesis there is a factorization $b = p_1 \ldots p_k$ into irreducibles; then $a_0 = p_1 \ldots p_k p$, contradiction. □

ACC(ideals) turns out to be the key hypothesis in the structure theory of commutative rings, in view of Exercises 19, 20, and 21; for example, for any primitive root ρ of 1, the important ring $\mathbb{Z}[\rho]$ is Noetherian, even though $\mathbb{Z}[\rho]$ need not be a UFD. Easier examples of this phenomenon are given in Exercise 6. In general, ideals have replaced elements in the structure theory. Prime ideals have turned out to be the "correct" generalization of prime elements, even for non-UFDs.

It is not so easy to construct an example of a ring that is *not* Noetherian. One approach is given in Exercise 24; however, I prefer the following example, since it illustrates how to transfer "bad" properties from groups to rings.

Example 19. Let $R = \{f(x) \in \mathbb{Q}[x] : f(0) \in \mathbb{Z}\}$. Thus $f \in R$ iff $f = m + xg$, where $m \in \mathbb{Z}$ and $g \in \mathbb{Q}[x]$. Clearly, R is a ring, and for any subgroup G of $(\mathbb{Q}, +)$ we have the corresponding ideal $Gx + x^2\mathbb{Q}[x]$ of R. Now let $G_i = \{\frac{m}{i} : m \in \mathbb{Z}\}$, a subgroup of $(\mathbb{Q}, +)$, and let A_i denote the corresponding ideal of R. The infinite ascending chain of subgroups

$$G_2 \subset G_4 \subset G_8 \subset \ldots$$

translates to an ascending chain of ideals $A_2 \subset A_4 \subset A_8 \subset \ldots$ in R.

We review different kinds of rings in increasing generality from left to right:

$$\begin{array}{ccccc} & & & \text{Noetherian} & \\ & & \nearrow & & \\ \text{Euclidean domain} & \rightarrow PID \rightarrow & & UFD \rightarrow & \text{Integral domain} \end{array}$$

Aside 20. Ironically, the conclusion of Corollary 17 holds without any hypothesis, by "Zorn's lemma," one of the pillars of ring theory:

Zorn's Lemma. Suppose A is any set and $\mathcal{S}$ is a collection of subsets of A, such that whenever $A_1 \subset A_2 \subset \ldots$ with each A_i in $\mathcal{S}$ we have $\cup A_i \in \mathcal{S}$. Then $\mathcal{S}$ contains some maximal member.

The set of proper ideals of an arbitrary ring R satisfies the condition of Zorn's lemma, since if $A_1 \subset A_2 \subset \ldots$ are proper ideals of R, then, clearly, $1 \notin \cup A_i$, implying $\cup A_i$ is a proper ideal. Zorn's lemma thus shows that R has maximal (proper) ideals. This application of Zorn's lemma requires the existence of a unit element of R.

The proof of Zorn's lemma relies on the "axiom of choice" from set theory, but this axiom is normally accepted by algebraists (because they do not want to give up Zorn's lemma!).

Exercises

1. In the "dictionary" add the following entry: the monoid of nonzero principal ideals of an integral domain R (with R as the neutral element) corresponds to the monoid of equivalence classes of nonzero elements of R (two elements being equivalent when they are associates). Show in the cases $R = \mathbb{Z}$ and $R = F[x]$ that this monoid can be identified with a suitable submonoid of R, namely $\mathbb{N}^+$ and {the monic polynomials} respectively. For this reason one sometimes deals more generally with unique factorization monoids instead of unique factorization domains. Fact 12 says that the monoid of nonzero ideals in a PID is a unique factorization monoid.
2. If $d = \gcd(a, b)$ in a UFD, then one can find r, s with r relatively prime to a, such that $d = ra + sb$. (*Hint*: Write $a = \hat{a}d$ and $b = \hat{b}d$. Then $1 = r\hat{a} + s\hat{b}$. Let q be the product of those primes dividing a that do *not* divide r, and let $r' = r + \hat{b}q$, $s' = s - \hat{a}q$. Then $\gcd(r', a) = 1$ since r', a have no common prime divisor; on the other hand, $r'a + s'b = d$.)

Counterexamples

3. Call an integral domain R *quasi-Euclidean* if R has a function $d: R \to \mathbb{N}$ satisfying the following property: If $d(x) \geq d(y)$, then either $y|x$ or there are $z, w \in R$ for which $d(xz - yw) < d(y)$. Modifying the proof of Theorem 16.11, show that any quasi-Euclidean domain is a PID.
4. Given any ring R and $S \subset R$, define $\hat{S} = \{s \in S : a + Rs \subseteq S$ for some a in $S\}$. Define $R_0 = R \setminus \{0\}$ and inductively $R_{i+1} = \hat{R}_i$ for each $i \geq 1$. Note that $R_1 = \{$noninvertible elements of $R_0\}$. Show that if R is Euclidean, then $\cap_{i \in \mathbb{N}} R_i = \emptyset$. (*Hint*: By induction, $d(s) \geq n$ for all s in R_n.)
5. Let $R = \left\{a + b\frac{\sqrt{-19}}{2} : a, b \in \mathbb{Z}\right\} \subset \mathbb{C}$. Show R is quasi-Euclidean (and thus a PID) but not Euclidean. (*Hint*: Given x, y relatively prime, we want to find z and w for which $d(xz - yw) < 1$. Write $\frac{x}{y} = \frac{a+b\sqrt{-19}}{c}$, where $c > 1$ and $(a, b, c) = 1$. For $c \geq 5$ take numbers d, e, f, q, r, for which $ae + bd + cf = 1$, $ad - 19be = cq + r$, $|r| \leq \frac{c}{2}$. Put $z = d + e\sqrt{-19}$ and $w = q - f\sqrt{-19}$. If $c = 2$, then a is odd [or else $y|x$], so take $w = \frac{a-1}{2} + \frac{b\sqrt{-19}}{2}$, $z = 1$. For $c = 3, 4$ take $z = a - b\sqrt{-19}$, and work out w. Hence R is quasi-Euclidean. To show R is not Euclidean it suffices to show $R_1 = R_2$ in the notation of Exercise 4. But $R_1 = \{R \setminus \{0, \pm 1\}$; if $s \in R_1 \setminus R_2$, then for each a in R_1 there is b in R for which $a + bs \notin R_1$, *i.e.*, $-bs \in \{a - 1,\ a,\ a + 1\}$. Taking $a = 2$, show that $-bs \in \{1, 2, 3\}$ so $s = \pm 2$ or ± 3. Now get a contradiction by taking $a = \frac{1+\sqrt{-19}}{2}$.)
6. $\mathbb{Z}[\sqrt{10}]$ is not a UFD. (*Hint*: 9 is a product of two primes in two inequivalent ways). Similarly, $\mathbb{Z}[\sqrt{-6}]$ is not a UFD. Using the fact that $\mathbb{Z}[x]$ is a UFD, to be shown in Chapter 20, conclude that an integral domain that is the homomorphic image of a UFD need not itself be a UFD.
7. Using the idea of Example 10, show that, for any n in $\mathbb{Z}$, the ideal of $\mathbb{Z}[x]$ generated by $x - n$ is prime but not maximal. (You might want to appeal to lemma 18.2.)
8. Explicit illustrations of Examples 10 and 11: $R = \mathbb{Z}[x]$ has the nonprincipal ideal $Rx + 2R$; $R = F[x_1, x_2]$ has the nonprincipal ideal $Rx_1 + Rx_2$.

Consequences of Zorn's Lemma

9. State the analog of Zorn's lemma, using descending chains instead of ascending chains. Nevertheless, many rings (such as $\mathbb{Z}$) fail to have minimal nonzero ideals. What goes wrong?

10. Suppose S is any submonoid of $R\setminus\{0\}$. Then R has a prime ideal P that is disjoint from S. (*Hint*: By Zorn's lemma there is an ideal of R that is maximal with respect to being disjoint from S; prove it is a prime ideal.)
11. An element a of a ring R is called *nilpotent* if $a^n = 0$ for suitable $n > 0$. An integral domain cannot contain any nonzero nilpotent elements. In an arbitrary commutative ring R, prove that the intersection of all the prime ideals of R consists precisely of the nilpotent elements of R. (*Hint*: If $a \in R$ is nilpotent, then a is contained in every prime ideal P since its image in R/P is 0; if a is not nilpotent, then by Exercise 10 there is a prime ideal disjoint from $\{a^n : n \in \mathbb{N}\}$.)
12. As in Remark 2′, one can define a *common multiple* of a and b; we write $c = \operatorname{lcm}(a, b)$, called the *least common multiple*, if c is a common multiple which divides every other common multiple. Show $c = \operatorname{lcm}(a, b)$ if $Ra \cap Rb = Rc$. Such c exists in a PID and is unique up to associate.
13. ("The resultant polynomial") Two polynomials $f(x) = \sum_{i=0}^{m} a_i x^i$ and $g(x) = \sum_{j=0}^{n} b_j x^j$ over a field F are relatively prime iff the following matrix is nonsingular:

$$A = \begin{pmatrix} a_m & a_{m-1} & \dots & a_0 & 0 & 0 & \dots & 0 \\ 0 & a_m & a_{m-1} & \dots & a_0 & 0 & \dots & 0 \\ \vdots & \vdots & \dots & \vdots & \vdots & \vdots & \dots & \vdots \\ 0 & 0 & \dots & \dots & \dots & \dots & \dots & a_0 \\ b_n & b_{n-1} & \dots & \dots & b_0 & 0 & \dots & 0 \\ 0 & b_n & \dots & \dots & b_1 & b_0 & \dots & 0 \\ \vdots & \vdots & \dots & \vdots & \vdots & \vdots & \dots & \vdots \\ 0 & 0 & \dots & \dots & \dots & \dots & \dots & b_0 \end{pmatrix}$$

(*Hint*: Let $B = \begin{pmatrix} x^{n+m-1} & 0 & 0 & \dots & & 0 \\ x^{n+m-2} & 1 & 0 & \dots & & 0 \\ x^{n+m-3} & 0 & 1 & \dots & & 0 \\ \vdots & & & \ddots & & \\ x & 0 & 0 & \dots & 1 & 0 \\ 1 & 0 & 0 & \dots & & 1 \end{pmatrix}$. Then

$$|A| x^{n+m-1} = |AB| = f(x)h(x) + g(x)k(x),$$

where $\deg h \le n-1$, $\deg k \le m-1$. If $|A| \ne 0$ and f, g have a nonconstant factor $r(x)$, then x divides r, so $a_0 = 0 = b_0$, implying $|A| = 0$ after all. Conversely, if $|A| = 0$, then $fh = -gk$; since $\deg f > \deg k$, conclude that some irreducible factor of f divides g.)

UFDs

(see also Exercise 16.21 and Exercises 20.3 through 20.7)

14. Define the ACC on principal ideals in anology to Definition 13. The ACC on principal ideals implies factorization (not necessarily unique) into irreducible elements.
15. Show that R is a UFD, iff R satisfies the ACC on principal ideals and every irreducible element is prime. (*Hint*: See Exercise 16.21)
16. If R is a UFD, then the intersection of any two principal ideals is principal; furthermore, any two elements of R have both a gcd and an lcm, and $\gcd(a,b)\text{lcm}(a,b) = ab$.
17. R is a UFD, iff every nonzero prime ideal of R contains a prime element. (*Hint*: Let $S = \{\text{finite products of prime elements}\}$. If $r \notin S$ then Exercise 10 implies that r is contained in a prime ideal P of R disjoint from S, contrary to the condition.)
18. In a UFD every *minimal* nonzero prime ideal (*i.e.*, does not contain any other nonzero prime ideal) is principal.

Noetherian Rings

19. We say an ideal L of a commutative ring R is *finitely generated* iff $L = Ra_1 + \cdots + Ra_t$ for suitable elements $a_1, \dots, a_t$ of R (and suitable t). Prove R is Noetherian iff every ideal of R is finitely generated. (*Hint*: ($\Rightarrow$) by the proof of Proposition 15.)
20. The homomorphic image of any Noetherian ring is Noetherian.
21. "Hilbert Basis Theorem" If R is a Noetherian ring then the polynomial ring $R[x]$ is Noetherian. (*Hint*: Given $L \triangleleft R[x]$, define $L_k = 0 \cup \{a \in R : a \text{ is the leading coefficient of some polynomial of degree} \leq k \text{ in } L\} \triangleleft R$, for each $k \geq 0$; write $L_k = \sum_{i=1}^{t(k)} Ra_{ik}$ for suitable a_{ik} in R, and take $f_{ik} \in L$ of degree $\leq k$ having leading coefficient a_{ik}. Note that $L_1 \subseteq L_2 \subseteq \dots$. Taking n such that $L_n = L_{n+1} = \dots$, conclude $L_k = \sum_{k=1}^{n} \sum_{i=1}^{t(k)} Rf_{ik}$.)
22. In a Noetherian ring, any ideal contains a product of prime ideals. (*Hint*: From the proof of Theorem 18.)
23. The rings of Exercise 6 are Noetherian, although they are not UFDs.
24. Generalizing Example 11, define inductively the polynomial ring $F[x_1, x_2, \dots, x_t]$, as $F[x_1, x_2, \dots, x_{t-1}][x_t]$, and show it is a Noetherian integral domain. Viewing $F[x_1, x_2, \dots, x_{t-1}] \subset F[x_1, x_2, \dots, x_t]$ in the natural way, define $R = \cup_{t=1}^{\infty} F[x_1, x_2, \dots, x_t]$. Note that

$$Rx_1 \subset Rx_1 + Rx_2 \subset Rx_1 + Rx_2 + Rx_3 \subset \dots$$

is an infinite ascending chain of ideals in R, so R is not Noetherian; however, R is a UFD, *cf.* Exercise 20.7.

Chapter 18. Roots of Polynomials

This chapter (and much of our subsequent material) grows out of the following innocuous application of the Euclidean algorithm for polynomials:

Remark 1. (Notation as in Proposition 16.7.) When R is commutative and $g = x - c$ with $c \in R$, we can calculate r directly, by means of "substituting" c for x; since r is constant and $g(c) = 0$,

$$r = f(c) - q(c)g(c) = f(c) \tag{1}$$

Although the reader might be willing to accept this Remark without further ado, let us be careful and justify the substitution we have just made, by proving

LEMMA 2. *"Substitution Lemma." Suppose R is commutative. For any given $c \in R$, there is a "substitution" homomorphism $R[x] \to R$ defined by $\sum a_i x^i \mapsto \sum a_i c^i$. (In other words we "substitute" c for x.)*

Actually the following more general result will also be useful (One recovers Lemma 2 by taking $C = R$ and $\psi = 1_R$).

LEMMA 3. *Suppose C is a commutative ring, and $c \in C$. Any given ring homomorphism $\psi: R \to C$ extends uniquely to a ring homomorphism $\varphi: R[x] \to C$, such that $\varphi(x) = c$ and $\varphi(a) = \psi(a)$ for all a in R.*

Proof. If such φ exists, we must have

$$\varphi(\sum a_i x^i) = \sum \varphi(a_i)\varphi(x)^i = \sum \psi(a_i)c^i, \tag{2}$$

proving uniqueness. It remains to show that φ as defined by (2) is a homomorphism.

$$\begin{aligned}
\varphi(\sum a_i x^i + \sum b_i x^i) &= \varphi(\sum (a_i + b_i)x^i) = \sum \psi(a_i + b_i)c^i \\
&= \sum \psi(a_i)c^i + \sum \psi(b_i)c^i \\
&= \varphi(\sum a_i x^i) + \varphi(\sum b_i x^i); \\
\varphi((\sum_i a_i x^i)(\sum_j b_j x^j)) &= \varphi(\sum_u \sum_{i=0}^{u} a_i b_{u-i} x^u) \\
&= \sum_u \sum_{i=0}^{u} \psi(a_i)\psi(b_{u-i})c^u \\
&= (\sum_i \psi(a_i)c^i)(\sum_j \psi(b_j)c^j) \\
&= \varphi(\sum_i a_i x^i)\varphi(\sum_j b_j x^j).
\end{aligned}$$

(Here the hypothesis C commutative was needed to move c past $\psi(b_j)$.) Finally, $\varphi(1) = \psi(1) = 1$. □

The careful reader may have spotted an application of a special case of Lemma 2 in Example 17.10. One can weaken the hypotheses a bit more, *cf.* Exercise 1.

COROLLARY 4. *Suppose R is commutative, and $c \in R$. For any polynomial $f(x)$ in $R[x]$, $f(c) = 0$ iff $x - c$ divides f in $R[x]$.*

Proof. Write $f = qg + r$, where $g = x - c$. Remark 1 yields $r = f(c)$, which is 0 iff $(x - c)|f$. □

THEOREM 5. *(Easy part of the Fundamental Theorem of Algebra.) Suppose R is an integral domain, and $a_1, \ldots, a_t$ are distinct roots of f. Then $t \le \deg f$, and $f = (x - a_1)\ldots(x - a_t)h$ for suitable $h \in R[x]$, where $\deg h = \deg f - t$.*

Proof. Induction on t, the case $t = 1$ being Corollary 4. In general write $f = (x - a_1)g$. Note $\deg g = \deg f - 1$. But $a_2, \ldots, a_t$ are roots of g; indeed, for each $i > 1$ we see $0 = f(a_i) = (a_i - a_1)g(a_i)$ implying $g(a_i) = 0$. By induction $t - 1 \le \deg g$ and $g = (x - a_2)\ldots(x - a_t)h$, where $\deg h = \deg g - (t - 1) = \deg f - t$, implying $f = (x - a_1)g = (x - a_1)\ldots(x - a_t)h$. □

COROLLARY 6. *If R is an integral domain and the polynomial $f \in R[x]$ of degree n has n distinct roots $a_1, \ldots, a_n$ in R, then*

$$f = c(x - a_1)\ldots(x - a_n),$$

where c is the leading coefficient of f.

Proof. Write $f = c(x - a_1)\ldots(x - a_n)$ where $c \in R[x]$; then $\deg c = n - n = 0$ so c is constant; clearly, the leading term of f is cx^n. □

There is a very nice application to fields.

Finite Subgroups of Fields

THEOREM 7. *Any finite multiplicative subgroup G of a field is cyclic.*

Proof. Let $n = |G|$ and $m = \exp G$. By remark 7.12′ it suffices to prove $n = m$. Clearly, $n \ge m$. On the other hand, each of the n elements of G is a root of the polynomial $x^m - 1$, so $n \le m$ by Theorem 5. □

This was so easy that we might wrongly be tempted to slight its impact. So let us quickly give a deep consequence.

COROLLARY 8. *Euler(p) is cyclic, for every prime number p.*

Proof. Apply Theorem 7 to the field $\mathbb{Z}_p$. □

However, I challenge the reader to find a constructive proof, *i.e.*, to determine systematically a generator of $(\mathbb{Z}_p \setminus \{0\}, \cdot)$. Concerning the structure of Euler(m) for m *not prime*, recall that Euler(8) $\approx (\mathbb{Z}_2, +) \times (\mathbb{Z}_2, +)$ is not cyclic; the general situation is described in Exercise 4.

Primitive Roots of 1

Consider an arbitrary finite subgroup G of $\mathbb{C}$. By Theorem 7, G is cyclic, so we write $G = \langle a \rangle$. Then $a^n = 1$ for $n = |G|$, so a is a root of the polynomial $x^n - 1$, one of the most important polynomials we shall encounter. Let us take a few moments to study its roots in $\mathbb{C}$. The most straightforward approach is by means of the unit circle $C = \{x + iy : x^2 + y^2 = 1\} = \{\cos\theta + i\sin\theta : \theta \in \mathbb{R}\}$ on the complex plane.

Remark 9. We can understand C better by defining $e^{i\theta} = \cos\theta + i\sin\theta$. Then $e^{2\pi i} = 1$, and

$$\begin{aligned} e^{i(\theta_1+\theta_2)} &= \cos(\theta_1 + \theta_2) + i(\sin\theta_1 + \theta_2) \\ &= \cos\theta_1\cos\theta_2 - \sin\theta_1\sin\theta_2 + i(\sin\theta_1\cos\theta_2 + \cos\theta_1\sin\theta_2) \\ &= (\cos\theta_1 + i\sin\theta_1)(\cos\theta_2 + i\sin\theta_2) \\ &= e^{i\theta_1}e^{i\theta_2}. \end{aligned}$$

Consequently, C is a multiplicative subgroup of $\mathbb{C}$, and there is a natural group homomorphism $\mathbb{R} \to C$ given by $\theta \mapsto e^{i\theta}$.

Now we see that the roots of $x^n - 1$ are $\{e^{2\pi i/n}, e^{4\pi i/n}, \ldots, e^{2n\pi i/n} = 1\}$, so $x^n - 1$ has precisely n roots in $\mathbb{C}$. Certain of these roots play a special role.

Definition 10. ρ is a *primitive* nth root of 1 if $\rho^n = 1$ but $\rho^m \neq 1$ for all $1 \leq m < n$.

If ρ is a primitive nth root of 1, then $\{\rho^k : 0 \leq k < n\}$ are distinct, so we have

Remark 11. $\{n\text{th roots of } 1\}$ is a cyclic group (under multiplication) which is isomorphic to $(\mathbb{Z}_n, +)$ and is generated by any of the primitive nth roots of 1. Consequently $e^{2k\pi i/n}$ is a primitive nth root of 1 iff $(k, n) = 1$; the number of primitive nth roots of 1 is the Euler number $\varphi(n)$.

Remark 12. If $\rho \neq 1$ is an nth root of 1 then $\sum_{j=1}^{n-1} \rho^j = -1$. (Indeed ρ is a root of $x^n - 1 = (x^{n-1} + \cdots + 1)(x - 1)$, so ρ is a root of $x^{n-1} + \cdots + 1$.)

These observations will elevate the Euler group to a key tool in field theory. However, let us first give a quick application to number theory.

Remark 13. $n = \sum_{d|n} \varphi(d)$, for any n. Indeed, any nth root of 1 is a primitive dth root of 1 for some d dividing n, so the equality is obtained merely by calculating the number of nth roots of 1 in this manner.

Roots of 1 also are used in factoring other polynomials.

PROPOSITION 14. *If F is a field containing a primitive nth root ρ of 1 then*

$$x^n - b^n = (x - b)(x - \rho b)\dots(x - \rho^{n-1}b)$$

for any b in F.

Proof. $b, \rho b, \dots, \rho^{n-1}b$ are all distinct, and $(\rho^k b)^n = \rho^{kn}b^n = (\rho^n)^k b^n = b^n$ implies each $\rho^k b$ is a root of $x^n - b^n$. Hence we are done by Theorem 5. □

COROLLARY 15. *If F is a field containing a primitive nth root ρ of 1 then for all a, b in F we have*

$$a^n - b^n = (a - b)(a - \rho b)\dots(a - \rho^{n-1}b)$$

Proof. Substitute a for x in the Proposition. □

Exercises

1. (A noncommutative generalization of Lemma 3): Suppose T is any ring, and suppose $c \in T$ commutes with all elements of T. Any homomorphism $\psi: R \to T$ extends to a unique homomorphism $\varphi: R[x] \to T$, such that $\varphi(r) = \psi(r)$ for all r in R and $\varphi(x) = c$.

The Structure of Euler(n)

2. If $8|n$, then Euler(n) is *not* cyclic. (*Hint:* ± 1 and $\pm(1 + \frac{n}{2})$ are four distinct elements whose squares are 1.)
3. For any $t \geq 1$, the number of solutions of $a^p = 1$ in Euler(p^t) is at most p. (*Hint:* Prove by induction on $t \geq 2$ that if $a^p \equiv 1 \pmod{p^t}$, then $a \equiv 1 \pmod{p^{t-1}}$. Indeed, for $t > 2$, by induction $a \equiv 1 \pmod{p^{t-2}}$, so write $a = 1 + kp^{t-2}$ and take p powers.)
4. If 8 does not divide n then Euler(n) is cyclic; if 8 divides n, then Euler(n) $\approx \mathbb{Z}_2 \times \mathbb{Z}_{\varphi(n)/2}$. (*Hint:* In view of Exercises 6.5 and 6.6, one may assume $n = p^t$ for some prime p. Let $G =$ Euler(n). Then $|G| = \varphi(n) = p^{t-1}(p-1)$.

   CASE I. $p \neq 2$. There is a natural surjection $\psi: G \to$ Euler(p), given as in exercise 4.8, so G has an element a of order $p - 1 = \varphi(p)$, by exercise 4.9. On the other hand, $\ker\psi$ has order p^{t-1}, so G is an internal direct product of $\ker\psi$ and $\langle a\rangle$. It suffices to prove $\ker\psi$ is cyclic, which follows from Exercises 3 and 7.2(i).

CASE II. $p = 2$. One may assume $t \geq 3$. G is not cyclic, by Exercise 2, so it suffices to find an element b of order $2^{t-2} = \frac{n}{4}$. Show via induction on t that [5] has order $> 2^{t-3}$ in Euler(2^t); hence take $b = [5]$.)

5. How many roots does the polynomial $x^5 - 1$ have in each of the following fields: $\mathbb{Z}_2, \mathbb{Z}_3, \mathbb{Z}_5$? In general how many roots does $x^n - 1$ have in $\mathbb{Z}_p$?
6. $x^m - 1 | x^n - 1$ iff $m|n$. (*Hint:* $x^m - 1$ divides $x^n - 1$, iff $x^m - 1$ divides $x^n - x^m = x^m(x^{n-m} - 1)$.)
7. For any a, b in a field containing a primitive nth root ρ of 1 show

$$\sum_{j=0}^{n-1} a^j b^{n-1-j} = (a - \rho b)\dots(a - \rho^{n-1}b);$$

$$1 + b + \cdots + b^{n-1} = (1 - \rho b)\dots(1 - \rho^{n-1}b);$$

$$n = (1 - \rho)\dots(1 - \rho^{n-1}).$$

(*Hint:* Cancel $x - b$ from each side of Corollary 14.)
8. $(\mathbb{R}/\mathbb{Z}, +) \approx C$, where C denotes the unit circle.

We are now ready to tackle number theory in Euclidean rings other than $\mathbb{Z}$. Our motivation is that polynomials such as

$$x^n - 1 = (x-1)(x^{n-1} + x^{n-2} + \cdots + 1)$$

factor more completely over extension rings of $\mathbb{Z}$ and thereby may enable us to solve equations more readily by means of these larger rings. We shall illustrate this fact in two famous results related to the famous French mathematician Pierre Fermat. Actually Fermat was a jurist by profession and never published his proofs; nevertheless, we believe that he knew the proofs, since there is only one mistake he is known to have made in his mathematical career.

A Theorem of Fermat

The first result concerns the following question about natural numbers: Which prime numbers p can be written as a sum of two square integers, *i.e.*, $a^2 + b^2$? Since a sum of two squares is clearly positive, we shall assume throughout that $p > 0$. It helps to look at the primes under 50.

p:	2	3	5	7	11	13	17	19
Sum:	$1+1$	NO	$4+1$	NO	NO	$9+4$	$16+1$	NO

p:	23	29	31	37	41	43	47
Sum:	NO	$25+4$	NO	$36+1$	$25+16$	NO	NO

Looking at these examples, we can conjecture Fermat's a result: A prime number $p \neq 2$ has the form $a^2 + b^2$ for suitable a, b in $\mathbb{N}$ iff $p \equiv 1 \pmod 4$.

One direction can be proved rather easily.

Remark 1. If a is odd, then $a^2 \equiv 1 \pmod 4$; if a is even then $a^2 \equiv 0 \pmod 4$. (Indeed, if $a = 2k+1$ then $a^2 = 4k^2 + 4k + 1 \equiv 1 \pmod 4$.)

Let us use these simple computations, assuming $p = a^2 + b^2$. Excluding the prime $p = 2$, we have p odd. Then, clearly, either a or b must be odd; we may assume a is odd, and then, clearly, b is even. Thus,

$$p = a^2 + b^2 \equiv 1 + 0 \pmod 4 \equiv 1 \pmod 4.$$

Hence there are no solutions for $p \equiv 3 \pmod 4$.

The remainder of our discussion focuses on proving conversely that every prime $p \equiv 1 \pmod 4$ does have a solution $p = a^2 + b^2$. A very short proof of this fact is given in Exercise 1, but that proof is not easy to reconstruct. A more intuitive proof is motivated by the observation

$$a^2 + b^2 = (a + bi)(a - bi) = y\bar{y},$$

where $y = a + bi$, $i = \sqrt{-1}$, and $^-$ is complex conjugation in $\mathbb{C}$. However, the ring $\mathbb{C}$ is too large for our purposes. Since we are only interested in integral solutions, we note that this factorization also holds in the subring $\mathbb{Z}[i] = \{m+ni : m,n \in \mathbb{Z}\}$ of $\mathbb{C}$, called the ring of *Gaussian integers*. Thus, any integer of the form $a^2 + b^2$ is not prime in $\mathbb{Z}[i]$, thereby leading us towards the following assertion, where we write $N(y)$ for $y\bar{y}$:

THEOREM 2. *The following assertions are equivalent, for a prime $p \neq 2$ in $\mathbb{N}$:*

(i) $p = a^2 + b^2$ *for suitable* a, b *in* $\mathbb{N}$;
(ii) $p \equiv 1 \pmod 4$;
(iii) $p = N(y)$ *for some* y *in* $\mathbb{Z}[i]$;
(iv) p *is not prime in* $\mathbb{Z}[i]$.

Proof. We have just seen the implications (i) $\Rightarrow$ (ii) and (i) $\Rightarrow$ (iv); furthermore, (i) $\Leftrightarrow$ (iii) is immediate, since $N(a + bi) = a^2 + b^2$. We shall demonstrate (ii) $\Rightarrow$ (iv) $\Rightarrow$ (iii) to complete the proof. But first we need to study arithmetic in $\mathbb{Z}[i]$. Note that $N(yz) = yz\overline{yz} = y\bar{y}z\bar{z} = N(y)N(z)$ for all y, z in $\mathbb{Z}[i]$.

THEOREM 3. $\mathbb{Z}[i]$ *is a Euclidean domain, with degree function* $d(y) = N(y)$.

Proof. We verify Definition 16.8. Property (1) is clear since $N(r) \geq 1$ for every $0 \neq r$ in $\mathbb{Z}[i]$. It remains to show, given y, z in $\mathbb{Z}[i] \setminus \{0\}$, that $y = qz + r$ for q, r in $\mathbb{Z}[i]$ satisfying $N(r) < N(z)$.

Since $\mathbb{C} = \mathbb{R}[i]$ is a field, we have $\frac{y}{z} \in \mathbb{C}$, *i.e.*, $\frac{y}{z} = u_1 + u_2 i$ for suitable u_1, u_2 in $\mathbb{R}$. Thus, $y = (u_1 + u_2 i)z$. For $j = 1, 2$, let q_j be the integer closest to u_j, and let $v_j = u_j - q_j$, so $|v_j| \leq \frac{1}{2}$. Then

$$N(v_1 + v_2 i) = v_1^2 + v_2^2 \leq \frac{1}{4} + \frac{1}{4} < 1;$$

letting $r = y - (q_1 + q_2 i)z = (v_1 + v_2 i)z$, we see that

$$N(r) = N(v_1 + v_2 i)N(z) < N(z),$$

so we conclude by writing $y = (q_1 + q_2 i)z + r$. $\square$

We need a few more facts, in order to perform arithmetic in $\mathbb{Z}[i]$.

Remark 4. (i) If $y|z$ in $\mathbb{Z}[i]$, then $\bar{y}|\bar{z}$. (Indeed, if $z = yq$, then $\bar{z} = \bar{y}\bar{q}$.)

(ii) If $c, d \in \mathbb{Z}$ and $c|d$ in $\mathbb{Z}[i]$, then $c|d$ in $\mathbb{Z}$ (since writing $d = c(a+bi) = ca + cbi$ for $a, b \in \mathbb{Z}$ we match real parts to get $d = ca$).

(iii) $y \in \mathbb{Z}[i]$ is invertible iff $N(y) = 1$. (This follows at once from the fact that $(a+bi)^{-1} = \frac{a-bi}{N(a+bi)}$.)

Proof of (iv) ⇒ (iii) in Theorem 2. Take a non-invertible, proper divisor $a+bi$ of p in $\mathbb{Z}[i]$. Then $N(a+bi)$ divides $N(p) = p^2$ in $\mathbb{Z}$, but $N(a+bi) \neq 1$ and $N(a+bi) \neq p^2$, both by Corollary 16.13. Hence $N(a+bi) = p$.

Proof of (ii) ⇒ (iv) in Theorem 2. This is more number-theoretic. Recalling that Fermat's Little Theorem can be obtained as an immediate application of Lagrange's Theorem to the group Euler(p), we are led to try to exploit the extra information that Euler(p) is cyclic (Corollary 18.8). Let c be a generator of Euler(p), and write $p = 4m + 1$ for suitable m. Then $o(c) = |\text{Euler}(p)| = p - 1 = 4m$, so $c^{4m} \equiv 1 \pmod p$ but $c^{2m} \not\equiv 1 \pmod p$. Taking $u = c^{2m}$ we see $u \neq 1$ but $u^2 = 1$ in $\mathbb{Z}_p$. But the field $\mathbb{Z}_p$ has only the two solutions ± 1 to the equation $x^2 - 1 = 0$; hence $u \equiv -1 \pmod p$. Taking $d = c^m$ yields $d^2 = u \equiv -1 \pmod p$, so

$$p \text{ divides } (d^2 + 1) = (d+i)(d-i). \tag{1}$$

Now suppose p were prime in $\mathbb{Z}[i]$. Then p divides $d + i$ or $d - i$, so Remark 4(i) implies p divides both $d + i$ and $d - i$, and so

$$p|((d+i) - (d-i)) = 2i,$$

i.e., $p|2$, contradiction. Thus, p is not prime in $\mathbb{Z}[i]$, as desired. □

Addendum: "Fermat's Last Theorem"

The raison d'être for much of the subject of "modern algebras" lies in a certain problem in number theory, which we shall discuss here. In a stimulating book, *The Last Problem*, (Gollantz: London, 1962), E.T. Bell argues that this question lies at the foundation of human knowledge and endeavor. Certainly it has inspired generations of mathematicians and has inspired many outstanding developments in algebra and algebraic geometry. The ancient Egyptians were aware that $3^2 + 4^2 = 5^2$, and it is not difficult to find an infinitude of solution of $a^2 + b^2 = c^2$ for a, b, c positive integers, merely by taking $u, v \in \mathbb{N}$ arbitrarily, and putting

$$a = u^2 - v^2, \quad b = 2uv, \quad c = u^2 + v^2$$

These solutions are called "Pythagorean triples," after the famous Greek mathematician and philosopher Pythagoras, but may well have been known previously by the Babylonians some 3500 years ago, *cf.* Edwards' book *Fermat's Last Theorem*, Springer-Verlag, 1977. It is somewhat harder to show that all solutions with a, b, c relatively prime can be obtained in this way, but the proof is still "elementary" and was known by the great classical Greek mathematician Diophantus, *cf.* Exercise 3.

The obvious continuation is, "Are there solutions of $a^3 + b^3 = c^3$ for nonzero integers a, b, c?", or more generally, "Are there integers $n > 2$, and $a, b, c \neq 0$ such that $a^n + b^n = c^n$?" This problem was brought to the attention of the mathematical world by Fermat, who jotted in the margin of a book (Bachet's translation of Diophantus' magnum opus, *Arithmetic*) that he had discovered a truly marvelous demonstration for proving there are no solutions. As usual, Fermat neglected to write down his proof, and since then the greatest mathematicians of the world have tried to rediscover his proof (or find new ones). Recently (June, 1993) a solution to Fermat's last theorem has been announced and outlined by Andrew Wiles. His proof builds on some of the most difficult mathematics of recent years, and contains some gaps.

Possibly by the time this book has appeared, the proof set forth by Wiles will have been found to be complete. This might cause a let-down, because one of the great quests of mathematicians would be complete. However, one should view as a great tribute to the human spirit, that the final solution of Fermat's Last Theorem would come before World War III or any other final solutions.

Let us see what is available via the methods of this course. Even the case $n = 3$ is no pushover. Attempting to generalize the approach used for $n = 2$ in Exercise 3, let us write

$$b^3 = c^3 - a^3 = (c - a)(c^2 + ca + a^2).$$

Unfortunately, the right-hand side is not yet factored sufficiently to carry out the argument. However, working instead in any ring that contains a primitive cube root ρ of 1, we can write

$$b^3 = (c - a)(c - \rho a)(c - \rho^2 a).$$

This implies the three right-hand factors are "almost" cubes, but instead of reaching a solution analogous to Pythagorean triples, one instead concludes by induction that no solution exists! Before proceeding with the details, let us develop some properties of ρ. We define $\mathbb{Z}[\rho] = \{a + b\rho : a, b \in \mathbb{Z}\}$.

Remark 6. (i) $\rho^2 = -\rho - 1$ by Remark 18.12, implying $\mathbb{Z}[\rho]$ is a ring. (Indeed,

$$(a + b\rho)(c + d\rho) = ac + bd\rho^2 + (ad + bc)\rho = (ac - bd) + (ad + bc - bd)\rho,$$

proving $\mathbb{Z}[\rho]$ is closed under multiplication.)

(ii) Since ρ^2 is the other primitive cube root of 1 we see by the quadratic formula (applied to (i)) that $\rho^2 = \bar{\rho}$. In particular, letting $N(y) = y\bar{y}$ we have $N(a + b\rho) = (a + b\rho)(a + b\rho^2) = a^2 + b^2 - ab \in \mathbb{N}$.

(iii) $\mathbb{Z}[\rho]$ is a Euclidean domain, again with degree function $N(\ )$ (*cf.* Exercise 4), and thus a UFD.

(iv) If $\gcd(a, b) = 1$ in $\mathbb{Z}$ then $\gcd(a + b\rho, a + b\rho^2)$ divides $1 - \rho$ in $\mathbb{Z}[\rho]$. (Indeed, let $d = \gcd(a + b\rho, a + b\rho^2)$. Then d divides $(a + b\rho)\rho - (a + b\rho^2) = a(\rho - 1)$, and d also divides $(a + b\rho) - (a + b\rho^2) = b(\rho - \rho^2) = b\rho(1 - \rho)$, and thus d divides $\gcd(a(1 - \rho), b\rho(1 - \rho)) = 1 - \rho$.)

(v) If $z \in \mathbb{Z}[\rho]$ and $N(z) = 1$, then z is invertible; consequently if $N(z)$ is prime in $\mathbb{Z}$ then z is prime in $\mathbb{Z}[\rho]$. (For if $z = yy'$ then $N(z) = N(y)N(y')$ in $\mathbb{N}$, so $N(y)$ or $N(y')$ is 1, implying y or y' is invertible.)

(vi) Let $q = 1 - \rho$. $N(q) = 3$, so q is prime by (v). Note $\rho \equiv 1 \pmod{q}$, so $a + b\rho \equiv a + b \pmod{q}$.

(vii) Every element of $\mathbb{Z}[\rho]$ is congruent to one of $\{0, 1, 2\} \pmod{q}$. (Indeed, take $z \in \mathbb{Z}[\rho]$. z is congruent to some integer m by (vi), which in turn is congruent to one of $\{0, 1, 2\} \pmod{q}$, since $q|3$.)

THEOREM 7. *(Fermat's Last Theorem for $n = 3$.) There are no integers $a, b, c \neq 0$ such that $a^3 + b^3 = c^3$.*

Proof. We shall rely heavily on the prime $q = 1 - \rho$ of Remark 6(vi) and the calculation (for $k \leq j$)

$$\begin{aligned}(y - \rho^j z) - (y - \rho^k z) &= \rho^k z - \rho^j z = \rho^k z(1 - \rho^{j-k}) \\ &= \rho^k z(1 - \rho)(1 + \rho + \cdots + \rho^{j-k-1}).\end{aligned} \tag{2}$$

In particular, $y - \rho^j z \equiv y - \rho^k z \pmod{q}$.

Claim 1. *If $y \equiv z \pmod{q}$ and $q \nmid z$, then $y^3 \equiv z^3 \pmod{q^4}$.*

Proof of Claim 1. $y^3 - z^3 = (y - z)(y - \rho z)(y - \rho^2 z)$ by Corollary 18.15. By hypothesis $q|(y - z)$ so (2) implies $q|(y - \rho^j z)$ for each j . Writing $y - \rho^j z = qw_j$, we see that $w_{j+1} - w_j = \rho^j z \equiv z \pmod{q}$, so in particular, w_1, w_2, w_3 are all distinct modulo q. Hence some $w_j \equiv 0 \pmod{q}$, by Remark 6(vii), so $q \cdot q \cdot q^2 = q^4$ divides $(y - z)(y - \rho z)(y - \rho^2 z) = y^3 - z^3$, as desired.

Let us return to the main proof. Replacing c by $-c$ we may rewrite the equation as

$$a^3 + b^3 + c^3 = 0. \tag{3}$$

We shall prove there is no solution for $a, b, c \neq 0$ in $\mathbb{Z}[\rho]$. Otherwise take such an example with $N(c)$ minimal. Then a, b, c are pairwise relatively prime; indeed, if, for example, some prime p divided a, b then p divides $c^3 = -a^3 - b^3$, and thus $p|c$, so we could divide a, b, c each by p and have a smaller solution to (3).

Claim 2. *a, b, or c is divisible by q.*

Proof of Claim 2. Otherwise, by Remark 6(vii) each of a, b, c is congruent to $\pm 1 \pmod{q}$, so Claim 1 implies each of a^3, b^3, c^3 is congruent to $\pm 1^3 = \pm 1 \pmod{q^4}$. But then $0 = a^3 + b^3 + c^3$ is congruent to ± 1 or $\pm 3 (\bmod q^4)$, which is impossible since $q^4 = ((1-\rho)^2)^2 = (-3\rho)^2 = 9\rho^2$ is an associate of 9.

We may thus assume $q|a$, and conclude by proving more generally

Claim 3. *It is impossible that $uq^{3m}a^3 + b^3 + c^3 = 0$, for u invertible in $\mathbb{Z}[i]$ and a, b, c in $\mathbb{Z}[i]$ prime to ρ.*

The proof is straightforward but involves lengthy computations, so we sketch it in Exercise 5. Thus Fermat's Last Theorem is proved for $n = 3$.

To prove Fermat's Last Theorem in general, it would suffice suffice to assume $n = 4$ or n is an odd prime (Exercise 7). The proof of $n = 4$ is easier than the above proof (*cf.* Exercise 6), so we are left with the case where n is prime. What is special about 3 in the above proof? The reader who traces through the proof will spot only two places where 3 cannot be replaced at once by an arbitrary odd prime number n:

(i) $\mathbb{Z}[\rho]$ needs to be a UFD, for ρ a primitive nth root of 1; and

(ii) The argument concerning invertible elements at the end of Exercise 5 is quite special, *cf.* Exercise 8.

The program of proving Fermat's Last Theorem along these lines was known by Lagrange and others, and who knows, possibly even by Fermat himself. Unfortunately, it turns out that $\mathbb{Z}[\rho]$ is not a UFD when $n = 37$. (However, it is Noetherian, of course.) Efforts to overcome this and other related questions in number theory led Kummer in the 1840s to a very careful investigation of $\mathbb{Z}[\rho]$, called the *ring of cyclotomic integers*, thereby laying the foundations for the subject of algebraic number theory.

The twentieth century ushered in the geometric approach to Fermat's Last Theorem. Clearly $a^n + b^n = c^n$ iff $(\frac{a}{c})^n + (\frac{b}{c})^n = 1$, so that the rational points of the graph $x^n + y^n = 1$ correspond precisely to the integral solutions of $a^n + b^n = c^n$. Thus, Fermat's Last Theorem is reduced to showing that certain algebraic curves have no nontrivial rational points. This has inspired rapid strides in algebraic geometry in the past fifteen years, and the ensuing powerful techniques are so much more advanced than the methods available even 30 years ago, that what Bell regarded in 1961 as "The Last Problem" of civilization is generally regarded as solvable. Recent work is far beyond the scope of this course.

A mathematical history of some of these developments is given in Edwards's book cited above. P. Ribenboim has written a stimulating set of lectures on Fermat's Last Theorem (Springer-Verlag, Berlin, 1979), and his Chapter 4 contains some striking results that have been obtained through elementary means, including an "elementary" proof for even n, discovered in this century. So one wonders, after all, might Fermat have had a proof?

Exercises

1. A quick proof of Theorem 2 ((*ii*) $\to$ (*i*)): Let $S = \{(a,b,c) \in \mathbb{N}^3 : a^2 + 4bc = p\}$. Define a map $\sigma: S \to S$ by

$$(a,b,c) \mapsto \begin{cases} (a+2c,\ c,\ b-a-c) & \text{if } a < b-c \\ (2b-a,\ b,\ a-b+c) & \text{if } b-c \le a \le 2b \\ (a-2b,\ a-b+c,\ b) & \text{if } a > 2b. \end{cases}$$

   Then $\sigma^2 = 1$, but σ has exactly one fixed point, *i.e.*, $\sigma(a,b,c) = (a,b,c)$ (since then one sees $b = a$ and thus $a(a+4c) = p$, implying $a = 1$ and $c = \frac{p-1}{4}$.) Hence, $|S|$ is odd. But then the map given by $(a,b,c) \mapsto (a,c,b)$ also has a fixed point, which is the solution to Fermat's theorem.
2. The primes of $\mathbb{Z}[i]$ are precisely those numbers of the following form: Primes of $\mathbb{N}$ congruent to 3 mod 4, and $a+bi : a^2+b^2$ is prime in $\mathbb{Z}$. (*Hint*: If $p \equiv 3 \pmod 4$ and $p = yz$ in $\mathbb{Z}[i]$, then $N(p) = N(y)N(z)$ so $N(y) = p$, impossible.) Show how the prime factorization of $N(y)$ in $\mathbb{N}$ leads to the prime factorization of y in $\mathbb{Z}[i]$.
3. Prove the that every solution in integers of $a^2 + b^2 = c^2$ is given by Pythagorean triples. (*Hint*: It is enough to show this for a, b, c pairwise relatively prime. In particular, a or b is odd. Assume a is odd. Then b is even by Remark 1. Hence c is odd, and

$$b^2 = c^2 - a^2 = (c+a)(c-a).$$

   Write $c + a = 2y$ and $c - a = 2z$. Then $c = y + z$ and $a = y - z$; hence y, z are relatively prime. But $yz = (\frac{b}{2})^2$ implies $y = u^2$ and $z = v^2$ where $uv = \frac{b}{2}$. Hence, $c = u^2 + v^2$ and $a = u^2 - v^2$.)
4. If ρ is a primitive cube root of 1 then $\mathbb{Z}[\rho]$ satisfies the Euclidean algorithm, with the degree function $N(\ )$. (The proof is analogous

to Theorem 3). What goes wrong with this proof for other primitive roots of 1?

5. Prove Claim 3, by induction on m. First note that $m \geq 2$, using Claim 1. But
$$-uq^{3m}a^3 = b^3 + c^3 = (b+c)(b+\rho c)(b+\rho^2 c),$$
so q^2 divides at least one of the factors; assume $q^2|(b+c)$. Then, as in the proof of Claim 1, $q^2 \nmid (b+\rho^j c)$ for $j = 1,2$, implying $q^{3m-2}|(b+c)$, *i.e.*,
$$b+c = u_1 q^{3m-2} f^3, \quad b+\rho c = u_2 q g^3, \quad b+\rho^2 c = u_3 q h^3$$
for suitable f, g, h relatively prime in $\mathbb{Z}[i]$ and u_1, u_2, u_3 invertible. But
$$\begin{aligned} 0 &= (b+c) + \rho(b+\rho c) + \rho^2(b+\rho^2 c) \\ &= u_1 q^{3m-2} f^3 + \rho u_2 q g^3 + \rho^2 u_3 q h^3. \end{aligned}$$
Dividing through by $u_2 q$ yields
$$0 = u' q^{3m-3} f^3 + g^3 + u'' h^3$$
with u', u'' are invertible. Hence $g^3 \equiv -u'' h^3 \pmod q$, implying $g \equiv \pm h \pmod q$ and thus $g^3 \equiv \pm h^3 \pmod{q^4}$, so $u'' \equiv \pm 1 \pmod{q^4}$. If one can prove $u'' = \pm 1$; then replacing h by $u''h$ would yield $0 = u' q^{3m-3} f^3 + g^3 + h^3$, contrary to the induction hypothesis.

Thus, it remains only to show if u'' is invertible and $u'' \equiv \pm 1 \pmod{q^4}$, then $u'' = \pm 1$. Write $u'' = m + n\rho$ for suitable m, n in $\mathbb{Z}$. Then $1 = N(u'') = m^2 + n^2 - mn$; thus, $4 = 4m^2 + 4n^2 - 4mn = (2m-n)^2 + 3n^2$, implying $|n| \leq 1$; the possibilities for u'' are $\pm 1, \pm\rho$, and $\pm(1+\rho)$. (Note $1+\rho = -\rho^2$.) Of these clearly ± 1 are the only ones congruent to $\pm 1 \pmod{q^4}$, as desired.

6. Prove Fermat's Last Theorem for n = 4, by reducing to Pythagorean triples. This method enables one to prove more generally that there is no solution to $a^4 + b^4 = c^2$.

7. If $n = st$ and $a^n + b^n = c^n$ then $(a^s)^t + (b^s)^t = (c^s)^t$; conclude from Exercise 6 that to prove Fermat's Last Theorem it suffices to prove the case for n an odd prime.

8. If ρ is a primitive fifth root of 1, then $\mathbb{Z}[\rho]$ has an infinite number of invertible elements.

9. For any primitive nth root ρ of 1, define $\mathbb{Z}[\rho] = \{\sum_{i=0}^{n-2} a_i \rho^i : a_i \in \mathbb{Z}\}$. Show that $\mathbb{Z}[\rho]$ is closed under multiplication and thus is a subring of $\mathbb{C}$. There is an onto ring homomorphism $\mathbb{Z}[x] \to \mathbb{Z}[\rho]$ given by $x \mapsto \rho$; what is the kernel? Conclude that $\mathbb{Z}[\rho]$ is Noetherian, even though it is known that $\mathbb{Z}[\rho]$ need not be a UFD.

CHAPTER 20. IRREDUCIBLE POLYNOMIALS

Having seen that irreducible polynomials play a special role in the polynomial ring $F[x]$, we should like to be able to determine which polynomials are irreducible. Later on we shall see that every irreducible polynomial gives rise to a field, thereby enhancing our interest in irreducible polynomials.

It is convenient to say a polynomial p is "over" a ring R when its coefficients all lie in R, *i.e.*, when $p \in R[x]$. Clearly, every polynomial of degree 1 over an arbitrary field F is irreducible; such polynomials have the form $ax + b$ for $a, b \in F$, and are called *linear*. In case $F = \mathbb{C}$, all irreducible polynomials are linear, by the celebrated Fundamental Theorem of Algebra (Theorem 25.30). However, the situation is completely different for $F = \mathbb{Q}$, where it is virtually hopeless to try to classify the irreducible polynomials. Instead we shall give several criteria for irreducibility, and from them we shall derive several interesting classes of irreducible polynomials over $\mathbb{Q}$, including those that arise from roots of 1.

First note that, dividing through by the leading coefficient, we may assume that our polynomial f is monic. Also recall Corollary 18.4: f has a factor $x - a$ iff a is a root of f. This is enough for us to check irreducibility of any polynomial f of degree ≤ 3; if $f = gh$ with g, h nonconstant, then g or h is linear. We shall also consider larger degrees.

To determine that a polynomial f over $\mathbb{Z}$ is irreducible over $\mathbb{Q}$, it often turns out to be easier to check irreducibility of f over $\mathbb{Z}$ and then to appeal to some general theory that we are about to develop. However, over $\mathbb{Z}$ we cannot pass automatically to monics, and we are confronted with the difficulty that $2x^2 + 2$ is irreducible over $\mathbb{Q}$ but has the proper factorization $2(x^2 + 1)$ over $\mathbb{Z}$. To isolate this problem we introduce the following concept, applicable more generally to any UFD that is not a field.

Definition 1. The *content* $c(f)$ of a polynomial f (over R) is the "greatest common divisor" (gcd) in R of the coefficients of f (*cf.* Remark 16.25).

Strictly speaking, the gcd is defined only up to associate; for example, if $R = \mathbb{Z}$ and $f = 2x^3 - 2x + 6$, then $c(f) = \pm 2$. We say f is *primitive*, and write $c(f) = 1$, if its coefficients are relatively prime, in the sense that no noninvertible element of R divides all the coefficients of f. Clearly, any monic polynomial is primitive. In general, we see

Remark 2. If R is a UFD, then any f in $R[x]$ can be written as cf_1, where $c = c(f)$ and f_1 is a primitive polynomial. Conversely, if $f = cf_1$ with f_1 primitive, then $c = c(f)$. (Alternatively, if $r \in R$ and $r | f$ in $R[x]$, then $r | c(f)$ in R.)

Remark 3. If $f = gh$ in $R[x]$, then $f(0) = g(0)h(0)$, and the leading coefficient of f is the product of the leading coefficients of g and h.

Our main goal is to show in general that factorization in $R[x]$ boils down to factorization in R coupled with factorization into primitive polynomials. We need a structural preliminary.

Remark 4. There is a natural homomorphism $\varphi: R[x] \to (R/Rp)[x]$, given by $\sum c_i x^i \mapsto \sum (c_i + Rp)x^i$. (Indeed, let $\psi: R \to (R/Rp)[x]$ be defined as the composition of the natural homomorphism $R \to R/Rp$ with the natural injection $R/Rp \to (R/Rp)[x]$. By Lemma 18.3, ψ extends to a homomorphism $\varphi: R[x] \to (R/Rp)[x]$ satisfying $\varphi(x) = x$.)

Now $\ker \varphi = \{\sum c_i x^i : \text{ each } c_i \in Rp\} = R[x]p$; thus, $R[x]/R[x]p \approx (R/Rp)[x]$, by the first Noether isomorphism theorem.

LEMMA 5. *If R is an integral domain, then any prime p of R is also a prime of $R[x]$.*

Proof. By Proposition 17.6, we need to show that if R/Rp is an integral domain, then so is $R[x]/R[x]p$. But $(R/Rp)[x]$ is certainly an integral domain, which is isomorphic to $R[x]/R[x]p$ by Remark 4. □

Polynomials over UFDs

The groundwork is now laid for a major result. Throughout, we assume that R is a UFD.

LEMMA 6. *(Gauss's lemma.) If f, g are primitive polynomials over R, then fg is also primitive over R.*

Proof. Otherwise $c(fg)$ has some irreducible factor p, so $p|fg$. But p is prime, by Remark 16.25. By Lemma 5 we see $p|f$ or $p|g$, *i.e.*, $p|c(f)$ or $p|c(g)$, contrary to f and g primitive. □

COROLLARY 6′. $c(fg) = c(f)c(g)$.

Proof. Let $c = c(f)$ and $d = c(g)$. Then $f = cf_1$ and $g = dg_1$ for suitable primitive polynomials f_1, g_1; thus, $fg = (cd)f_1g_1$ implying $cd = c(fg)$ by Remark 2. □

COROLLARY 6″. *Suppose $f, g \in R[x]$, and write $f = c(f)f_0$ and $g = c(g)g_0$. Then $f|g$, iff $c(f)|c(g)$ in R and $f_0|g_0$ in $R[x]$.*

Proof. ($\Leftarrow$) is clear. ($\Rightarrow$) Write $g = hf$. Then $c(g) = c(h)c(f)$, and canceling the contents from each side yields $g_0 = h_0 f_0$. □

To proceed further we pass to the field of fractions F of the integral domain R, by means of the following observation.

LEMMA 7. *For any f in $F[x]$ there is some $s \neq 0$ in R for which $sf \in R[x]$.*

Proof. Write $f = \sum_{i=0}^{t} \frac{c_i}{d_i} x^i$ where $c_i, d_i \in R$ and $d_i \neq 0$. Take $s = d_1 \dots d_t$; evidently all the coefficients of sf are in R. □

THEOREM 8. *Suppose R is a UFD whose field of fractions is F. If $f = gh$ for $f, g \in R[x]$ with g primitive and h in $F[x]$, then $h \in R[x]$.*

Proof. $sh \in R[x]$ for some s in R, and

$$sc(f) = c(sf) = c(gsh) = c(g)c(sh) = c(sh),$$

since g is primitive. Thus, $s|c(sh)$ so each coefficient of sh is divisible by s, implying $h \in R[x]$. □

COROLLARY 9. *Suppose R is a UFD, and $f \in R[x]$ is primitive. Then f is irreducible in $F[x]$ iff f is irreducible in $R[x]$.*

Proof. ($\Rightarrow$) If $f = gh$ in $R[x]$ then g or h is constant; assuming g is constant we see g divides $c(f) = 1$.

($\Leftarrow$) Suppose $f = gh$ in $F[x]$. Taking s in R for which $sg \in R[x]$ we can write $sg = c_1 g_1$ where $c_1 \in R$ and $g_1 \in R[x]$ is primitive; then $f = g_1 h_1$ where $h_1 = s^{-1} c_1 h \in F[x]$. But then h_1 is in $R[x]$ by Theorem 8, and is primitive, by corollary 6′. Hence g_1 or h_1 is invertible and thus constant, implying that g or h is constant. □

Note 9′. Actually we showed that for $f \in R[x]$, any factorization $f = gh$ in $F[x]$ has a factorization $f = g_1 h_1$ in $R[x]$, with g_1 primitive and equivalent to g in $F[x]$.It follows at once by induction that any factorization of f in $F[x]$ has an equivalent factorization with each factor in $R[x]$.

Example 9″. Let us pause for a moment to consider rational roots of polynomials. Suppose $\frac{m}{n}$ is a root of a polynomial $f \in \mathbb{Q}[x]$, with m, n relatively prime. Then $nx - m$ divides f over $\mathbb{Q}$, and thus over $\mathbb{Z}$. But $nx - m$ is a primitive polynomial, implying n divides the leading coefficient of f, and $m|f(0)$, *cf.* Remark 3. In particular, this gives us a finite procedure for finding all rational roots of f.

Theorem 8 implies that the irreducible elements of $R[x]$ are precisely the irreducible primitive polynomials and the irreducible constants in R. Although not the focus of the present discussion, the following result is now rather straightforward.

THEOREM 10. *If R is a UFD, then $R[x]$ is a UFD.*

Proof. The existence of a factorization for any f in $R[x]$ is clear: Writing $f = c(f) f_0$ in $R[x]$ with f_0 primitive, we factor $c(f)$ into irreducibles in R, and, by note 9′, we can factor f_0 into irreducible polynomials in $R[x]$ which, in view of Gauss's lemma, are primitive. In view of Proposition 16.20 it remains to show that this is a factorization into primes. On the one hand, any irreducible element of R is prime in R and thus prime in $R[x]$, by Lemma 5. On the other hand, any primitive irreducible polynomial is irreducible in $F[x]$ and thus is prime. □

For example, $6x^2-6$ factors in $\mathbb{Z}[x]$ as $2\cdot 3\cdot (x-1)(x+1)$, where the only possible modification is multiplication by ± 1. Theorem 10 can be pushed a bit further, *cf.* Exercise 3. Applications are given in Exercises 7ff.

As mentioned earlier, Noetherian rings have replaced unique factorization domains as the focus of commutative ring theory, and the analog of Theorem 10 for Noetherian rings (*cf.* Exercise 17.21) is much more important in current research.

Eisenstein's Criterion

By Corollary 9, irreducibility of polynomials over $\mathbb{Z}$ leads to irreducibility over $\mathbb{Q}$. However, we do not yet have general techniques to establish irreducibility of polynomials of degree ≥ 4. An easy structural argument yields an infinite class of irreducible polynomials over $\mathbb{Q}$, of arbitrary degree.

THEOREM 11. *(Eisenstein's Criterion.) Suppose $f \in \mathbb{Z}[x]$. If some prime p in $\mathbb{Z}$ does not divide the leading coefficient of f but divides all other coefficients of f, and if $p^2 \nmid f(0)$, then f is irreducible.*

Proof. Otherwise factor $f = gh$, where $\deg g = m > 0$ and $\deg h = n > 0$. Then $\deg f = m + n$. Writing $^-$ for the image in $\mathbb{Z}_p[x]$ obtained by taking each coefficient mod p (*cf.* Remark 4), we have

$$ax^{m+n} = \bar{f} = \bar{g}\bar{h},$$

where a is invertible in $\mathbb{Z}_p$. Then

$$m + n = \deg \bar{g} + \deg \bar{h} \leq \deg g + \deg h = m + n;$$

thus equality holds at each stage, implying $\deg \bar{g} = m$ and $\deg \bar{h} = n$. Thus, by unique factorization of the polynomial x^{m+n} over the field $\mathbb{Z}_p$, we have $\bar{g} = ux^m$ and $\bar{h} = vx^n$ for suitable $u, v \neq 0$ in $\mathbb{Z}_p$. But this means $\bar{g}(0) = 0 = \bar{h}(0)$, so p divides $g(0)$ and $h(0)$, and thus p^2 divides $g(0)h(0) = f(0)$, contradiction. Hence f is indeed irreducible. □

Note that this criterion does not make sense in $\mathbb{Z}_p$ and is useless there; for example, $x^2 + 3 = (x-2)(x+2)$ in $\mathbb{Z}_7$. Thus, in the remainder of this discussion, we restrict our attention to polynomials over $\mathbb{Q}$.

Example 12. (i) For any prime p in $\mathbb{Z}$ and any $t > 0$, the polynomials $x^t - p$ and $x^t + p$ are irreducible.

(ii) Of course, $x^2 - p^2 = (x+p)(x-p)$ is reducible; likewise $x^t - p^t$ is reducible for each t.

(iii) Although it does not satisfy Eisenstein's criterion, $x^2 + p^2$ is irreducible in $\mathbb{Q}[x]$, since it has no roots in $\mathbb{Q}$. What about $x^4 + p^2$? Let us turn to $\mathbb{Z}[i]$ for help. $x^4 + 4 = (x^2 - 2i)(x^2 + 2i)$. But $2i = (1+i)^2$, and $-2i = (i(1+i))^2 = (-1+i)^2$, so

$$\begin{aligned} x^4 + 4 &= (x + (1+i))(x - (1+i))(x + (-1+i))(x - (-1+i)) \\ &= (x+1+i)(x+1-i)(x-1+i)(x-1-i) \\ &= ((x+1)^2+1)((x-1)^2+1) = (x^2+2x+2)(x^2-2x+2). \end{aligned}$$

This factorization exists over $\mathbb{Q}$ because of the fluke that $2i$ is a square in $\mathbb{Q}[i]$. A similar analysis for $p \neq 2$ actually shows that $x^4 + p^2$ is irreducible for all $p \neq 2$ (*cf.* Exercise 2).

Our next application is quite cute. We saw that every primitive nth root ρ of 1 satisfies the polynomial $x^n - 1$. But $x^n - 1 = (x-1)(x^{n-1} + \cdots + 1)$ is reducible, and we should like to find some irreducible factor that ρ satisfies. The obvious candidate works, for n prime.

THEOREM 13. *For p prime the polynomial $f = x^{p-1} + x^{p-2} + \cdots + 1$ is irreducible.*

Proof. We would like to be able to apply Eisenstein's criterion. If $f = gh$ then $f(x+1) = g(x+1)h(x+1)$, as seen by the substitution lemma (18.2). Thus it suffices to prove $f(x+1)$ is irreducible. But $(x-1)f(x) = x^p - 1$. Applying $x \mapsto x+1$ yields

$$xf(x+1) = (x+1)^p - 1 = x^p + px^{p-1} + \binom{p}{2}x^{p-2} + \cdots + px,$$

so

$$f(x+1) = x^{p-1} + px^{p-2} + \binom{p}{2}x^{p-3} + \cdots + p.$$

To prove Eisenstein is applicable it suffices merely to show $p | \binom{p}{i}$ for all $1 \leq i < p$. But this is obvious since in $\binom{p}{i} = \frac{p!}{i!(p-i)!}$ the prime p appears in the factorization of the numerator but not in the factorization of the denominator. □

Here is another instance of the same trick.

Example 14. The polynomial $f = 8x^3 - 6x^2 + 1$ is irreducible over $\mathbb{Q}$. Indeed

$$f(x+1) = 8(x+1)^3 - 6(x+1)^2 + 1 = 8x^3 + 18x^2 + 12x + 3$$

is irreducible, by Eisenstein. Likewise $g = 8x^3 - 6x^2 - 1$ is irreducible. (Why?) A less elegant way to check that these polynomials are irreducible would be to show they have no rational roots, by means of Example 9″.

Exercises

1. Suppose $f \in \mathbb{C}[x]$. If z is a nonreal root of f, then $(x-z)(x-\bar{z}) = x^2 - (z+\bar{z})x + z\bar{z}$ is an irreducible factor of f in $\mathbb{R}[x]$.
2. Let $f = x^4 + p^2$ for $p > 2$. Factor f completely over $\mathbb{C}$, and matching complex conjugates, factor f as a product of two irreducible quadratics over $\mathbb{R}$. Since these factors are not in $\mathbb{Z}[x]$, conclude that f is irreducible over $\mathbb{Q}$.

Nagata's Theorem and Its Applications

3. (Nagata's Theorem.) Suppose R is an integral domain, and S is a multiplicative subset consisting of products of prime elements, such that $R[S^{-1}]$ is a UFD. If every element of R can be written as a finite product of irreducible elements then R is a UFD. (Hint: An irreducible of R either divides an element of S and so is prime, or does not and stays irreducible in $S^{-1}R$ and thus is prime.)
4. Reprove Theorem 10 using Nagata's Theorem.
5. The "circle ring" $F[x,y,z]/\langle x^2 + y^2 + z^2 - 1\rangle$ is a UFD. (Hint: Noting $x^2 + y^2 = (1+z)(1-z)$, localize at $1-z$ and apply Nagata's theorem.)
6. State and prove Eisenstein's criterion for an arbitrary UFD.

The Ring $\mathbb{Z}[x_1, \ldots, x_n]$ and the Generic Method

7. Suppose C is a UFD. Then the polynomial ring $C[x_1, \ldots, x_n]$ in n indeterminates is also a UFD; likewise, the polynomial ring in an infinite number of commutative indeterminates over C is a UFD.
8. (Substitution in several indeterminates.) Prove, for any commutative ring C and elements $\{c_1, \ldots, c_n\}$ in C, that there is a unique ring homomorphism $f: \mathbb{Z}[x_1, \ldots, x_n] \to C$, such that $f(x_i) = c_i$ for $1 \le i \le n$.
9. Over $\mathbb{Z}[x_1, \ldots x_n]$ the determinant d of the matrix

$$A = \begin{pmatrix} 1 & 1 & \cdots & 1 \\ x_1 & x_2 & \cdots & x_n \\ \vdots & \vdots & \ddots & \vdots \\ x_1^{n-1} & x_2^{n-1} & \cdots & x_n^{n-1} \end{pmatrix}$$

equals $\prod_{1\le j<i\le n}(x_i - x_j)$. (Hint: Subtracting the jth row from the ith row of A yields the row

$$x_i - x_j \quad x_i^2 - x_j^2 \quad \dots \quad x_i^n - x_j^n,$$

each term of which is divisible by $(x_i - x_j)$. By unique factorization,

$$\prod_{1\le j<i\le n} (x_i - x_j)|d;$$

prove equality by matching degrees and the coefficient of the monomial $x_2 x_3^2 x_4^3 \dots x_{n-1}^n$.)

10. (The Vandermonde Determinant.) For any commutative ring C, and any $c_1, \dots, c_n$ in C, the matrix

$$\begin{pmatrix} 1 & 1 & \dots & 1 \\ c_1 & c_2 & \dots & c_n \\ \vdots & \vdots & \ddots & \vdots \\ c_1^{n-1} & c_2^{n-1} & \dots & c_n^{n-1} \end{pmatrix}$$

has determinant $\prod_{1\le j<i\le n}(c_i - c_j)$. (Hint: Apply exercise 8 to exercise 9.)

11. (Symmetric polynomials.) Any permutation π of $\{1\ 2\ \dots\ n\}$ induces a homomorphism $\varphi_\pi\colon \mathbb{Z}[x_1, \dots x_n] \to \mathbb{Z}[x_1, \dots x_n]$ given by $\varphi_\pi(x_i) = x_{\pi i}$. We say $f \in \mathbb{Z}[x_1, \dots x_n]$ is a *symmetric* polynomial if $\varphi_\pi(f) = f$ for every permutation π. For example, one can write

$$(x - x_1)\dots(x - x_n) = x^n - s_1 x^{n-1} + s_2 x^{n-2} - \dots \pm s_n$$

where s_i are the following symmetric polynomials:

$$s_1 = x_1 + \dots + x_n,$$
$$s_2 = x_1x_2 + x_1x_3 + x_2x_3 + \dots,$$
$$\dots$$
$$s_n = x_1 \dots x_n.$$

$s_1, \dots, s_n$ are called the *elementary symmetric polynomials* in n indeterminates. Prove that any symmetric polynomial f can be written $f = h(s_1, \dots, s_n)$, for a suitable polynomial h. (Hint: Induction on n and on the total degree of f. Write $\bar{f}$ for $f(x_1, \dots, x_{n-1}, 0)$ in $\mathbb{Z}[x_1, \dots x_{n-1}]$. If f is symmetric, then $\bar{f}$ is symmetric in $n - 1$ indeterminates, and in fact the elementary symmetric polynomials in $n-1$ indeterminates are $\bar{s}_1, \ldots, \bar{s}_{n-1}$. (Note $\bar{s}_n = 0$.) By induction, $\bar{f} = \bar{g}(\bar{s}_1, \ldots, \bar{s}_{n-1})$ for suitable g. Then $f - g(s_1, \ldots, s_{n-1})$ is divisible by x_n and thus by each x_i. Write $f - g(s_1, \ldots, s_{n-1}) = s_n q$ and apply induction to q.)

12. Given n, denote the symmetric polynomial $x_1^t + x_2^t + \cdots + x_n^t$ by f_t. Verify the following formulas:

$$f_2 = s_1^2 - 2s_2; \quad f_3 = s_1^3 - 3s_1s_2 + 3s_3;$$

$$f_4 = s_1^4 - 4s_1^2 s_2 + 4s_1s_3 + 2s_2^2 - 4s_4.$$

More generally, prove the following recursive formulas, known as *Newton's formulas*:

$$\begin{aligned} f_k =& s_1 f_{k-1} - s_2 f_{k-2} \pm \cdots + (-1)^k s_{k-1} f_1 + (-1)^{k+1} k s_k \\ & \quad (\text{for} \quad k \leq n); \\ f_{n+k} =& s_1 f_{n+k-1} \pm \cdots + (-1)^{n+1} s_n f_k. \end{aligned}$$

(Hint: Working in $F[x_1, \ldots, x_n]$, take another indeterminate y and let $q = \prod_{i=1}^n (1 - x_i y) = \sum_{k=0}^n (-1)^k s_k y^k$. The formal logarithmic derivative developed as a formal power series yields

$$q' q^{-1} = \sum_{i=1}^{n} -x_i (1 - x_i y)^{-1} = -\sum_{k=0}^{\infty} f_{k+1} y^k,$$

cf. Exercises 16.8ff, 16.12, and 16.15. Thus, $q' = -\sum_{k=0}^{\infty} q f_{k+1} y^k$; match coefficients of y^{k-1}.)

13. If d is the Vandermonde determinant of exercise 10, prove d^2 is the determinant of the matrix

$$B = \begin{pmatrix} n & f_1 & \cdots & f_{n-1} \\ f_1 & f_2 & \cdots & f_n \\ \vdots & \vdots & \ddots & \vdots \\ f_{n-1} & f_n & \cdots & f_{2n-2} \end{pmatrix}.$$

(Hint: $|B| = |AA^t| = |A|^2$.) Coupled with exercises 11,12, this gives an explicit computation of d^2 in terms of the elementary symmetric polynomials.

Review Exercises for Part II

1. Prove or give a counterexample: The union of two ideals is an ideal.
2. List the ideals of $\mathbb{Z}_{100}$.
3. Find an integral domain which is not a field, but which contains $\mathbb{C}$.
4. Find rings $R_1 \subset R_2 \subset R_3 \subset \mathbb{Q}$.
5. Prove $\mathbb{Z}[\frac{1}{6}] \approx (\mathbb{Z}[\frac{1}{2}])[\frac{1}{3}]$.
6. Find (with proof) all the subrings of $\mathbb{Z}[\frac{1}{6}]$.
7. If $p = a^2 + b^2 = c^2 + d^2$ is prime with $a, b, c, d \in \mathbb{N}$ and $a \leq b$ and $c \leq d$, then $a = c$ and $b = d$.
8. Define "prime" and "irreducible" (elements), and show that every prime element is irreducible.
9. Give an example of a prime nonzero ideal which is not a maximal ideal.
10. Show $f = 8x^3 - 6x^2 + 1$ is irreducible over $\mathbb{Q}$, without using Eisenstein's criterion.
11. Give an example of a UFD which is not a PID.
12. Show that 9 has two inequivalent factorizations in $\mathbb{Z}[\sqrt{10}]$.
13. Prove that the ring of Gaussian integers $\mathbb{Z}[i]$ is a UFD.
14. Prove that every natural prime number of the form $4k + 1$ is a sum of two squares.
15. State and prove Gauss' lemma.
16. Determine the primes of $\mathbb{Z}[i]$.
17. Prove the formula $\sum_{d|n} \varphi(d) = n$, where φ is the Euler function.
18. In the ring $\mathbb{Z}[\rho]$, where ρ is a primitive cube root of 1, show that: (i) $1 - \rho$ is irreducible; (ii) $1 - \rho$ is prime; (iii) 3 is reducible.
19. Prove $(1 - \rho) \dots (1 - \rho^{n-1}) = n$, for any primitive nth root ρ of 1.
20. Factor the following polynomials over $\mathbb{Q}$:
    $x^4 + 1,\ x^4 + 2,\ x^4 + 4,\ x^4 + 6,\ x^4 + 8,\ x^4 + 9.$

PART III — FIELDS

Our previous applications of abstract algebra have been to discover properties of integers and rational numbers, using the minimum possible computation. We turn now to the main goal of this course, which is to study irrational numbers, in particular, roots of polynomial equations. Since this will involve some rather difficult and abstract theory, we pause to survey the history of (polynomial) equations, preceding the revolutionary work of Galois.

Historical Background

Solving equations has been one of the principal occupations of mathematicians since the time of the Pythagorean school, although their methods were based on geometric reasoning. The name algebra itself derives from the famous treatise on equations, called Al-jabr wal muqabala, written in 830 by Mohammed ibn Musa al-Khowarizmi.

It is natural to start with polynomial equations whose coefficients are integers. The simplest case is the linear equation $ax + b = 0$, which has the root $x = -\frac{b}{a} \in \mathbb{Q}$. Note that any linear equation with coefficients in a field F has its root in F, so we are led to consider equations with coefficients in a given field F. In general, since any root of a polynomial $f(x)$ is also a root of the polynomial obtained by dividing through by the leading coefficient of f, we might as well assume from the onset that f is monic.

The degree 2 case is not quite as trivial. Every high-school student has had to struggle with the "quadratic formula," which says that the roots of the equation $x^2 + bx + c = 0$ are $\frac{-b \pm \sqrt{b^2-4c}}{2}$. In other words, any quadratic equation over a field can be "solved" by taking one square root. The well-known method of obtaining this solution is by "completing the square";

i.e., letting $y = x + \frac{b}{2}$, one sees $y^2 = -c + \frac{b^2}{4} = \frac{b^2-4c}{4}$, so one solves by taking square roots.

For perhaps a thousand years or more, one of the major research questions in mathematics was to solve the general cubic equation

$$x^3 + bx^2 + cx + d = 0.$$

In analogy to the quadratic case, one can obtain the simpler equation

$$y^3 + py + q = 0,$$

by taking $y = x + \frac{b}{3}$, but further reductions are rather tricky. The famous poet Omar Khayyam found a geometric solution for a wide range of cubic equations, but a general method was not discovered until the sixteenth century. In 1515, Scipione del Ferro apparently solved the cubic (for $p > 0$ and $q < 0$) , but only disclosed his findings to a select group of pupils. Challanged by a former student of del Ferro, Niccolo Tartaglia discovered the solution to the cubic but also kept his results secret. He finally divulged them in confidence to Cardano, who published the general solution in his major work Ars Magna sive de Reglis Algebraicis ("The Great Art — or the Rules of Algebra"). Clearly, it is enough to give one root, for then one is left with solving the other two roots in a quadratic equation. The root of $y^3 + py + q = 0$ given in Cardano's book is

$$y = \sqrt[3]{\frac{-q}{2} + \sqrt{\left(\frac{p}{3}\right)^3 + \left(\frac{q}{2}\right)^2}} + \sqrt[3]{\frac{-q}{2} - \sqrt{\left(\frac{p}{3}\right)^3 + \left(\frac{q}{2}\right)^2}}.$$

Note that for $q = 0$, one gets $y = 0$; for $p = 0$, one gets $y = -\sqrt[3]{q}$, as one would expect. However, the conscientious reader will find some difficulties with this formula, since it often gives an unexpected or unrecognizable root. (Try solving $x^3 + x = 2$ or $x^3 - 15x = 4$.)

The quartic equation was solved by Ferrari, a student of Cardano, using a method reminiscent of completing the square: To solve $y^4+py^2+qy+r = 0$, one notes

$$\left(y^2 + \frac{p}{2} + u\right)^2 = -qy - r + \left(\frac{p}{2}\right)^2 + 2uy^2 + pu + u^2,$$

with u to be determined. Considering the terms in y and y^2 in the right-hand side, one sees this can be a square of a polynomial in y, if and only if it is the square of $\sqrt{2u}y - \frac{q}{2\sqrt{2u}}$, in which case we could then take square roots and solve. But for this to happen, we must have

$$-r + \left(\frac{p}{2}\right)^2 + pu + u^2 = \frac{q^2}{8u}, \quad i.e.,$$

$$8u^3 + 8pu^2 + \left(2p^2 - 8r\right) u - q^2 = 0, \tag{1}$$

which can be solved via Cardano's formula for cubics. Thus the quartic is solved by means of (1), which is called the "resolvent cubic."

For some time, it was thought that an inductive procedure could be established to solve an equation of arbitrary degree n in terms of a "resolvent" polynomial of degree $< n$; in fact a number of papers proposing solutions to the quintic were submitted, but each of them was found eventually to contain some error or another. (Unfortunately, the resolvent polynomial of the quintic turns out to have degree 6.) Finally in 1799 Ruffini published a proof that there is no general formula to solve equations of degree 5. His proof required more than 500 pages, and was met with skepticism; in 1824 the brilliant Norwegian mathematician Niels Henrik Abel settled the question (negatively) to everyone's satisfaction, by finding certain clearcut algebraic criteria which would follow from a formula, but which are violated by the "generic" equation of degree ≥ 5.

In 1831, a year before being killed in a duel (at the age of 21), Evariste Galois carved his niche for mathematical posterity by sketching his theory of solving polynomial equations, thereby clarifying the formulas for polynomials of degree 3 and 4, as well as reproving the theorem of Ruffini and Abel.[1]

One of the triumphs of early 20th-century-algebra was to explain Galois theory in terms of the theory of fields, groups, rings, and vector spaces. We start with a polynomial f over a field F and take a larger field K which contains a root of f. Thus the study of *field extensions* $K \supseteq F$ is basic to the theory, and a superficial discussion of properties of vector spaces suffices to dispose of various famous geometric questions considered by the Greeks, as will be seen in Chapter 22. (However, to answer one of these questions we need to know that π is transcendental, and this deep theorem is deferred until Appendix A.)

Our next step, in Chapter 23, is to utilize ring-theoretic tools from Part II, and consider a field E that contains *all* of the roots of f; the theory is applied to determine all finite fields (Chapter 24). The set of isomorphisms from E to E that fix F turns out to be a group, with respect to composition of functions, called the Galois group; its subgroups determine

[1]Unfortunately, Galois's memoir was rejected by the French Academy, but, fortunately, the prominent French mathematician Liouville went over the details of Galois's memoir and arranged for it to be published posthumously in 1846. A fascinating account of this subject, including an analysis of all major contributions leading to Galois's crowning achievement, including the material sketched here, can be found in *Galois' Theory of Algebraic Equations*, by Jean-Pierre Tignol, Longman Scientific and Technical, Essex, England (1988).

all fields between E and F, via the celebrated "Galois correspondence" of Chapter 25. A host of applications follows in Chapter 26, culminating in a determination in Chapter 27 of precisely which polynomials can be solved in terms of explicit formulas. In particular, one can show with a minimum of computation that every equation of degree 4 can be solved in such a way, whereas there are equations of degree 5 and larger that cannot be so solved. Appendix A contains a complete proof that π is transcendental, and Appendix B contains a brief discussion of how fields, polynomials, and other topics of this text generalize to noncommutative algebras.

We start with a given base field F, which is to contain all the coefficients of the equation to be studied, and build up from there.

Definition 1. An *F-field* is a field K that contains F; we also say that K is a *field extension* of F, denoted as K/F (not to be confused with the other uses of this notation).

Example 2. Some examples of field extensions: $\mathbb{R}/\mathbb{Q}$, $\mathbb{C}/\mathbb{Q}$, and $\mathbb{C}/\mathbb{R}$.

Actually, although K will almost always be a field, it could be any integral domain containing the field F, and we assume this throughout the discussion. (We shall need this greater generality for technical reasons, but at the end we reduce to the field extension case, in remark 13 (ii).)

Definition 3. If $a \in K$ and $f = \sum \alpha_i x^i \in F[x]$, we write $f(a)$ for the evaluation $\sum \alpha_i a^i$ in K. Let $\mathcal{I}_a = \{f \in F[x] : f(a) = 0\}$, easily seen to be an ideal of $F[x]$; these f are called the polynomials *satisfied by* a. We say a is *algebraic over* F when $\mathcal{I}_a \neq 0$; we say a is *transcendental over* F when $\mathcal{I}_a = 0$.

Whereas an algebraic number often can be described rather concretely, such as $\sqrt{2}$, it is not easy to pinpoint any single transcendental number. The most important examples of transcendental numbers (over $\mathbb{Q}$) are π and e (the base of the natural logarithms), although the proofs are rather involved, *cf.* Appendix A. The easiest explicit number to prove transcendental is given in Exercise 1, but by far the most straightforward way to prove that transcendental numbers exist is to show that "most" real numbers are transcendental, by means of set theory (*cf.* Exercises 2ff)!

Algebraic Elements

For the remainder of this discussion, we turn to the algebraic case ($\mathcal{I}_a \neq 0$). Write $\langle f \rangle$ for the ideal $F[x]f$. Since $F[x]$ is a PID, we see $\mathcal{I}_a = \langle f_a \rangle$ for a suitable polynomial $f_a \neq 0$. Actually we know already from Remark 16.10$'$ that if also $\mathcal{I}_a = \langle g \rangle$ then $g = \alpha f_a$ for suitable $\alpha \neq 0$ in F (since all invertible elements of $F[x]$ are constants). In particular, dividing through by the leading coefficient we have a unique monic f_a, called the *minimal polynomial of* a. Note that if $a \neq 0$, then f_a is nonconstant so $\deg f_a \geq 1$. If $n = \deg f_a$ we say a is *algebraic of degree* n.

Remark 4. $\deg f_a$ is the least degree among all polynomials satisfied by a.

Let us pause for one familiar example, now, and postpone the others until later. $\mathbb{C}$ is an $\mathbb{R}$-field. Let $i = \sqrt{-1} \in \mathbb{C}$. The minimal polynomial of i is $x^2 + 1$. Every element of $\mathbb{C}$ has the form $x + yi$ for x, y in $\mathbb{R}$.

In this example, we realize at once that x^2+1 is the minimal polynomial since its degree is only 2. In general, it is useful to have a structural criterion for the minimal polynomial.

Remark 5. Suppose $a \in K$. If $f = gh \in F[x]$ then $f(a) = g(a)h(a)$ (as can be seen directly or by Remark 18.2).

PROPOSITION 6. *Suppose $a \in K$, and f is a monic polynomial satisfied by a. Then f is the minimal polynomial of a, iff f is irreducible.*

Proof. ($\Rightarrow$) Suppose $f = gh$. Then $0 = f(a) = g(a)h(a)$ implying $g(a) = 0$ or $h(a) = 0$. But f has minimal degree in $\mathcal{I}_a$, so either $\deg f = \deg g$ or $\deg f = \deg h$; we conclude f is irreducible.

($\Leftarrow$) Let f_a be the minimal polynomial of a. Then $f \in \langle f_a \rangle$ so $f = gf_a$ for some g in $F[x]$. But $\deg f_a \geq 1$ so g must be constant, *i.e.*, f is associate to f_a, implying $f = f_a$, since both are monic. □

Given a field extension K/F, we want to compare the structure of the fields K and F. There are two main methods to do this: The observations of this chapter will be obtained by viewing K as a vector space over F; later, we shall obtain deeper results by means of ring homomorphisms. So first we review some linear algebra.

A *vector space* over a field F is an Abelian group $(V, +)$ endowed with *scalar multiplication* $F \times V \to V$ satisfying the following axioms, for all α_i in F and v_i in V:

$$1v = v$$
$$(\alpha_1 + \alpha_2)v = \alpha_1 v + \alpha_2 v$$
$$\alpha(v_1 + v_2) = \alpha v_1 + \alpha v_2$$
$$(\alpha_1\alpha_2)v = \alpha_1(\alpha_2 v).$$

A subset $S \subset V$ is called *linearly independent*, or *F-independent*, if $\sum \alpha_i s_i = 0$ (for distinct $s_i \in S$) necessarily implies each $\alpha_i = 0$. A *base* of the vector space V is a linearly independent set B that spans V, *i.e.*, every element of V is a linear combination of elements of B. The number of elements of a base B is called the *dimension* of V, denoted here as $[V : F]$, and is known to be independent of the particular choice of B. If B is an infinite set we write $[V : F] = \infty$.

Any ring $K \supseteq F$ can be viewed as a vector space over the field F, with respect to the given addition and multiplication in K. (The vector space axioms then follow as a special case of the ring axioms). Nevertheless, we maintain our additional assumption that K is an integral domain, in order to avoid contending with situations such as in Exercise 18.

Given a in K, let $F[a]$ denote the image of $F[x]$ in K, under the substitution $x \mapsto a$, *i.e.*,

$$F[a] = \{\sum_{i=0}^{t} \alpha_i a^i : t \in \mathbb{N},\ \alpha_i \in F\}.$$

Clearly $F[a]$ is an integral domain, being a subring of the integral domain K. Our immediate interest in $F[a]$ is as a subspace of K as vector space over F.

Remark 7. $F[a]$ is spanned over F by $B = \{1, a, a^2, \dots\}$, so some subset of B must be a base. B is itself a base iff a is transcendental over F.

PROPOSITION 8. *Suppose $a \in K$ is algebraic of degree n over F. Then $[F[a] : F] = n$. In fact $1, a, a^2, \dots, a^{n-1}$ are a base of $F[a]$ over F.*

Proof. Independence of $\{1, a, a^2, \dots, a^{n-1}\}$ is clear, for if $\sum_{i=0}^{n-1} \alpha_i a^i = 0$, then a satisfies the polynomial $\sum_{i=0}^{n-1} \alpha_i x^i$, which must be 0 since its degree is less than that of f_a; hence each $\alpha_i = 0$.

It remains to show that $1, a, \dots, a^{n-1}$ span $F[a]$, *i.e.*, that $F[a] = V$ where $V = \sum_{i=0}^{n-1} Fa^i$. We must show a typical element $b = \sum_{i=0}^{m} \alpha_i a^i$ of $F[a]$ belongs to V. This follows from the Euclidean algorithm for $F[x]$. Indeed $b = g(a)$, for $g = \sum_{i=0}^{m} \alpha_i x^i$. Writing $g(x) = q(x)f_a(x) + r(x)$, where $\deg r < \deg f_a$ or $r = 0$, we see that $r(a) = g(a) = b$, but by inspection $r(a) \in V$. □

The converse is also true.

COROLLARY 9. *If $[F[a] : F] = n$, then a is algebraic of degree n over F.*

Proof. $1, a, \dots, a^n$ are dependent over F (since these are $n+1$ elements); thus $\sum_{i=0}^{n} \alpha_i a^i = 0$ for suitable α_i in F, not all zero, implying a satisfies the polynomial $f = \sum_{i=0}^{n} \alpha_i x^i$. Now Proposition 8 implies $[F[a] : F]$ is the degree of a. □

Actually, it is important to know that $F[a]$ is *always* a field whenever a is algebraic. This is seen by a surprising application of the structure theory.

Remark 9′. If $f \neq 0$ is an irreducible polynomial of $F[x]$, then $\langle f \rangle$ is a maximal ideal of the principal ideal domain $F[x]$, so $F[x]/\langle f \rangle$ is a field.

THEOREM 10. *Suppose $a \in K$ is algebraic over F. Then*
(i) *$F[x]/\mathcal{I}_a \approx F[a]$, under the substitution homomorphism $(x \mapsto a)$; and*
(ii) *$F[a]$ is a field.*

Proof. (i) The substitution homomorphism $\psi_a : F[x] \to F[a]$ is onto, with kernel $\mathcal{I}_a$, so $\psi_a : F[x]/\mathcal{I}_a \to F[a]$ is the desired isomorphism, by Noether I.

(ii) $\mathcal{I}_a = \langle f_a \rangle$, and f_a is irreducible by Proposition 6. By Remark 9′, $F[x]/\mathcal{I}_a$ is a field, which by (i) is isomorphic to $F[a]$. □

Summary 11. The following statements are equivalent, for any element a in K:

(i) a is algebraic over F;

(ii) $[F[a] : F] < \infty$; and

(iii) a is contained in some field that is finite dimensional over F. (In fact this field could be taken to be $F[a]$.)

Let us look at some familiar examples. Although one can demonstrate directly that $F[a]$ is a field, direct computations sometimes would be quite intricate.

Example 12. (i) The special case $a \in F$. Then $F[a] = F$, so $[F[a] : F] = 1$, the base being $\{1\}$. Note that $f_a = x - a$ is linear. In general, $[K : F] = 1$ iff $K = F$. This example is rather trivial.

(ii) $\mathbb{C} = \mathbb{R}[i]$, where $i = \sqrt{-1}$ satisfies the polynomial $x^2 + 1$. $[\mathbb{C} : \mathbb{R}] = 2$, the base being $\{1, i\}$.

(iii) $a = \sqrt{2}$ satisfies the polynomial $x^2 - 2$, which is irreducible and thus must be the minimal polynomial of a, implying $\mathbb{Q}[\sqrt{2}]$ is two-dimensional over $\mathbb{Q}$, with base $1, \sqrt{2}$.

(iv) In general, we say K/F is a *quadratic* extension if $[K : F] = 2$. If $\alpha \in F$ then $F[\sqrt{\alpha}]/F$ is a quadratic extension, with base $\{1, \sqrt{\alpha}\}$, and the minimal polynomial of $\sqrt{\alpha}$ is $x^2 - \alpha$. Conversely, suppose F is a field in which $1+1 \neq 0$. We claim that any quadratic extension K/F can be written $K = F[\sqrt{\alpha}]$ for suitable α in F. This is seen by the familiar argument from high school of "completing the square." Indeed, take any a in $K \setminus F$. Then $1 < [F[a] : F] \leq [K : F] = 2$, implying $[F[a] : F] = 2$; hence $F[a] = K$, and the minimal polynomial of a has the form $x^2 + \alpha_1 x + \alpha_0$. Thus $a^2 + \alpha_1 a + \alpha_0 = 0$, implying

$$(a + \frac{\alpha_1}{2})^2 = a^2 + \alpha_1 a + \frac{\alpha_1^2}{4} = -\alpha_0 + \frac{\alpha_1^2}{4} \in F.$$

Thus, letting $\alpha = -\alpha_0 + \frac{\alpha_1^2}{4}$, we see $K = F[\sqrt{\alpha}]$. Note that we required $1+1 \neq 0$ in order that $\frac{1}{2}$ make sense; this idea will be pursued in Chapter 23 when we study the characteristic of a field.

(v) Generalizing (iii) in another direction, we see that the polynomial $x^n - p$ is irreducible over $\mathbb{Q}$ for any prime number p, by Eisenstein's criterion, implying $\mathbb{Q}[\sqrt[n]{p}]$ is an n-dimensional field extension of $\mathbb{Q}$.

(vi) Let ρ be a primitive cube root of 1. Then the minimal polynomial of ρ is $x^2 + x + 1$. (Indeed $0 = \rho^3 - 1 = (\rho - 1)(\rho^2 + \rho + 1)$, implying that

ρ satisfies the polynomial x^2+x+1, which is irreducible by Theorem 20.13. Clearly, $[\mathbb{Q}[\rho]:\mathbb{Q}]=2$, and in fact $\mathbb{Q}[\rho]=\mathbb{Q}+\mathbb{Q}\rho$.)

(vii) More generally, if ρ is a primitive pth root of 1, for p prime, then $[\mathbb{Q}[\rho]:\mathbb{Q}]=p-1$; proof as in (vi). (The nonprime case will be handled in Example 25.12.)

Note that the cases $[K:F]=1,2$ have been described completely in (i) and (iv), except when $1+1=0$ in F.

Finite Field Extensions

Next we consider all the elements of K at once. We say K/F is *algebraic* if each element of K is algebraic over F; also, K/F is *finite* if $[K:F]<\infty$. (Of course, K/F can be finite even if F and K both are infinite as sets; *e.g.*, $\mathbb{C}/\mathbb{R}$ is finite, since it is quadratic.)

Remark 13. (i) Every finite extension is algebraic. More precisely, if $[K:F]=n$, then each element a of K has degree $\leq n$ over F, since $[F[a]:F]\leq[K:F]=n$.

(ii) If the integral domain K is algebraic over a field F, then K is itself a field. (Indeed, by Theorem 10, $F[a]$ is a field for any $a\neq 0$ in K, so $a^{-1}\in F[a]\subseteq K$.)

THEOREM 14. *If $F\subseteq K\subseteq L$ are fields, then $[L:F]=[L:K][K:F]$.*

Proof. Let $\{a_i : i\in I\}$ and $\{b_j : j\in J\}$ be respective bases of K over F and of L over K. We shall show that $\{a_ib_j : i\in I,\ j\in J\}$ is a base of L over F; the result then follows at once. Clearly,

$$L=\sum_{j\in J}Kb_j=\sum_{j\in J}(\sum_{i\in I}Fa_i)b_j=\sum_{i,j}Fa_ib_j,$$

so the a_ib_j span L over F. It remains to show the a_ib_j are F-independent. If $\sum_{i,j}\alpha_{ij}a_ib_j=0$, then rewriting this as $\sum_j(\sum_i\alpha_{ij}a_i)b_j=0$ we see each $\sum_i\alpha_{ij}a_i=0$, since the b_j are K-independent. But by assumption the a_i are F-independent, so each $\alpha_{ij}=0$. □

This basic result has many consequences, some of which are given in Exercises 10 ff.

COROLLARY 15. *Notation as above, $[K:F]$ divides $[L:F]$; in particular, if $[L:F]$ is prime, then $K=L$ or $K=F$.*

Proof. The first assertion follows at once from the theorem, and the second assertion is an obvious consequence. □

Given a field extension L/F and $a,b\in L$ algebraic over F, we define $F[a,b]=(F[a])[b]$. Note that $K=F[a]$ is a field, and thus so is $F[a,b]$.

Furthermore the minimal polynomial of b over K divides the minimal polynomial f_b of b over F, so $[F[a,b]:K] \leq \deg f_b = \deg b$. Consequently

$$[F[a,b]:F] = [F[a,b]:K][K:F] \leq \deg b \deg a.$$

One can iterate this procedure, to define $F[a_1,\ldots,a_t]$ for any t, whenever $a_1,\ldots,a_t$ are algebraic over F. (Also see Exercise 10.)

COROLLARY 16. *If a and b are algebraic elements of an F-field K, then $a \pm b$, ab, and $\frac{a}{b}$ (for $b \neq 0$) are algebraic.*

Proof. All of these elements are in $F[a,b]$, which is finite over F and thus algebraic over F. □

COROLLARY 17. *For any F-field K, the set of elements of K algebraic over F is a field, called the algebraic closure of F in K.*

Occasionally we shall need the following generalization of $F[a,b]$. Given subfields K, L of a field W, we define the *compositum* KL to be the intersection of all subfields of W containing both K and L.

PROPOSITION 18. *Suppose $F \subseteq K, L \subseteq W$ are fields, with $[K:F]$ and $[L:F]$ finite. Then $KL = \{\sum_{\text{finite}} a_i b_i : a_i \in K,\ b_i \in L\}$.*

Proof. Let $E = \{\sum_{\text{finite}} a_i b_i : a_i \in K,\ b_i \in L\}$. Any ring that contains both K and L clearly contains E, so $E \subseteq KL$. But $[E:F]$ is finite, since one gets a finite spanning set by multiplying a base of K with a base of L, as in the proof of Theorem 14. Hence E is a field by Remark 13 (ii), implying $KL = E$. □

Remark 19. Obviously $[KL:L] \leq [K:F]$, since any base of K over F also spans KL over L. In Exercise 25.2 we shall see that $[KL:L]$ divides $[K:F]$.

Remark 20. Of course, this construction could be generalized to the compositum of any number of subfields; and if $K_1,\ldots,K_t$ are subfields of W all finite dimensional over F, then $K_1\ldots K_t = \{\sum_{\text{finite}} a_{i_1}\ldots a_{i_t} : a_{i_j} \in K_j\}$.

Exercises

1. Any real number between 0 and 1 can be described as a string of digits following a decimal point. Define *Liouville's number*

   $$a = .110001000000000000000001000000000000000000\ldots,$$

   which is given by 1 in the $n!$-position (for each n) and 0 elsewhere. Determine the only possible nonzero positions in a^m for each m, and by showing that these do not match, conclude that a is transcendental. A concise proof using calculus is given in Appendix A.

Countability and Transcendental Numbers

Recall that two sets have the same cardinality iff there is a 1:1 correspondence between them, and a set S is called *countable* if it is in 1:1 correspondence with $\mathbb{N}$. (In other words one can list the elements of S as $\{s_1, s_2, \dots\}$.) S is called *uncountable* if S is not countable.

2. The union of a countable number of countable sets is countable. (*Hint*: If $S_u = \{s_{1u},\ s_{2u}, \dots\}$ and $S = \cup S_u$ then

$$S = \{s_{11},\ s_{21},\ s_{12},\ s_{22},\ s_{31},\ s_{22},\ s_{13}, \dots\}. \quad)$$

3. The set of complex numbers that are algebraic over $\mathbb{Q}$ is countable. (*Hint*: Since any polynomial has a finite number of zeros, it is enough to show that there are a countable number of polynomials over $\mathbb{Z}$. For any polynomial f, define its "index" to be $\deg f$ plus the sum of the absolute values of the coefficients. There are only a finite number of polynomials with a given index, so the set of polynomials is countable.)
4. (Kantor's Diagonalization Trick) The set S of real numbers between 0 and 1 is uncountable. (*Hint*: Any real number between 0 and 1 can be written in the form $.d_1d_2d_3\dots$ where each $d_i \in \{0, 1, \dots, 9\}$. Suppose S were countable, and write the nth real number as $.d_{1n}d_{2n}\dots$. Construct a number $.d'_1d'_2\dots$ that formally does not appear on the list, *i.e.*, $d'_i \neq d_{ii}$ for each i. (One has to be a bit careful, since, for example, $.19999\dots = .20000\dots$)
5. There are uncountably many transcendental real numbers between 0 and 1, by Exercises 3 and 4.
6. Generalizing the argument of Exercise 3, show that if $F \subset K$ are fields with F infinite, then F has the same cardinality as its algebraic closure in K.

Algebraic Extensions

7. Give an example of an algebraic extension that is not finite.
8. Using the fact that every polynomial over $\mathbb{C}$ has a root (to be proved in Chapter 26), prove that $\mathbb{C}$ has no proper algebraic field extensions.
9. Find a proper field extension of $\mathbb{C}$, bearing in mind that it cannot be algebraic, by Exercise 8. (*Hint*: $\mathbb{C}[x]$ is an integral domain containing $\mathbb{C}$.)
10. Show that $F[a, b] = (F[b])[a]$. Prove that if $m = \deg a$ and $n = \deg b$ are relatively prime then $F[a, b]$ is a field of dimension mn over F.

($Hint$: $F[a,b]$ contains both $F[a]$ and $F[b]$, and thus $[F[a,b]:F]$ is divisible both by m and n.)

11. $\sqrt{3}+\sqrt{7}$ has degree 4 over $\mathbb{Q}$. (*Hint*: $[\mathbb{Q}[\sqrt{3},\sqrt{7}]:\mathbb{Q}]=4$. But $\sqrt{3}+\sqrt{7}$ is not quadratic over $\mathbb{Q}$.) For a more systematic method of computation, see Exercise 15.
12. $[\mathbb{Q}[\sqrt{p_1},\dots,\sqrt{p_t}]:\mathbb{Q}]=2^t$, for distinct positive prime numbers $p_1,\dots,p_t$.
13. Suppose a is a root of the polynomial $\sum_{i=0}^t m_i x^i$ for m_i in $\mathbb{Z}$. Then $m_0 a$ is a root of a suitable monic polynomial (over $\mathbb{Z}$) of degree t. (*Hint*: $0 = m_t^{t-1}\sum m_i a^i = (m_t a)^t + m_{t-1}(m_t a)^{t-1} + \dots$.)
14. Suppose V is a vector space of dimension n over a field F. Write $\mathrm{Hom}_F(V,V)$ for the set of linear transformations of V, taken naturally as a subring of $\mathrm{Hom}(V,V)$ (viewing V as an Abelian group). Taking a base $\{b_1,\dots,b_n\}$ of V over F, show that any linear transformation $T\colon V\to V$ over F is defined uniquely by the action on the base $b_i \mapsto \sum_{j=1}^n \alpha_{ij}b_j$, where each $\alpha_{ij}\in F$. Thus $T\mapsto(\alpha_{ij})$ defines a 1:1 correspondence $\mathrm{Hom}_F(V,V)\to M_n(F)$, which is an "anti-isomorphism" of rings — it preserves addition but reverses the order of multiplication.
15. (The regular representation revisited, *cf.* Exercise 14.4) Suppose A is an algebra of dimension n over a field F. The composite sending a to the right multiplication map and then to the corresponding matrix yields an injection of A into $M_n(F)$. (*Hint*: Multiplication is reversed twice.)
16. View $\mathbb{C}$ explicitly as a subfield of $M_2(\mathbb{R})$.
17. Reprove Remark 13 by means of Exercise 15 and the characteristic polynomial. This proof is "constructive," in the sense that one can actually compute a polynomial satisfied by an element; as an application, solve Exercise 11 by this method.
18. The direct product $R=\mathbb{C}\times\mathbb{C}$ contains the subfield $\{(\alpha,\alpha):\alpha\in\mathbb{C}\}$ which is a field isomorphic to $\mathbb{C}$. Show that $[R:\mathbb{C}]=2$; why does this not contradict the fact that $\mathbb{C}$ is algebraically closed?

In the eyes of the ancient Greeks, there was a deep connection between mathematics and life, and numbers entered into every facet of philosophy, science, and art, such as the "golden ratio" ($\frac{1+\sqrt{5}}{2}$), an important guide in architecture. The only real numbers of interest to them could be constructed, using a compass and straight edge, by means of certain well-defined rules, which in modern terms could be described as follows:

Construction by Straight Edge and Compass

Definition 1. We work in the Euclidean plane, and construct real numbers in a sequence of steps. We start at Step 1 with two points which we call $\hat{0}$ and $\hat{1}$, and with the line L_0 defined by these two points (which we identify as the "real line"), but with no other lines or circles. By convention we identify L_0 with the X-axis in the plane, and we shall define any real number a in terms of its corresponding point $\hat{a}$ on L_0. Thus the numbers 0 and 1 are defined by the points $\hat{0}$ and $\hat{1}$.

Inductively, suppose we have defined various lines, circles, points, and numbers at Step $i-1$. In addition to carrying these over to Step i, we proceed to define new lines, circles, points, and numbers at Step i, as follows:

1. A point is defined at Step i if it lies in the intersection of a line and a line, a line and a circle, or a circle and a circle, which already have been defined at Step $i-1$. (Note that the intersection of the circle with a line or another circle will normally define two points, so there may be a certain ambiguity in describing these constructions. There are various ways of avoiding this ambiguity, but they are irrelevant to the present discussion.)
2. A number a is defined at Step i if the point $\hat{a}$ has been defined at Step i.
3. A line is defined at Step i if it passes through two points defined at Step i. The line defined by points P and Q will be denoted $L(P,Q)$.
4. A circle is defined at Step i if its center is a point defined at Step i and if its radius is the distance between two points defined at Step i. The circle defined by center P and radius r will be denoted $C(P;r)$. (Intuitively, we construct the circle by opening the compass to the distance between the two given points, and then placing the center of the compass at the prescribed center of the circle.) Although we do not specify *a priori* that the radius be defined, it turns out to be a constructible number, by Remark 2 below.

A number that is defined at some step is called a *constructible number*.

Thus, at Step 1 we may define the two circles $C(\hat{0};1)$ and $C(\hat{1};1)$. These intersect L_0 also at the points $-\hat{1}$ and $\hat{2}$ respectively, which thus are defined at Step 2. Hence the numbers -1 and 2 are defined at Step 2.

The two points of intersection of $C(\hat{0};1)$ and $C(\hat{1};1)$ are defined at Step 2, and thus the line L_1 connecting these two points is defined at Step 2. Hence the point $L_1 \cap L_0 = \hat{\frac{1}{2}}$ is defined at Step 3, thereby yielding the number $\frac{1}{2}$, the first noninteger to be constructed.

Remark 2. The radius r of a definable circle is a constructible number; indeed, $\hat{r} \in C(\hat{0};r) \cap L_0$.

We shall use the following familiar constructions:

(1) A perpendicular to a given line, passing through a defined point off the line;
(2) A perpendicular to a given line, passing through a defined point on the line; and
(3) A line passing through a defined point P and parallel to a defined line L. (Actually this is an immediate consequence of (1) and (2); we drop the perpendicular L_1 from P to L, and then take another perpendicular from L_1 at P.)

Diagram 1 illustrates constructions (1) and (2) respectively.

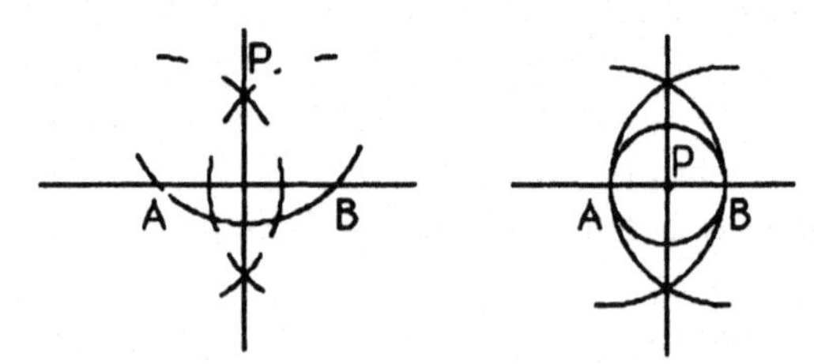

Diagram 1
Construction of perpendiculars.

One question of utmost importance to the ancient Greeks was, "Which numbers are constructible?" More explicitly, they asked:

1. "Doubling the cube." Can one construct a cube having double the volume of a given cube?
2. "Squaring the circle." Can one construct a square of the same area as a given circle?
3. "Trisecting an angle." Can one construct an angle one-third as wide as a given angle? (There is a well-known procedure for bisecting an angle.)

4. "Constructing the n-gon." For which n can one construct a regular polygon of n sides, also called the regular n-gon?

Our objective is to utilize the field theory developed so far, to handle these questions. Questions 1,2, and 3 will be shown to have negative answers, whereas the solution of question 4 is started in Exercises 8ff (although the full solution is given only in Chapter 26.)

Remark 3. Let us rephrase these questions in terms of constructibility of numbers.

1. If the edge of the given cube is 1, then the edge of the second cube is $\sqrt[3]{2}$, so we want to construct $\sqrt[3]{2}$.
2. If the radius of the circle is 1, then its area is π, so we want to construct $\sqrt{\pi}$.
3. If the given angle is 3θ then we want to construct θ.

   Let us consider further the construction of angles. Putting the vertex of the angle at $\hat{0}$ and taking one of the sides of the angle to be L_0, we let P be the intersection of the other side with $C(\hat{0}; 1)$. Dropping a perpendicular from P to L_0 yields the number $\cos\theta$; conversely, given $\cos\theta$ we can reverse the steps, raising a perpendicular to the unit circle and thereby reconstructing the angle θ. In this way, constructing an angle is tantamount to constructing its cosine. Thus, our question is, "If $\cos(3\theta)$ is constructible then is $\cos\theta$ constructible?" Clearly, $\cos\frac{\pi}{3} = .5$ is constructible, so a special case of this question is whether $\cos\frac{\pi}{9}$ is constructible.
4. We need to construct the angle $\frac{2\pi}{n}$, since then we could subdivide the circle into n equal parts, thereby enabling us to inscribe the regular n-gon in the circle. As above, this is equivalent to constructing $\cos\frac{2\pi}{n}$, although we shall find it much more instructive to pass to primitive roots of 1.

Algebraic Description of Constructibility

All of these questions thereby boil down to verifying whether or not certain explicit numbers are constructible, and we shall solve these problems by finding an algebraic criterion for a number to be constructible. It turns out that $\sqrt[3]{2}, \sqrt{\pi}$, and $\cos\frac{\pi}{9}$ all fail this criterion, and so the answer to each of these questions is "No." Our task is to relate constructibility to the previous material about fields. In order to apply that theory we consider all constructible numbers at once.

PROPOSITION 4. *The set C of constructible real numbers is a subfield of $\mathbb{R}$, satisfying the property that if $a \in C$ then $\sqrt{a} \in C$.*

Proof. First we verify the field axioms for $\mathcal{C}$. If $a, b \in \mathcal{C}$, then $b \pm a \in \mathcal{C}$, seen by taking the intersections of $C(\hat{b}; a)$ with L_0. Thus, $(\mathcal{C}, +)$ is a group.

Given $a, b \in \mathcal{C}$ with $a > 0$ and $b \neq 0$ it remains to construct $\frac{a}{b}$ and $\sqrt{a}$, which is done by means of Diagrams 2 and 3:

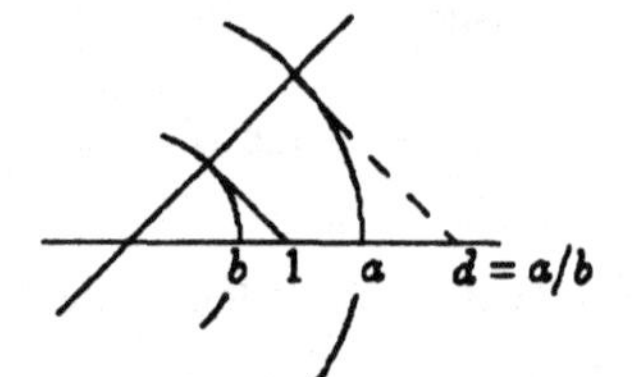

Diagram 2
Construction of the quotient.

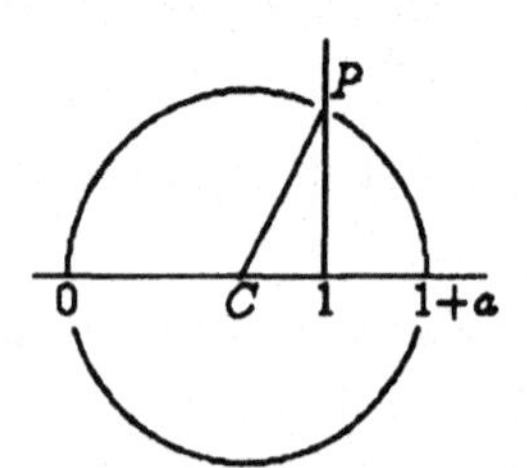

Diagram 3
The square root.

Note. Although one could construct $\sqrt{a}$ rather easily by means of similar triangles, the construction here uses an elegant application of Pythagoras' theorem applied to the right triangle $C\hat{1}P$. Note that C corresponds to $\frac{1+a}{2}$; the distance along the perpendicular down from P to $\hat{1}$ is

$$\sqrt{\overline{CP}^2 - \overline{C\hat{1}}^2} = \sqrt{\left(\frac{1+a}{2}\right)^2 - \left(1 - \frac{1+a}{2}\right)^2} = \sqrt{a}.$$

(Note that we assumed $a < 1$, but this can always be done if we replace a by $\frac{1}{a}$ when $a > 1$.) □

Let us call a set S of real numbers *constructible* if each element of S is constructible.

COROLLARY 5. *If $K = F[\sqrt{\alpha}]$ is a field, where F is constructible and $0 < \alpha \in F$, then K is also constructible.*

In view of this result, we would like to see what it means to adjoin the square root of an element to a field. In preparation for the main result, we say a field $K \subset \mathbb{C}$ is *quadratically defined* if there is a chain of fields

$$\mathbb{Q} = F_0 \subset F_1 \subset F_2 \cdots \subset F_t = K \tag{1}$$

for suitable t, such that each $[F_{i+1} : F_i] = 2$.

The next result could be viewed as an introduction to the deeper theory of Chapter 27. Recall the definition of the compositum KL from Proposition 21.18.

LEMMA 6. (i) *If K and L are quadratically defined fields then KL is a quadratically defined field.*

(ii) *Any subfield L of a quadratically defined field K is quadratically defined.*

Proof. (i) If there are chains of fields

$$\mathbb{Q} = F_0 \subset F_1 \subset F_2 \cdots \subset F_m = K,$$
$$\mathbb{Q} = E_0 \subset E_1 \subset E_2 \cdots \subset E_n = L,$$

for suitable m, n, such that each $[F_{i+1} : F_i] = 2$ and each $[E_{i+1} : E_i] = 2$, then

$$\mathbb{Q} = F_0 \subset F_1 \subset F_2 \cdots \subset F_m = K \subseteq KE_1 \subseteq KE_2 \subseteq \cdots \subseteq KE_n = KL;$$

each successive extension has degree ≤ 2, by Remark 21.19.

(ii) Take the chain (1). Letting $F_i' = F_i \cap L$, we see (by extending a base of F_i' to a base of F_i) that $[F_{i+1}' : F_i'] \leq [F_{i+1} : F_i] = 2$, so the F_i' produce the desired chain from $\mathbb{Q}$ to L. □

THEOREM 7. *The following assertions are equivalent for a real number a:*

(i) *a is constructible;*

(ii) *a is contained in a quadratically defined subfield of $\mathbb{R}$; and*

(iii) *$\mathbb{Q}[a]$ itself is a quadratically defined subfield of $\mathbb{R}$.*

Proof. (ii) ⇒ (iii) by Lemma 6(ii).

(*iii*) ⇒ (*i*) Take a chain $\mathbb{Q} = F_0 \subset F_1 \subset F_2 \cdots \subset F_m = K$, such that each $[F_{i+1} : F_i] = 2$. By Example 21.12(iv) we can write $F_{i+1} = F_i[\sqrt{a_i}]$ for $a_i \in F_i$. Clearly $\mathbb{Q}$ is constructible, by Proposition 4. But by Corollary 5, if F_i is constructible, then F_{i+1} is constructible. Hence F_t is constructible, by induction.

(i) ⇒ (ii) We shall prove for any constructible point P, viewed in terms of its Cartesian coordinates (x_P, y_P), that there is a chain (1) such that $x_P, y_P \in F_t$. This is the same as proving the assertion of the theorem, since P is a constructible point iff x_P and y_P are constructible, *cf.* Exercise 4. However, it is easier to set up the induction in this manner, on the Step i at which P is defined. Let S be the set of (X- and Y-) coordinates of the points defined before Step i. Since S is finite, we appeal to induction and Lemma 6 to obtain a chain

$$\mathbb{Q} = F_0 \subset F_1 \subset F_2 \cdots \subset F_n = L$$

for which $S \subset L$, with each $[F_{i+1} : F_i] = 2$. Thus, it suffices to prove that $[L[x_P, y_P] : L] \leq 2$ (since then $K = [L[x_P, y_P] : L]$ is quadratically defined.) This turns out to be an exercise in analytic geometry, as we shall see now.

The equation of the line passing through points (x_1, y_1) and (x_2, y_2) is

$$y - y_1 = m(x - x_1) \quad \text{where} \quad m = \frac{y_2 - y_1}{x_2 - x_1}, \tag{2}$$

and the equation of the circle with center (x_0, y_0) and radius equal to the distance between points (x_3, y_3) and (x_4, y_4) is

$$(x - x_0)^2 + (y - y_0)^2 = s \quad \text{where} \quad s = (x_4 - x_3)^2 + (y_4 - y_3)^2. \tag{3}$$

By definition, (x_P, y_P) is a simultaneous solution of two equations, each having the form (2) and (3), where all $x_i, y_i \in L$.

We treat each case in turn. The intersection of two lines involves solving two simultaneous linear equations each of the form (2), and thus has its solution in L, see Exercise 6.

The intersection of a circle and a line involves the solution of (2) and (3). But using (2) we could substitute $m(x - x_1) + y_1$ for y in (3), so (3) turns into a quadratic equation in x, which has its solution in some quadratic extension K of L, and $y \in K$ by (2).

Finally, the intersection of two circles involves solving (3) and

$$(x - x_0')^2 + (y - y_0')^2 = s'^2, \tag{4}$$

But subtracting (4) from (3) cancels out the x^2 and y^2 terms and thus leaves a linear equation with coefficients in L, so we continue as in the case of the intersection of a circle and a line. (Geometrically, our equation describes the line passing through the points of intersection of the two circles.) □

COROLLARY 8. *If a is a constructible number, then a is algebraic and its degree is a power of 2.*

Proof. By the theorem $\mathbb{Q}[a]$ is a quadratically defined field. Thus, applying theorem 21.14 to (1), we see $[\mathbb{Q}[a] : \mathbb{Q}] = 2^t$ for some t; hence $\deg a = 2^t$. □

Solution of the Problems of Antiquity

We are now ready to solve problems 1,2, and 3 given before Remark 3.

COROLLARY 9. *(Compare with Remark 3.) The numbers $\sqrt[3]{2}$, $\sqrt{\pi}$, and $\cos\frac{\pi}{9}$ are not constructible.*

Proof. We shall show that none of these are algebraic of degree a power of 2, and thus we shall be done.

1. $\sqrt[3]{2}$ has degree 3, *cf.* Example 21.12(v).
2. Since $\sqrt{\pi}$ is quadratic over π, we may pass to π, which is transcendental. However, this is a rather deep theorem, whose proof is given in Appendix A.
3. Let us determine the minimal polynomial f of $\cos\frac{\pi}{9}$. More generally, we want to determine $\cos\theta$ in terms of $\cos 3\theta$. To streamline the computations we appeal to the mathematics of the complex unit circle; the point corresponding to the angle θ is $e^{\theta i} = \cos\theta + i\sin\theta$. Then

$$\begin{aligned} e^{3\theta i} &= (e^{\theta i})^3 \\ &= (\cos\theta + i\sin\theta)^3 \\ &= (\cos\theta)^3 + 3i(\cos\theta)^2\sin\theta + 3i^2\cos\theta(\sin\theta)^2 + i^3(\sin\theta)^3 \\ &= (\cos^3\theta - 3\cos\theta\sin^2\theta) + i(3\cos^2\theta\sin\theta - \sin^3\theta). \end{aligned}$$

Matching real parts and substituting $1 - \cos^2\theta$ for $\sin^2\theta$ yields

$$\begin{aligned} \cos 3\theta &= \cos^3\theta - 3\cos\theta\sin^2\theta \\ &= 4\cos^3\theta - 3\cos\theta \end{aligned}$$

Now putting $\theta = \frac{\pi}{9}$, we see $\cos 3\theta = \cos\frac{\pi}{3} = \frac{1}{2}$, so $\cos\theta$ is a root of the polynomial $4x^3 - 3x - \frac{1}{2}$, or equivalently, of $f(x) = 8x^3 - 6x - 1$. But f is irreducible, by Example 20.14. □

Remark 10. The regular n-gon: Initial results. The same line of reasoning can be used to study the constructibility of the regular n-gon, or equivalently, constructing the angle $\frac{2\pi}{n}$. But interpreting the unit circle as in the complex plane, this is tantamount to constructing the point corresponding to a primitive nth root ρ_n of 1. Although we have focused on constructible real numbers, one could just as easily have defined constructible complex numbers, by saying $a + bi$ is constructible if both a and b are constructible real numbers. The theory goes through in this greater generality, as is seen in Exercise 4; the proofs of Theorem 7 and Corollary 8 also apply to complex numbers, so ρ_n can be constructed only if its minimal polynomial f_n has degree a power of 2. When n is prime, we have seen $f_n = x^{n-1} + \cdots + 1$, which has degree $n - 1$, so $n - 1$ must be a power of 2. In particular, one *cannot* construct the regular 7-gon, 11-gon, or 13-gon. Furthermore, one cannot construct the regular 9-gon, in view of corollary 9(iii). Further insight is provided in Exercises 9ff, in which we see the regular 5-gon can be constructed. Thus, the smallest nontrivial case is $n = 17 (= 2^{2^2} + 1)$.

In a tour de force, the young Euler constructed the 17-gon; his result follows from Theorem 26.7 below, which determines the constructibility of all regular n-gons, modulo a famous question in number theory.

Exercises

Constructibility

1. Describe rigorously the construction of a perpendicular to a given line, at a given point on the line.
2. Describe rigorously the construction of a perpendicular to a given line, at a given point off the line.
3. Define the Y-axis of the plane to be the line perpendicular to the X-axis at $\hat{0}$. (Thus $\hat{0}$ is the origin, and we have the axes of the Cartesian plane.) Written in Cartesian coordinates, the point $P = (x_P, y_P)$ is defined iff x_P and y_P are constructible numbers. (This is seen by projecting down onto the X- and Y-axes.)
4. The number $a+bi$ is quadratically defined iff the point (a,b) on the plane (with respect to the axes of Exercise 3) is constructible. (*Hint*: a,b are quadratically defined, as is $i=\sqrt{-1}$, so apply Lemma 6(i).)
5. Describe rigorously the construction of the bisector of an angle. How can one infer the existence of the angle bisector using algebra instead of geometry?
6. Solve explicitly equation (2) (from the proof of Theorem 7) and
$$y - y' = m'(x - x')$$
to get $x = \frac{(mx_1-y_1)-(m'x_1'-y_1')}{m-m'} \in L$ and $y \in L$.
7. Suppose a is algebraic of degree 4. Show that a is constructible iff $\mathbb{Q}[a]$ contains a proper F-subfield $\neq F$.

Constructing a Regular n-gon

8. If a regular p-gon can be constructed, for p prime, then p has the form $2^{2^t}+1$. (*Hint*: $p = 2^m+1$ for some m. But if m has an odd factor k, then $2^{m/k}+1$ divides p, a contradiction.) Construct a regular 3-gon.
9. Any primitive fifth root ρ of 1 is constructible. (*Hint*: This is sometimes proved directly in geometry courses, but is not difficult to show algebraically; show that $\rho+\rho^{-1}$ satisfies $x^2+x=1$. Conclude that the regular 5-gon is constructible.
10. If the regular n-gon is constructible, then so is the regular $2n$-gon.
11. Determine the constructibility of all regular n-gons for $n \leq 16$. (*Hint*: No for $n = 7, 9, 11, 13$; yes for all others. Note, for example, that if ρ_n is a primitive n-th root of 1 then $\rho_{15} \in \mathbb{Q}[\rho_5, \rho_3]$.)

CHAPTER 23. ADJOINING ROOTS TO POLYNOMIALS: SPLITTING FIELDS

In the previous sections we studied algebraic elements in terms of the polynomials that they satisfy. Now we reverse the question and ask, how can we determine the roots of a polynomial? This question has several possible interpretations, and there exists a highly developed theory of approximating roots of a polynomial in a given field. However, we are interested in an exact algebraic solution, and thus our first interpretation is, "Given $f \in F[x]$, can we find a field K that contains all the roots of f?" First we locate one root. To understand the question, let us examine a familiar situation.

Suppose we met someone who does not know what the complex numbers are. We might very possibly describe them as the field $\mathbb{R}[\sqrt{-1}]$. That person would then be looking for an $\mathbb{R}$-field that contained an element i satisfying $i^2 = -1$, or equivalently $i^2 + 1 = 0$; in other words he or she would want to adjoin a root of the polynomial $x^2 + 1$ to the field $\mathbb{R}$. There is a formal construction that works for any irreducible polynomial f. To obtain a root x of f we declare formally $f(x) = 0$; the easiest way to do this is by taking the polynomial ring $F[x]$ and factoring out the ideal $\langle f(x) \rangle$ (thereby "forcing" $f(x)$ to become 0). Thus, we shall be combining aspects of ring theory with the vector space applications used in Chapter 21, and to this end we need to reconsider the definition of F-fields, or more generally about inclusions of rings.

What do we really mean when we say $\mathbb{Z} \subset \mathbb{Q}$? When defining the rational numbers, as in Chapter 15, we do not actually include $\mathbb{Z}$ itself, but rather the set $\{\frac{n}{1} : n \in \mathbb{Z}\}$, which is naturally identified with $\mathbb{Z}$ under the correspondence $\frac{n}{1} \mapsto n$. In the language of algebra, $\mathbb{Q}$ contains a subring which is isomorphic to $\mathbb{Z}$, or equivalently, there is a ring injection $\psi: \mathbb{Z} \to \mathbb{Q}$ sending $n \mapsto \frac{n}{1}$. Thus, we could think of an F-field K as a field K together with a ring injection $\psi: F \to K$. We can weaken this formally, in view of the following straightforward but useful observation.

Remark 1. If F is a field and $\psi: F \to R$ is a ring homomorphism, then ψ is an injection; indeed, $\ker \psi$ is a proper ideal of F and thus is 0.

The way has been cleared to introduce the following structure.

Definition 2. Suppose F is a given field. A commutative ring R is called an *F-algebra* if there is a ring homomorphism $\psi: F \to R$.

Then ψ will be an injection, by Remark 1, so we identify F with its image under ψ, called the "canonical image" of F; in this way we view F as a subfield of R. In particular, an F-algebra K that happens to be a field is an F-field.

Example 2′. Any subfield F of $\mathbb{C}$ is a $\mathbb{Q}$-algebra, and thus a $\mathbb{Q}$-field. Indeed, we define the injection $\psi: \mathbb{Q} \to F$ by $\psi(\frac{m}{n}) = \frac{m \cdot 1}{n \cdot 1}$, *cf.* Example 15.9.

By being precise, we have introduced an ambiguity: An F-algebra R could be an F-algebra in several different ways, depending on the choice of the homomorphism $\psi: F \to R$, cf. Exercise 1. Thus one should keep the particular injection ψ in mind. Nevertheless, in this treatment we shall tend to view $F \subseteq K$, and we shall downplay the role of ψ. This ambiguity motivates the development of the theory in Chapter 25.

If R_1, R_2 are F-algebras, then, viewing both $F \subseteq R_1$ and $F \subseteq R_2$, we define an *F-homomorphism* to be a ring homomorphism $\varphi: R_1 \to R_2$ that fixes F, *i.e.*, $\varphi(\alpha) = \alpha$ for all α in F. An *F-isomorphism* is an F-homomorphism that is an isomorphism.

One very important example: $F[x]$ is clearly an F-algebra, where the map $\psi: F \to F[x]$ sends an element α of F to the constant polynomial α. For any F-algebra R, one can extend ψ to the substitution homomorphism $\psi_a: F[x] \to R$ given by $x \mapsto a$, *cf.* Lemma 18.2, and this is an F-homomorphism. Note that the image of $F[x]$ under ψ_a is $F[a]$.

Here is another important example.

PROPOSITION 3. *Suppose F is a field, and $f \in F[x]$ is monic irreducible. Let $L = F[x]/\langle f \rangle$. Then L is an F-field, and (identifying F with its canonical image in L), $L = F[\bar{x}]$ where $\bar{x}$ is the image of x in L. Furthermore, f is the minimal polynomial of $\bar{x}$ over F, and $[L : F] = \deg f$.*

Proof. L is a field, by Remark 21.9′, so is an F-field. Let $^-$ denote the canonical homomorphism from $F[x]$ onto $L = F[x]/\langle f \rangle$. Any element of L has the form $\overline{\sum \alpha_i x^i} = \sum \overline{\alpha_i} \bar{x}^i$, so, identifying F with $\bar{F}$ we have $L = F[\bar{x}]$. Furthermore, if $f = \sum_{i=0}^{n} \beta_i x^i$, then $0 = \bar{f} = \sum_{i=0}^{n} \beta_i \bar{x}^i = f(\bar{x})$; since f is irreducible we conclude from Proposition 21.6 that f is the minimal polynomial of $\bar{x}$. Hence $[L : F] = \deg f$, by Proposition 21.8. □

Note. In case we already have found an F-field $K = F[a]$, where a is a root of f, then $F[a] \approx F[x]/\langle f \rangle$, by Theorem 21.10. For example, $\mathbb{C} = \mathbb{R}[i] \approx \mathbb{R}[x]/\langle x^2 + 1 \rangle$. Let us continue this reasoning.

PROPOSITION 4. *Suppose K_1, K_2 are F-fields, and $a_1 \in K_1$, $a_2 \in K_2$ are both roots of the same irreducible polynomial $f \in F[x]$. Then there is an F-isomorphism $F[a_1] \to F[a_2]$ sending $a_1 \mapsto a_2$.*

Proof. Apply Theorem 21.10 twice (once in reverse) to get the isomorphism $F[a_1] \to F[x]/\langle f \rangle \to F[a_2]$ sending $a_1 \mapsto x \mapsto a_2$. □

Sometimes both of these roots are in the same field.

Example 5. (i) The roots of $x^2 + 1$ in $\mathbb{C}$ are $\pm i$. Therefore, there is an F-isomorphism $\mathbb{C} \to \mathbb{C}$ given by $i \mapsto -i$; this is complex conjugation.

(ii) the roots of $x^2 - 2$ in $\mathbb{Q}[\sqrt{2}]$ are $\pm\sqrt{2}$. Hence there is a $\mathbb{Q}$- isomorphism $\mathbb{Q}[\sqrt{2}] \to \mathbb{Q}[\sqrt{2}]$ given by $\sqrt{2} \mapsto -\sqrt{2}$.

Let us digress briefly, to obtain an important (although easy) converse.

PROPOSITION 6. *Suppose K is an F-field, $f(x) \in F[x]$, and $a \in K$ is a root of f. Then $\sigma(a)$ is also a root of f, for any F-isomorphism $\sigma: K \to K$.*

Proof. Write $f = \sum \alpha_i x^i$ for suitable α_i in F. Then

$$\begin{aligned} f(\sigma(a)) = \sum \alpha_i \sigma(a)^i &= \sum \sigma(\alpha_i)\sigma(a^i) = \sigma(\sum \alpha_i a^i) \\ &= \sigma(f(a)) = \sigma(0) = 0. \quad \square \end{aligned}$$

Splitting Fields

We return to our main quest. Having found one root, we now want them all, in a suitable extension field K of F; we also want such K to be uniquely defined (up to F-isomorphism). This task might seem unmanageable at the outset, but becomes much easier if we recall from Theorem 18.5 that f can have at most n roots in K, where $n = \deg f$. Thus, our strategy is to use Proposition 3 to adjoin one root at a time, until we are done (after at most n steps). Let us describe this procedure more precisely.

We say a polynomial f *splits* over K if f can be written in $K[x]$ in the form

$$\gamma(x - a_1) \dots (x - a_n), \quad a_i \in K.$$

(Here γ is the leading coefficient of f and thus is in K; without loss of generality, one may take f to be monic, so that $\gamma = 1$.)

Note that this condition implies that $a_1, \dots, a_n$ are roots of f and actually are *all* the roots of f in any field extension of K, since any other root b would satisfy

$$0 = f(b) = (b - a_1) \dots (b - a_n),$$

which is a contradiction.

Definition 7. A *splitting field* of a polynomial f (over the field F) is a field $E \supseteq F$ satisfying the following two conditions:

(i) f splits over E, and

(ii) f does not split over K for any $F \subseteq K \subset E$.

Remark 8. Another way of saying that E is a splitting field of a monic polynomial f is that $E = F[a_1, \dots, a_n]$, with $f = (x - a_1) \dots (x - a_n)$

in $E[x]$. Intuitively, E is the field we get by adjoining the roots of f. (In particular, $[E : F] < \infty$ since E is obtained by adjoining at most n roots.)

This point of view is very intuitive and enables us to work in any field L over which f splits as $f = (x-a_1)\dots(x-a_n)$, since then $F[a_1,\dots,a_n] \subseteq L$. Then L cannot contain any other splitting field of f; in fact f has no other roots in L, as noted above. (Most commonly, $L = \mathbb{C}$.) Other immediate consequences of Remark 8:

Remark 9. Suppose E is a splitting field of f over F.

(i) If $F \subseteq K \subseteq E$, then E is also a splitting field of f over K.

(ii) Any factor g of f in $F[x]$ also splits over E. (Indeed, by unique factorization in $E[x]$, the factorization of g is part of the factorization of f into linear factors.)

(iii) If $F \subseteq K \subseteq E$ and $f = (x-a)g$ in $K[x]$, then E is a splitting field of g over K. (Indeed, E is generated over K by the roots of g.)

(iv) Conversely to (iii), if $K = F[a]$ and $f = (x-a)g$, then any splitting field L of g over K is also a splitting field of f over E. (Indeed, writing $g = (x-a_2)\dots(x-a_t)$ in $K[x]$, we see $L = K[a_2,\dots,a_t] = F[a,a_2,\dots,a_t]$.)

Example 10. (based on Example 21.12.) In each case, E will denote the splitting field of f over F.

(i) If $f = x - a$ for $a \in F$, then $E = F$.

(ii) $\mathbb{C}$ is the splitting field of x^2+1 over $\mathbb{R}$.

(iii) $\mathbb{C}$ contains the splitting field of x^2+1 over $\mathbb{Q}$, which is $\mathbb{Q}[i]$. In general, if $f \in F[x]$ has a root in $K \setminus F$ with $[K : F] = 2$, then $E = K$. (This can be seen either via Example 21.12(iv) or as a special case of Theorem 11 below.)

(iv) If $f = x^n - 1$ and $F = \mathbb{Q}$, then $E = \mathbb{Q}[\rho]$, where ρ is a primitive nth root of 1. Indeed, $f = (x-1)(x-\rho)\dots(x-\rho^{n-1})$, by Proposition 18.14.

(v) If $f = x^{n-1} + \dots + 1$ and $F = \mathbb{Q}$, then, by Remark 9(iii), $E = \mathbb{Q}[\rho]$, where ρ is a primitive nth root of 1; indeed, $x^n - 1 = (x-1)f$.

(vi) If $f = x^n - p$ for p prime, and $F = \mathbb{Q}$, then $E = \mathbb{Q}[\sqrt[n]{p},\rho]$, where ρ is a primitive n-th root of 1. (Indeed, by Proposition 18.14, $E = \mathbb{Q}[a,\rho a,\dots,\rho^{n-1}a]$, where $a = \sqrt[n]{p}$. But then $\rho = (\rho a)a^{-1} \in E$, implying $E = \mathbb{Q}[a,\rho]$.)

If one wants to start with an arbitrary base field F and develop the theory from scratch, one must confront the following questions, for any polynomial f in $F[x]$:

(1) Does f have splitting fields?

(2) Is the splitting field unique?

(3) How many distinct roots does f have in its splitting field (since perhaps a root repeats)?

We start by answering question (1).

THEOREM 11. *Any polynomial f over F has a splitting field E for which $[E : F] \leq n!$, where $n = \deg f$.*

Proof. By Proposition 3, we can find $K \supseteq F$ containing a root a of f, with $[K : F] \leq n$. But then $f = (x - a)g$ in $K[x]$, by Corollary 18.4, so $\deg g = n - 1$. By induction g has a splitting field E over K, for which $[E : K] \leq (n-1)!$, so $[E : F] = [E : K][K : F] \leq n!$; Remark 9(iv) shows that E is the splitting field of f. □

Remark 12. In Theorem 11, if f is irreducible, then $n | [E : F]$. (Indeed, if a is a root of f, then $[F[a] : F] = n$ divides $[E : F]$.)

We can answer (2) by proceeding formally. Suppose $\varphi: F_1 \to F_2$ is an F-homomorphism of F-fields. Then φ extends to an F-homomorphism $\tilde{\varphi}: F_1[x] \to F_2[x]$, given by $\sum a_i x^i \mapsto \sum \varphi(a_i) x^i$.

THEOREM 13. *Suppose $\varphi: F_1 \to F_2$ is an F-isomorphism, and $f_2 = \tilde{\varphi}(f_1)$. Let E_i be a splitting field of f_i over F_i, for $i = 1, 2$. Then φ extends to a natural F-isomorphism $E_1 \to E_2$.*

Proof. Induction on $n = [E_1 : F_1]$. The idea of proof is to show that the method used in the proof of Theorem 11 works essentially the same way in all situations. Let g_1 be an irreducible factor of f_1, and let a_1 be any root of g_1 in E_1; let $g_2 = \tilde{\varphi}(g_1)$, and let a_2 be any root of g_2. The composition $F_1[x] \xrightarrow{\tilde{\varphi}} F_2[x] \to F_2[x]/\langle g_2 \rangle$ is an onto homomorphism whose kernel is $\langle g_1 \rangle$, thereby yielding an isomorphism $\phi: F_1[x]/\langle g_1 \rangle \to F_2[x]/\langle g_2 \rangle$ that extends φ. Now, by Proposition 4 we have a composition of isomorphisms

$$F_1[a_1] \approx F_1[x]/\langle g_1 \rangle \overset{\phi}{\approx} F_2[x]/\langle g_2 \rangle \approx F_2[a_2],$$

which we call φ_1. Certainly E_i is a splitting field of f_i over $F_i[a_i]$, for $i = 1, 2$, but $[E_1 : F_1[a_1]] < n$, so by induction φ_1 extends to the desired F-isomorphism $E_1 \to E_2$. □

Now taking $F_1 = F_2 = F$ in Theorem 13, we have

COROLLARY 13′. *The splitting field of a given polynomial over a given field F is unique up to F-isomorphism.*

Thus, we may speak of "the" splitting field. In Chapter 25 we shall need a more precise result concerning the number $n_{E/K;\,L} = n_{E/K;\,L}(\varphi)$ of F-isomorphisms extending any given F-homomorphism $\varphi\colon K \to L$ to an F-homomorphism $E \to L$, where $E = F[a_1, \ldots, a_u]$ is an F-field generated by various roots $a_1, \ldots, a_u$ of a given polynomial $f \in F[x]$, and K is an F-subfield of E. If $f = \sum \alpha_i x^i$, we write f_φ for $\sum \varphi(\alpha_i)x^i$, the polynomial naturally corresponding to f in $L[x]$.

THEOREM 14. *With notation as above, $n_{E/K;\,L} \le [E : K]$, equality holding if E is a splitting field (of f) and f_φ has $\deg(f)$ distinct roots in L.*

Proof. Induction on $n = [E : K]$. If $n = 1$, then $E = K$, so the only F-homomorphism extending φ must be φ itself.

In general, for $n > 1$ we take any root a of f in $E \setminus K$. Any extension of φ to an F-homomorphism $\sigma\colon E \to L$ can be viewed as an extension first to an F-homomorphism $\eta\colon K[a] \to L$ and then to σ. By induction, the number of F-homomorphisms $E \to L$ extending $\eta\colon K[a] \to L$ is at most $[E : K[a]]$, equality holding if f_η has $\deg(f)$ distinct roots in L. Hence

$$n_{E/K;\,L} \le n_{K[a]/K;\,L}[E : K[a]].$$

Note that $\eta\colon K[a] \to L$ is given by $\eta(\sum \alpha_i a^i) = \sum \varphi(\alpha_i)\eta(a)^i$; hence η is determined by $\eta(a)$, a root of f_φ since $f_\varphi(\eta(a)) = \eta(f(a)) = \eta(0) = 0$. On the other hand, as in Proposition 4, any root of f_φ gives rise to an F-homomorphism $\eta\colon K[a] \to L$. Thus, $n_{K[a]/K;\,L}$ is the number of distinct roots of g_φ in L, where g is the minimal polynomial of a over K, yielding

$$n_{K[a]/K;\,L} \le \deg g_\varphi = \deg g = [K[a] : K],$$

equality holding if f_φ has $\deg(f)$ distinct roots in L (since g_φ divides f_φ in $L[x]$). We conclude

$$n_{E/K;\,L} \le [K[a] : K][E : K[a]] = [E : K],$$

equality holding if f_φ has $\deg(f)$ distinct roots in L. □

Example 15. The splitting field of $f = x^4 - 2$ over $\mathbb{Q}$ is $E = \mathbb{Q}[\sqrt[4]{2}, i]$, by Example 10 (vi). Another way of constructing the splitting field is first to adjoin $\sqrt[4]{2}$ and then observe that f factors over $\mathbb{Q}[\sqrt[4]{2}]$ as

$$(x^2 - \sqrt{2})(x^2 + \sqrt{2}) = (x - \sqrt[4]{2})(x + \sqrt[4]{2})(x^2 + \sqrt{2}).$$

Thus, $(\mathbb{Q}[\sqrt[4]{2}])[x]/\langle x^2 + \sqrt{2}\rangle$ is another splitting field of f and, of course, is isomorphic to $\mathbb{Q}[\sqrt[4]{2}, i\sqrt[4]{2}] = \mathbb{Q}[\sqrt[4]{2}, i] = E$.

Separable Polynomials and Separable Extensions

In order to treat question (3) (posed after Example 10) we need a definition. We say a polynomial $f \in F[x]$ is *separable* if its roots (in a splitting field of f over F) are all distinct, *i.e.*, the number of distinct roots is $\deg f$. We say f is *inseparable* if f is not separable, *i.e.*, f has a repeated root. To avoid a vacuous situation, we shall always assume that $\deg f \geq 1$. Separable polynomials tie in at once to our previous result, which foreshadows the theory of Chapter 25:

COROLLARY 16. *Suppose E is the splitting field of a separable polynomial f over F. Then the number of F-isomorphisms from E to itself is precisely $[E : F]$.*

Proof. Just take $\varphi: F \to E$ to be the inclusion map. □

This result is so critical to the theory that we shall concentrate exclusively on separable polynomials for the rest of this course. The study of separable polynomials involves an algebraic approach to calculus. Given a polynomial $f = \sum a_i x^i$, we formally define its derivative f' to be $\sum i a_i x^{i-1}$; it is a simple matter to check that the usual rules of derivative hold, for all polynomials f, g:

$$\begin{aligned}(f+g)' &= f' + g'; \\ (\alpha f)' &= \alpha f' \quad \text{for } \alpha \text{ in } F; \\ (fg)' &= f'g + fg'.\end{aligned} \tag{1}$$

PROPOSITION 17. *Let E be a splitting field of f. The polynomial f is separable iff f and f' are relatively prime in $E[x]$.*

Proof. We prove, equivalently, that f has a repeated root iff f, f' are not relatively prime.

($\Rightarrow$) On the contrary suppose $(x-a)^2 | f$. Then $f = (x-a)^2 g$ so

$$f' = (x-a)'(x-a)g + (x-a)((x-a)g)',$$

which obviously is divisible by $x - a$.

($\Leftarrow$) Suppose f and f' are not relatively prime in $E[x]$. Then some irreducible factor $x - a$ of f also divides f'. Write $f = (x-a)g$; we know $x - a$ divides $f' = g + (x-a)g'$, so $x - a$ divides g, implying $(x-a)^2 | f$. □

It would be a shame to have to pass to $E[x]$ when applying this criterion; fortunately, we have the following fact.

PROPOSITION 18. *Suppose $F \subseteq K$ are fields, and $f, g \in F[x]$. Then $\gcd(f,g)$ is the same in $F[x]$ and in $K[x]$.*

Proof. Let d_F, d_K be the respective gcd in $F[x]$ and $K[x]$. *A fortiori*, d_F divides f and g in $K[x]$, so $d_F | d_K$. On the other hand, $\langle d_F \rangle = \langle f \rangle + \langle g \rangle$ in $F[x]$, so there are h_1, h_2 in $F[x]$ such that $d_F = h_1 f + h_2 g$; but d_K divides the right-hand side, and thus $d_K | d_F$. □

COROLLARY 19. *A polynomial f in $F[x]$ is separable iff f and f' are relatively prime in $F[x]$.*

COROLLARY 20. *An irreducible polynomial f is separable iff $f' \neq 0$.*

Proof. f, f' are not relatively prime, iff $\gcd(f, f') = f$ (since f is irreducible), iff $f | f'$, iff $f' = 0$ (since $\deg f' < f$).

How can $f' = 0$? From calculus (where $F \subseteq \mathbb{R}$) we learned this implies f is constant, which is ruled out since we are assuming $\deg f \geq 1$. Thus, we conclude for $F \subseteq \mathbb{R}$ that *every* irreducible polynomial over F is separable. Soon we shall improve this result, but first let us record a useful fact.

Remark 21. If $f = g_1 g_2$ in $F[x]$ is separable then g_1 and g_2 are relatively prime. (Indeed, let h be the gcd of g_1 and g_2. Then $h^2 | f$, implying any root of h is a multiple root of f.)

The Characteristic of a Field

In order to study fields that are not necessarily subfields of $\mathbb{R}$ or $\mathbb{C}$, we define $m{\cdot}1$ in a field F to be $1+1+\cdots+1$, taken m times. Note $(m{\cdot}1)(n{\cdot}1) = mn{\cdot}1$; the formal proof is an easy exercise in induction.

Definition 22. The *characteristic* of a field F, denoted $\text{char}(F)$, is the order of the element 1 in $(F, +)$, *i.e.*, $\text{char}(F) = m$ if $m \cdot 1 = 0$, for $m > 0$ minimal such. If $m \cdot 1 \neq 0$ for all $m > 0$, we say $\text{char}(F) = 0$.

Example 23. $\text{char}(\mathbb{C}) = 0 = \text{char}(\mathbb{R}) = \text{char}(\mathbb{Q})$. On the other hand, $\text{char}(F) > 0$ for any finite field F (since $(F, +)$ is a finite group), and, in particular, $\text{char}(\mathbb{Z}_p) = p$.

Along another line, a field F has characteristic $\neq 2$ iff $1 + 1 \neq 0$, iff 2 is invertible in F, *i.e.*, $\frac{1}{2} \in F$. The existence of $\frac{1}{2}$ often is very useful, for example, in Example 21.12 (iv).

Remark 24. Suppose $p = \text{char}(F) > 0$.

(i) p is a prime number. (Otherwise let $p = ab$ for $1 < a, b < p$. Then $0 = (p \cdot 1) = (a \cdot 1)(b \cdot 1)$ so $a \cdot 1 = 0$ or $b \cdot 1 = 0$, contrary to the minimality of p.)

(ii) $pa = (p \cdot 1)a = 0a = 0$, for all a in F.

We can view the characteristic more structurally. For any field F there is a ring homomorphism $\varphi: \mathbb{Z} \to F$ given by $\varphi(m) = m \cdot 1$.

Case I. φ is an injection. Then $\text{char}(F) = 0$. Note that F contains a copy of the field of fractions of $\mathbb{Z}$, which is $\mathbb{Q}$.

Case II. φ is not an injection. Since $\varphi(\mathbb{Z}) \subset F$ is an integral domain, we see that $\ker \varphi$ is a nonzero prime ideal of $\mathbb{Z}$, so $\ker \varphi = p\mathbb{Z}$ for some prime number p, implying $\text{char}(F) = p$. (This gives a "structural" proof that $\text{char}(F)$ is prime.) Also, Noether I yields an injection $\bar{\varphi}: \mathbb{Z}/p\mathbb{Z} \to F$.

Remark 25. Summarizing the two cases above, we have

(i) $\text{char}\, F = 0$ iff F contains an isomorphic copy of $\mathbb{Q}$.

(ii) $\text{char}(F) = p > 0$ iff F contains an isomorphic copy of $\mathbb{Z}_p$.

Let us return to separability.

Definition 26. A *separable* element over F is an element whose minimal polynomial over F is separable. A *separable extension* of F is an F-field K all of whose elements are separable over F. In this case, we also say the extension K/F is separable.

F is *perfect* if every finite extension is separable.

Example 27. $\mathbb{Q}$ is a perfect field; more generally, every field of characteristic 0 is perfect. (Indeed, in characteristic 0 the derivative $\sum i a_i x^{i-1}$ of a non-constant polynomial $f = \sum a_i x^i$ cannot be 0, implying every irreducible polynomial is separable.) We shall see in Chapter 24 that every field of finite order also is perfect.

Although perfect fields are very useful, our definition is unsatisfying since it depends on verifications for many polynomials; a more esthetic criterion is given in Exercise 16.

Exercises

1. $\mathbb{C}$ and $\mathbb{Q}[\sqrt{2}]$ each can be viewed as an algebra over itself in two different ways. (Hint: Either by the identity or by means of the homomorphism of example 5.)
2. Notation as in Theorem 14, show $n_{E/K;\,\varphi} < [E : K]$ if $\varphi(f)$ has fewer than $\deg(f)$ distinct roots in L.
3. (The Primitive Root Theorem.) If K is a separable, finite extension of an infinite field F, then $K = F[a]$ for some a in K. (*Hint:* By iteration, it is enough to show that, whenever $K = F[a, b]$, there is c in K for which $K = F[c]$. Let $f, g \in F[x]$ be the respective minimal polynomials of a, b. Passing to a splitting field E of fg, take the roots $a = a_1, \ldots, a_m$ and $b = b_1, \ldots, b_n$ of f and g (respectively),

and note that $a + \alpha b \neq a_i + \alpha b_j$ for all $j \neq 1$ and all i, for all but a finite number of α in F. Prove $F[a,b] = F[a+\alpha b]$ for any such α, as follows: Let $c = a+\alpha b$ and $L = F[c]$. Define $h(x) = f(c-\alpha x) \in L[x]$. Then $h(b) = f(a) = 0$, and $h(b_j) \neq 0$ for all $j > 1$, implying $x - b = \gcd(f(x), h(x))$, which can be taken in $L[x]$, implying $b \in L$. We shall improve this result in Chapter 26.)

4. $(\sqrt{2} + \sqrt{3})$ has degree 4 over $\mathbb{Q}$. (*Hint:* As in Exercise 3.)
5. Suppose $K = F[a]$ is a field, and $F \subseteq L \subseteq K$. If g is the minimal monic polynomial of a over L, then L is generated by the coefficients of g. (*Hint:* let L_1 be the subfield of L generated by the coefficients of g. Then g is also the minimal polynomial of a over L_1, implying $[K : L_1] = \deg g = [K : L]$, so $L = L_1$.)
6. Suppose $K = F[a]$. Then there are only finitely many fields between F and K. (*Hint:* let f be the minimal polynomial of a over F, and take $F \subseteq L \subseteq K$. By Exercise 5, L is generated by the coefficients of the minimal polynomial of a over L, and f has only a finite number of monic factors.)
7. (Steinitz' Theorem.) Suppose F is an infinite field, and K is an F-field. Then $K = F[a]$ for some a, iff there are only finitely many fields between F and K. (*Hint:* ($\Leftarrow$) As before, one needs only show that $F[a,b] = F[a+\alpha b]$ for some α in F. But $F[a+\alpha_1 b] = F[a+\alpha_2 b]$ for suitable $\alpha_1 \neq \alpha_2$ in F; this field contains $a + \alpha_1 b - (a + \alpha_2 b) = (\alpha_1 - \alpha_2)b$.)
8. If K/F is separable and finite, then there are only finitely many fields L between F and K, by Exercises 3 and 7. This result foreshadows the study of intermediate field extensions, and will also be seen as a consequence of the theory of Chapter 25.

The Roots of a Polynomial in Terms of Its Coefficients

9. Let $s_1, \ldots, s_n$ be the elementary symmetric polynomials in the indeterminates $x_1, \ldots, x_n$, *cf.* Exercise 20.11. Suppose $a_1, \ldots, a_n$ are the roots of a monic polynomial $f = \sum_{i=0}^{n} \alpha_i x^i \in \mathbb{Z}[x]$ in a suitable splitting field. Then $s_i(a_1, \ldots, a_n) = (-1)^i \alpha_{n-i}$, for $1 \leq i \leq n$. In particular, the sum of the roots is $-\alpha_1$, and the product of the roots is $(-1)^n \alpha_0$.
10. Suppose $f \in \mathbb{Z}[x]$ has leading coefficient m, and $a_1, \ldots, a_n$ are the roots of f in a suitable splitting field. If $h(x_1, \ldots, x_n)$ is an arbitrary symmetric polynomial, then $h(ma_1, \ldots, ma_n) \in \mathbb{Z}$. (*Hint:* By Exercise 21.13, $ma_1, \ldots, ma_n$ are the roots of suitable monic polynomial of degree n over $\mathbb{Z}[x]$, so each $s_i(ma_1, \ldots, ma_n) \in \mathbb{Z}$.)

Separability and the Characteristic

11. Give an example of an infinite field of positive characteristic. (*Hint:* Same idea as Exercise 21.9.)
12. If an irreducible polynomial f of $F[x]$ is inseparable, then $\text{char}(F) = p > 0$, and f has the form $\sum \alpha_i x^{pi}$; in other words, f has the form $g(x^p)$. Conversely, if $\text{char}(F) = p$ and $f = g(x^p)$, then f is inseparable. (*Hint:* $f' = 0$.)
13. Suppose $K = F[a]$, and f is the minimal polynomial of a. Suppose $\text{char}(F) = p > 0$. Then f is separable iff $F[a^p] = K$. (*Hint:* If f is inseparable, then write $f = g(x^p)$ and note a^p is a root of $g(x)$, implying
$$[F[a] : F[a^p]] = \frac{[F[a] : F]}{[F[a^p] : F]} = \frac{\deg f}{\deg g} = p > 1.$$
Conversely, if f is separable, then a is a common root of its minimal polynomial g over $F[a^p]$ and of $x^p - a^p = (x - a)^p$, implying $x - a$ is their gcd, and thus $a \in F[a^p]$.)
14. Suppose $\text{char}(F) = p > 0$, and $\alpha \in F$. Let $f = x^p - \alpha$. Either f is irreducible, or else α is a pth power in F, in which case $x^p - \alpha$ is also a pth power in $F[x]$. (*Hint:* Suppose that $f = gh$, and take a root b in a splitting field E of g. Then $g = (x - b)^k$ for some $k < p$; checking the degree of x^{k-1} shows $b \in F$.)
15. Give an example of a field extension K/F that is not separable. (*Hint:* F must be infinite, of positive characteristic.)
16. A field F of characteristic $p > 0$ is perfect iff every element of F is a pth power in F.

Calculus through the Looking Glass

17. Define a *derivation* of a field F to be a map $D: F \to F$ that satisfies equations (1) of the text. Let $F_0 = \{\alpha \in F : D(\alpha) = 0\}$. Show that $D(\sum_i \alpha_i a^i) = \sum_i i\alpha_i a^{i-1} D(a)$ for all α_i in F_0, a in F. In other words, $D(f(a)) = f'(a)D(a)$, for any f in $F_0[x]$. Conclude F_0 is a field; if $a \in F$ is separable over F_0, then $a \in F_0$. In particular, any derivation on a field of characteristic 0 takes on the value 0 on every element algebraic over $\mathbb{Q}$.
18. Generalizing Exercise 17, given a field extension K/F, suppose D is a derivation of F. If $a \in K$ is separable over F, then D has a unique extension to $F[a]$. (*Hint:* $D(\sum_i \alpha_i a^i) = \sum_i D(\alpha_i) a^i + \sum_i i\alpha_i a^{i-1} D(a)$. In particular, if f is the minimal polynomial of a then $0 = D(0) = D(f(a)) = g(a) + f'(a)D(a)$, where g is obtained

by applying D to each coefficient of f. This has a unique solution for $D(a)$.)

19. As in Exercise 18, except now suppose the minimal polynomial f is inseparable. Then taking f to be monic, show that D cannot be extended to K unless D sends every coefficient of f to 0, in which case $D(a)$ can be taken to be any arbitrary element of K.

CHAPTER 24. FINITE FIELDS

Since the beginning of computer technology the field $\mathbb{Z}_2$ has had important applications, with "1" denoting a closed circuit and "0" denoting an open circuit. Since computers use more than one circuit, a natural and important question is whether there are fields whose orders are higher powers of 2, or of other orders. So far we know $\mathbb{Z}_p$ is a field for p prime, so, in particular, there are fields of order 2,3,5,7, and so forth. What about 4 and 6? More generally, we want to tackle the following two questions, given n in $\mathbb{N}$:

(1) Is there a field of n elements?

(2) Are there two nonisomorphic fields of n elements?

The study of finite fields is an excellent example of reasoning from the converse. Instead of trying to construct new fields at the outset, we shall presume that we have found a finite field F and determine several of its properties. Having determined the basic properties, we shall be able to construct all finite fields and determine their structure. First let us study a very significant subfield.

The *characteristic subfield* F_0 of F is defined to be the subfield generated by 1; by Remark 23.25, $F_0 \approx \mathbb{Q}$ iff $\text{char}(F) = 0$, and $F_0 \approx \mathbb{Z}_p$ iff $\text{char}(F) = p$. Now $F \supset F_0$ is a field extension, and we can operate our machinery of the previous chapters. Suppose $|F| = n < \infty$. Then $\text{char}(F) = p$ for some prime number $0 < p \leq n$, and $F_0 \approx \mathbb{Z}_p$. Let $t = [F : F_0] < n$. Then F has a base $\{b_1, \ldots, b_t\}$ over F_0, so each element of F can be written uniquely in the form $\sum_{i=1}^{t} \alpha_i b_i$ for α_i in F_0.

Thus, an arbitrary element of F can be described uniquely by means of the vector $(\alpha_1, \ldots, \alpha_t)$, where each $\alpha_i \in F_0$. But F_0 has p elements, and there are t choices to be made, yielding p^t possibilities for $(\alpha_1, \ldots, \alpha_t)$. This proves $n = p^t$.

PROPOSITION 1. *If F is a finite field then $|F| = p^t$ where $p = \text{char}(F)$. Conversely if $f \in \mathbb{Z}_p[x]$ is irreducible of degree t, then $\mathbb{Z}_p[x]/\langle f \rangle$ is a field of order p^t.*

Proof. We just proved the first assertion; the second is clear, since taking $F = \mathbb{Z}_p[x]/\langle f \rangle$, we have $[F : \mathbb{Z}_p] = t$, so $|F| = p^t$ as just shown. □

Example 2. (i) There is no field of order 6, since 6 is not a prime power.

(ii) To find a field of order 4 we merely need to find an irreducible polynomial f of degree 2 over $\mathbb{Z}_2$; we may assume that f is monic, so $f = x^2 + \alpha_1 x + \alpha_0$. Note 0,1 cannot be roots of f, so

$$0 \neq f(0) = \alpha_0 \quad \text{implying} \quad \alpha_0 = 1;$$

$$0 \neq f(1) = 1 + \alpha_1 + \alpha_0 = 1 + \alpha_1 + 1 = \alpha_1, \quad \text{implying} \quad \alpha_1 = 1.$$

Thus, $f = x^2 + x + 1$ is irreducible, and $\mathbb{Z}_2[x]/\langle f\rangle$ is the desired field.

We might think that we are ready to answer question 1. Indeed, by Proposition 1, to obtain a field of arbitrary prime power order p^t, one merely need find an irreducible polynomial of degree t over $\mathbb{Z}_p$. This is not as easy one might expect, since our more sophisticated techniques (*e.g.*, Gauss's lemma and Eisenstein's criterion) only work in characteristic 0. There exist direct combinatorial proofs of the existence of irreducible polynomials of any degree over $\mathbb{Z}_p$, for p arbitrary, but these tend to be rather intricate. (One reasonable approach is via Exercises 3 and 4.) So let us turn first to question 2. There is a nice answer using splitting fields. This will not only provide us with the means of solving questions 2 and 1, but eventually will provide us with a precise determination of the number of irreducible polynomials of degree t over $\mathbb{Z}_p$, *cf.* Exercise 3.

Remark 3. If F is a field of order n, then $G = (F \setminus \{0\}, \cdot)$ is a group of order $n-1$, so any element of G satisfies $x^{n-1} = 1$, and thus $x^n = x$. But $0^n = 0$, so we see that every element of F satisfies the polynomial $x^n - x$. It follows at once that F is the splitting field of $x^n - x$ over the characteristic subfield of F. We have proved:

PROPOSITION 4. *Any field of order $n = p^t$ is the splitting field of the polynomial $x^n - x$ over $\mathbb{Z}_p$.*

COROLLARY 5. *Any two fields of the same order are isomorphic (being splitting fields of the same polynomial over the same base field).*

Proposition 4 also gives us the clue for attacking (1); let us take the splitting field of $x^n - x$ over $\mathbb{Z}_p$ and prove that it has order n. First we need an easy observation.

PROPOSITION 6. *Suppose R is any integral domain with* $\operatorname{char}(R) = p$. *Then there is an injection $\varphi: R \to R$ given by $a \mapsto a^p$.*

Proof. First we prove φ is a homomorphism. Clearly, $\varphi(ab) = (ab)^p = a^p b^p = \varphi(a)\varphi(b)$, and

$$\begin{aligned}\varphi(a+b) &= (a+b)^p = a^p + \binom{p}{1} a^{p-1}b + \binom{p}{2} a^{p-2}b^2 + \cdots + b^p = a^p + b^p \\ &= \varphi(a) + \varphi(b).\end{aligned}$$

Finally, $\varphi(-a) = (-a)^p = (-1)^p \varphi(a) = -a$. (This is clear if p is odd; on the other hand, $-1 = +1$ if $p = 2$.)

But φ is an injection, since $\ker \varphi = 0$. $\square$

This homomorphism, called the *Frobenius map*, is most useful.

COROLLARY 7. *With notation as in Proposition 6, for any t there is an injection $\psi: R \to R$ given by $a \mapsto a^{p^t}$.*

Proof. Take $\psi = \varphi \circ \cdots \circ \varphi$, taken t times, where φ is the Frobenius map. □

PROPOSITION 8. *Suppose F is a finite field. Then {elements of F satisfying the equation $x^n = x$} is a subfield containing $\mathbb{Z}_p$, whose order is precisely the number of roots in F of the polynomial $x^n - x$.*

Proof. First of all, let $S = \{a \in F : a^n = a\}$. In view of Corollary 7, if $a, b \in S$, then $(a \pm b)^n = a^n \pm b^n = a \pm b$, proving S is an additive subgroup of F; but $S \setminus \{0\} = \{a \in F : a^{n-1} = 1\}$ is clearly a subgroup of $(F \setminus \{0\}, \cdot)$, *cf.* Remark 7.0. Hence S is a field, of characteristic p, and so contains $\mathbb{Z}_p$. The last assertion is obvious. □

THEOREM 9. *There exists a field of $n = p^t$ elements, unique up to isomorphism, for any prime number p and any $t \neq 0$ in $\mathbb{N}$. Furthermore, this field satisfies $a^n = a$ for each element a.*

Proof. Let E be the splitting field of the polynomial $f = x^n - x$ over $\mathbb{Z}_p$. By Proposition 8, $E_0 = \{a \in E : a^n = a\}$ is a subfield that consists of the roots of f, so $E_0 = E$ by definition. It remains to show the roots of f are distinct (so that there are n of them.) But $f' = nx^{n-1} - 1 = -1$ (since p divides n), so f' is relatively prime to f, implying f is separable by Corollary 23.19. Uniqueness is by Corollary 5. □

The field of n elements is called $\mathrm{GF}(n)$. Note that although we put aside the task of finding irreducible polynomials over $\mathbb{Z}_p$, we now know that they exist, and also have some help in finding them.

Remark 10. Taking $n = p^t$, we see that $K = \mathrm{GF}(n)$ satisfies the following properties:

(1) $K = \{0, 1, a, a^2, \dots, a^{n-2}\}$ for some $a \in K$. (Indeed, $K \setminus \{0\}$ is a finite multiplicative subgroup of the field K, and thus is cyclic, by Theorem 18.7.) In particular, $K = F[a]$ for any field $F \subset K$, since all of the elements of K have been expressed algebraically in terms of a. (Compare with Exercise 23.3.)
(2) Taking a as in (1), let g be the minimal polynomial of a over $\mathbb{Z}_p$. Clearly, $K = \mathbb{Z}_p[a] \approx \mathbb{Z}_p/\langle g \rangle$, so $\deg g = [K : \mathbb{Z}_p] = t$, implying g is an irreducible polynomial of degree t over $\mathbb{Z}_p$.
(3) Notation as in (2), g divides $x^n - x$, by Proposition 4, and thus is separable.
(4) The same reasoning shows that any irreducible polynomial f over K divides $x^m - x$, where $m = |K|^{\deg f}$.

Remark 11. Every finite field K is perfect. (Indeed, by Remark 10(iv), any irreducible polynomial f over K divides $x^m - x$, which is separable since its derivative is -1; Hence f is separable.)

Further results along these lines are given in the exercises. This analysis is continued in Theorem 26.8 (as an application of the Galois correspondence) and Exercises 26.10, 26.11.

Reduction Modulo p

Finite fields have surprising applications to algebraicity over $\mathbb{Q}$. Already in Chapter 20 we proved the basic irreducibility results of Gauss and Eisenstein by passing modulo p to $\mathbb{Z}_p$. In Example 25.12 we shall proceed one step further, using this technique to obtain the minimal polynomial of any primitive nth root of 1.

Exercises

1. Writing $\mathrm{GF}(4) = \{0, 1, a, a^2\}$, show directly $1 + a = a^2$, proving there is a unique irreducible polynomial of degree 2 over GF(2).
2. Factor $x^8 - x$ into irreducible factors over $\mathbb{Z}_2$, and conclude that there are exactly two irreducible polynomials of degree 3 over $\mathbb{Z}_2$.
3. Let $n_p(t)$ denote the number of irreducible polynomials of degree t over $\mathbb{Z}_p$. Remark 10(4) implies that $n_p(t) \le \frac{p^t-2}{t}$. However, a closer examination shows $\sum_{d|t} n_p(d)d = p^t$; conclude by Möbius inversion (Exercise 2.12) that
$$n_p(t) = \frac{\sum_{d|t} \mu(d)p^d}{t}.$$
For example, $n_2(4) = \frac{16-4}{4} = 3$.
4. Using the formula from Exercise 3, show that $n_p(t) \neq 0$ for each t, thereby providing an independent proof of Theorem 9, which does not rely on splitting fields.
5. $\mathrm{GF}(p^s)$ is isomorphic to a subfield of $\mathrm{GF}(p^t)$, iff $s|t$.
6. Reprove Remark 11, via Exercise 23.16.
7. Let $n = p^t$. The field $K = \mathrm{GF}(p^t) = \{0, 1, a, \dots, a^{n-2}\}$ has $\varphi(n-1)$ multiplicative generators, and the minimal polynomial of each has t roots. If a is such a generator, then $a, a^p, \dots, a^{p^t-1}$ are the roots of the minimal polynomial of a, and this provides at least $\frac{\varphi(n-1)}{t}$ irreducible polynomials over GF(p) (and incidentally proves that $t|\varphi(n-1)$). (For the precise formula, see Exercise 3.) Display an irreducible polynomial of degree 4 over GF(2), none of whose roots is a multiplicative generator of the multiplicative group of GF(16).

One of the most beautiful achievements of mathematics is Galois's theory, linking field extensions to groups and thereby enabling one to apply group theory to study field extensions. Although this theory can fill a course by itself, we shall focus on the main results, in the following three chapters.

The Galois Group of Automorphisms of a Field Extension

Definition 1. An *automorphism* of a field K is an isomorphism $K \to K$. If K is an F-field, an *automorphism of* K/F is an F-automorphism of K, *i.e.*, an automorphism that fixes F elementwise. $\mathrm{Gal}(K/F)$ is the set of automorphisms of K/F and is called the *Galois group* of K over F.

Remark 2. $\mathrm{Gal}(K/F)$ is indeed a group, whose group operation is given by composition of functions.

This modest observation revolutionized mathematics, by enabling one to use group theory to study fields. But first some basic remarks.

LEMMA 3. *Suppose K/F is a finite extension. Then any homomorphism $\sigma: K \to K$ satisfying $\sigma(F) = F$ is an automorphism.*

Proof. σ is an injection, by Remark 23.1. On the other hand, taking a base B of K over F, we see that $\sigma(B)$ is a base of $\sigma(K)$ over $\sigma(F) = F$, so $[\sigma(K) : F] = [K : F]$; thus $\sigma(K) = K$, implying σ is onto. □

Remark 4. If $F \subseteq L \subseteq K$, then any automorphism of K/L is an automorphism of K/F; hence $\mathrm{Gal}(K/L) \subseteq \mathrm{Gal}(K/F)$.

Remark 5. Suppose $a \in K$, and let $a_1 = a,\ a_2, \dots, a_t$ be the roots (in K) of a polynomial $f \in F[x]$. By Proposition 23.6, any σ in $\mathrm{Gal}(K/F)$ satisfies $\sigma(a) = a_i$, for suitable i. In particular, if $K = F[a]$, then there is some permutation π_σ of $\{1, \dots, t\}$ such that $\sigma(a_i) = a_{\pi_\sigma(i)}$ for each i. In this way $\sigma \mapsto \pi_s$ describes a group injection $\mathrm{Gal}(K/F) \to S_t$.

On the other hand, Proposition 23.4 shows that if $K = F[a]$, then for each root a_i of f there is an automorphism σ of K/F given by $a \mapsto a_i$; thus in this case $|\mathrm{Gal}(K/F)|$ equals the number of roots of f in K. Before continuing this analysis, let us consider some examples.

Example 6.

(1) $F = \mathbb{Q}$ and $K = \mathbb{Q}[\sqrt{3}]$. The minimal polynomial of $\sqrt{3}$ is $x^2 - 3$, whose roots are $\pm\sqrt{3}$; hence, any automorphism σ of K/F is given by $\sigma(\sqrt{3}) = \pm\sqrt{3}$, and thus $\mathrm{Gal}(K/F) \approx S_2$. The same argument shows more generally that $\mathrm{Gal}(K/F) \approx S_2$ whenever $K = F[\sqrt{\alpha}]$,

where $\alpha \in F$ is not a square in F, provided $\operatorname{char}(F) \neq 2$ (to insure that $\sqrt{\alpha}$ and $-\sqrt{\alpha}$ are distinct). But, when $\operatorname{char}(F) \neq 2$, any quadratic extension has this form, by Example 21.12(iv). The characteristic 2 case is handled in exercise 9.

(2) $F = \mathbb{Q}$ and $K = \mathbb{Q}[\sqrt[4]{2}]$. Since the only other root of $x^4 - 2$ in K is $-\sqrt[4]{2}$, we see $\operatorname{Gal}(K/F) \approx S_2$.

(3) $F = \mathbb{Q}[i]$ and $K = \mathbb{Q}[i, \sqrt[4]{2}]$. Now there are automorphisms sending $\sqrt[4]{2}$ to $i^m \sqrt[4]{2}$, for $m = 0,1,2,3$, so $|\operatorname{Gal}(K/F)| = 4$, and, indeed, $\operatorname{Gal}(K/F) = \langle\sigma\rangle$, where σ is given by $\sqrt[4]{2} \mapsto i\sqrt[4]{2}$.

(4) $F = \mathbb{Q}$, $K = \mathbb{Q}[i, \sqrt[4]{2}]$. Any automorphism in $\operatorname{Gal}(K/F)$ sends $\sqrt[4]{2} \mapsto i^m \sqrt[4]{2}$ and $i \mapsto \pm i$, so there are at most $4 \cdot 2 = 8$ possible automorphisms. On the other hand, $\operatorname{Gal}(K/F)$ contains $\operatorname{Gal}(K/\mathbb{Q}[i])$ and $\operatorname{Gal}(K/\mathbb{Q}[\sqrt[4]{2}]$, so contains both the automorphisms σ (given in (3)) and τ given by complex conjugation; furthermore,

$$\tau\sigma\tau^{-1}: \sqrt[4]{2} \mapsto -i\sqrt[4]{2} = \sigma^3(\sqrt[4]{2}),$$
$$\tau\sigma\tau^{-1}: i \mapsto i = \sigma^3(i),$$

from which we see $\tau\sigma\tau^{-1} = \sigma^3$. Thus the generators σ, τ of $\operatorname{Gal}(K/F)$ satisfy the relations of the dihedral group D_4, and we conclude $\operatorname{Gal}(K/F) \approx D_4$.

Note: Once one observes that $\operatorname{Gal}(K/F)$ has a cyclic subgroup $\langle\sigma\rangle$ of order 4, and that $\sigma\tau \neq \tau\sigma$, it is obvious that $\operatorname{Gal}(K/F) \approx D_4$. Such short cuts are standard fare in the theory.

In all of these examples we note $\operatorname{Gal}(K/F) \leq [K : F]$, with equality often holding. Splitting fields play a key role.

THEOREM 7. *Suppose E is the splitting field of a polynomial f over F. Then $|\operatorname{Gal}(E/F)| \leq [E : F]$, equality holding if f is separable.*

Proof. In view of Lemma 3, this is just a restatement of Theorem 23.14. □

In order to extract as much as possible from this result we need a related notion. Suppose S is a set of automorphisms of K. Define the *fixed subfield* K^S to be $\{a \in K : \sigma(a) = a \text{ for all } \sigma \text{ in } S\}$.

Remark 8. K^S is a subfield of K, for any set of automorphisms S of K. (Indeed, if $\sigma(a) = a$ and $\sigma(b) = b$, then $\sigma(a \pm b) = \sigma(a) \pm \sigma(b) = a \pm b$, $\sigma(ab) = \sigma(a)\sigma(b) = ab$, and (for $a \neq 0$) $\sigma(a^{-1}) = \sigma(a)^{-1} = a^{-1}$.)

Remark 9. Let $G = \operatorname{Gal}(K/F)$. Then $K^G \supseteq F$, and $G = \operatorname{Gal}(K/K^G)$. (Indeed, G fixes F by definition, so $F \subseteq K^G$. Now $\operatorname{Gal}(K/K^G) \subseteq G$ by

Remark 4. On the other hand, any σ in G fixes K^G by definition, so lies in $\mathrm{Gal}(K/K^G)$.)

We already can make a striking observation.

PROPOSITION 10. *If E is the splitting field of a separable polynomial over F, then $E^G = F$ where $G = \mathrm{Gal}(E/F)$.*

Proof. $[E : F] = |G|$ by Theorem 7. But taking $F_1 = E^G$ we have $G = \mathrm{Gal}(E/F_1)$ by Remark 9, so also $[E : F_1] = |G|$. But $F \subseteq F_1 \subseteq E$ so we conclude $F_1 = F$ since $[F_1 : F] = \frac{[E:F]}{[E:F_1]} = 1$. □

To utilize this result we need a cute converse to Remark 5.

Remark 11. Suppose G is a set of automorphisms of K, and

$$f = (x - a_1)\dots(x - a_t) \in K[x].$$

Also suppose $\{a_1, \dots, a_t\}$ are distinct, and each σ in G acts as a permutation on $\{a_1, \dots, a_t\}$. (In other words, suppose for each σ in G that there is a suitable permutation π_σ of $\{1, \dots, t\}$ for which $\sigma(a_u) = a_{\pi_\sigma u}$, for each $1 \le u \le t$.) Then $f \in K^G[x]$. (Indeed, extending σ to an automorphism of $K[x]$ via $\sigma(x) = x$, we have

$$\sigma(f) = \sigma(x - a_1)\dots\sigma(x - a_t) = (x - \sigma(a_1))\dots(x - \sigma(a_t)) = f,$$

proving that σ fixes each coefficient of f, as desired.)

Example 12. Cyclotomic extensions. Let us apply these results to the the extension $E = \mathbb{Q}[\rho]$ of $\mathbb{Q}$, where ρ is a primitive nth root of 1. We know already for n prime that $x^{n-1} + x^{n-2} + \cdots + 1$ is the minimal polynomial of ρ. Our goal is to see what happens for n not necessarily prime. Define the polynomial $f_n = \prod_{j \in \mathrm{Euler}(n)} (x - \rho^j)$, called the nth *cyclotomic polynomial.* Recall $\mathrm{Euler}(n)$ is $\{k : 1 \le k < n,\ (k, n) = 1\}$. Thus $\{\rho^j : j \in \mathrm{Euler}(n)\}$ is precisely the set of primitive nth roots of 1. We shall prove for all n that $f_n \in \mathbb{Q}[x]$ is irreducible and thus is the minimal polynomial of ρ, and that $\mathrm{Gal}(E/\mathbb{Q}) \approx \mathrm{Euler}(n)$, so that $\deg f_n = |\,\mathrm{Gal}(E/\mathbb{Q})| = \varphi(n)$. In the process we shall also find an inductive formula that enables us to compute f_n in terms of $\{f_d :$ proper divisors d of $n\}$.

Indeed, let $G = \mathrm{Gal}(E/\mathbb{Q})$. Since E is the splitting field of $x^n - 1$, we see by Proposition 10 that $E^G = \mathbb{Q}$. But any automorphism σ of G merely permutes the primitive nth roots of 1 and thus, by Remark 11, $f_n \in E^G[x] = \mathbb{Q}[x]$.

We want to show that $f_n \in \mathbb{Z}[x]$. This follows by induction on n. Indeed, for any $d|n$ note that $\{(\rho^d)^j : j \in \text{Euler}(\frac{n}{d})\}$ is the set of primitive $\frac{n}{d}$-roots of 1, so

$$f_{n/d} = \prod_{j \in \text{Euler}(\frac{n}{d})} (x - (\rho^d)^j),$$

and thus,

$$x^n - 1 = \prod_{j=1}^{n}(x - \rho^j) = \prod_{d|n} f_{n/d} = \prod_{d|n} f_d = f_n g,$$

where $g = \prod_{d|n,\ d \neq n} f_d$. But $f_d \in \mathbb{Z}[x]$ for $d < n$, by induction, and is clearly monic, so $g \in \mathbb{Z}[x]$ and is monic. Hence $f_n \in \mathbb{Z}[x]$ by Theorem 20.8.

To prove f_n is irreducible we appeal to reduction modulo p. Suppose f_n were reducible. Then the minimal polynomial g of ρ would be a proper factor of f_n in $\mathbb{Z}[x]$, so some other primitive root ρ^k of 1 would *not* be a root of g, for suitable k prime to n. Write $k = p_1 \dots p_t$ as a product of prime numbers. Letting $\rho_0 = \rho$ and $\rho_i = \rho^{p_1 \cdots p_i}$, a primitive nth root of 1 for each i, we see that $\rho_{i+1} = \rho_i^{p_{i+1}}$. Furthermore, for suitable i, ρ_i is a root of g, and ρ_{i+1} is not a root of g. Replacing ρ by ρ_i and taking $p = p_{i+1}$, we may assume ρ is a root of g, but ρ^p is not a root of g. Note that p does not divide n.

Writing $f_n = gh$, we see that ρ^p is a root of h; hence ρ is a root of $h(x^p)$, implying that g and $h(x^p)$ are not relatively prime, and thus g divides $h(x^p)$.

Let $^-$ denote the canonical image in $\bar{\mathbb{Z}} = \mathbb{Z}/p\mathbb{Z}$. Then $x^n - 1 \in \bar{\mathbb{Z}}[x]$ is separable since it is relatively prime to its derivative nx^{n-1} (for p does not divide n). Thus $\bar{f}_n$ is separable. On the other hand, $\bar{f}_n = \bar{g}\bar{h}$; hence $\bar{g}$ and $\bar{h}$ are relatively prime, by Remark 23.21, implying $\bar{g}$ and $\bar{h}^p$ are relatively prime. But the Frobenius map fixes $\mathbb{Z}_p$; if $h = \sum_i m_i x^i$ then

$$\bar{h}^p = \sum \bar{m}_i^p x^{ip} = \sum \bar{m}_i x^{pi} = \bar{h}(x^p),$$

so by the previous paragraph $\bar{g}$ divides $\bar{h}^p$, contradiction.

Having concluded that f_n is irreducible, we see that E is the splitting field of f_n over $\mathbb{Q}$, and thus $|G| = \deg f_n = \varphi(n)$. It remains to show that $G \approx \text{Euler}(n)$. Note that for any k in Euler(n) there is σ_k in G given by $\rho \mapsto \rho^k$; since these automorphisms all act differently on ρ we have $\varphi(n)$ distinct automorphisms, so $G = \{\sigma_k : k \in \text{Euler}(n)\}$. Furthermore, $k \mapsto \sigma_k$ is a 1:1 correspondence from Euler(n) to G, which is a group homomorphism (and thus an isomorphism), since for any k, k' in Euler(n) we have

$$\sigma_k \sigma_{k'}(\rho) = \sigma_k(\rho^{k'}) = (\rho^{k'})^k = \rho^{kk'} = \sigma_{kk'}(\rho).$$

We already know the structure of Euler(n) from Exercise 18.4, and thus we know the Galois group of every cyclotomic extension.

In the course of this proof, we have shown that $x^n - 1 = \prod_{d|n} f_d$. For example, for $n = 6$ we see

$$\begin{aligned} f_1 f_2 f_3 f_6 &= x^6 - 1 = (x^3 - 1)(x^3 + 1) \\ &= (x-1)(x^2+x+1)(x+1)(x^2-x+1) \\ &= f_1 f_3 f_2 \cdot (x^2 - x + 1), \end{aligned}$$

so we conclude $f_6 = x^2 - x + 1$.

So far, we have obtained nice results when E is the splitting field of a separable polynomial f over F. However, we often are presented with the finite field extension E/F without the polynomial f, and thereby leading us to inquire whether this situation really depends on the polynomial. To this end we add two definitions to definition 23.26.

Definition 13. E/F is *normal* if for each element a of E the minimal polynomial of a (over F) splits (into linear factors over E).

E/F is a *Galois extension* if E/F is both normal and separable.

Note 14. If E/F is Galois, then E/L is Galois, for any fields $F \subseteq L \subseteq E$.

Although the definitions of separable and normal extensions are formally rather strong, holding for every irreducible polynomial with a root in E rather than merely for one particular polynomial, they are justified by the following important result.

THEOREM 15. *The following conditions are equivalent, for a finite field extension E/F:*

(i) *E is the splitting field of a separable polynomial f over F;*

(ii) *$E^G = F$ for a suitable group G of automorphisms of E;*

(iii) *E/F is Galois.*

Proof. (i) $\Rightarrow$ (ii) by Proposition 10.

(ii) $\Rightarrow$ (iii) For any a_1 in E, we must show that the minimal polynomial f of a_1 is separable and splits over E. Let $S = \{a_1, \ldots, a_t\}$ be the set of distinct roots of f in E; and let

$$g = (x - a_1)\ldots(x - a_t) \in E[x].$$

By inspection g is separable and splits, and $g|f$. But Remark 11 implies $g \in F[x]$ so $f = g$ (since f is irreducible in $F[x]$).

(iii) $\Rightarrow$ *(i)* Write $E = F[a_1, \ldots, a_t]$, let f_i be the minimal (monic) polynomial of a_i over F, for each i, and let f be the product of the f_i, discarding duplications. Obviously f splits and is separable, since each f_i splits and is separable irreducible. □

The really surprising part of Theorem 15 is that condition (ii) is enough to guarantee that E/F is Galois, and we shall use this criterion repeatedly.

COROLLARY 16. *E/F is Galois iff* $|\,\mathrm{Gal}(E/F)| = [E : F]$.

Proof. ($\Leftarrow$) Let $G = \mathrm{Gal}(E/F)$ and $F' = E^G \supseteq F$. Then $[E : F] = [E : F']$, implying $F' = F$. Thus E/F is Galois, by Theorem 15. □

We also want to show that if Theorem 15(ii) holds, then $G = \mathrm{Gal}(E/F)$. This requires the following fact.

LEMMA 17. *(E. Artin's Lemma.) Suppose G is any finite group of automorphisms of a field E. Then $[E : E^G] \leq |G|$. (In particular, $[E : E^G]$ is finite.)*

Proof. Write $G = \{\sigma_1, \ldots, \sigma_n\}$. We must show that any $m > n$ elements $\{a_1, \ldots, a_m\}$ of E are linearly dependent over E^G. Indeed, by linear algebra, there is a nontrivial solution $b_1, \ldots, b_m$ (in E) to the n equations

$$\sum_{j=1}^{m} \sigma_i(a_j)b_j = 0, \quad 1 \leq i \leq n. \tag{1}$$

Take such $b_1, \ldots, b_m$ with the smallest possible number of nonzero b_j; reordering the b_i if necessary, we may assume $b_1 \neq 0$; multiplying through by b_1^{-1} we may assume $b_1 = 1$. We shall show that each $b_j \in E^G$, thereby yielding the desired dependence (taking $i = 1$). It is enough to show that each b_j is fixed by every σ in G. But applying σ to (1) yields

$$\sum_{j=1}^{m} \sigma\sigma_i(a_j)\sigma(b_j) = 0, \quad 1 \leq i \leq n;$$

since also $G = \{\sigma\sigma_1, \ldots, \sigma\sigma_n\}$ one sees that $\sigma(b_1), \ldots, \sigma(b_m)$ is a solution of (1), and thus $\sigma(b_1) - b_1, \ldots, \sigma(b_m) - b_m$ is also a solution of (1). But $\sigma(b_1) - b_1 = 1 - 1 = 0$, so by hypothesis this solution must be trivial, *i.e.*, each $\sigma(b_j) - b_j = 0$, as desired. □

PROPOSITION 18. *If $F = E^G$, then $G = \mathrm{Gal}(E/F)$.*

Proof. E/F is Galois by Theorem 15, so $|\,\mathrm{Gal}(E/F)| = [E : F] \leq |G|$; since G is a subgroup of $\mathrm{Gal}(E/F)$, we conclude that $\mathrm{Gal}(E/F) = G$. □

COROLLARY 19. *If G is any subgroup of* $\mathrm{Gal}(E/F)$, *then the extension* E/E^G *is Galois, with Galois group* G.

Proof. Apply Proposition 18 to Theorem 15(ii). □

The Galois Group and Intermediate Fields

Let us fix a Galois extension E/F, and let $G = \mathrm{Gal}(E/F)$. An F-subfield L of E is called an *intermediate field, i.e.*, $F \subseteq L \subseteq E$. In Exercise 22.7 we saw that a number a of degree 4 is constructible iff $\mathbb{Q}[a]/\mathbb{Q}$ has an intermediate field (other than $\mathbb{Q}[a]$ and $\mathbb{Q}$). In general, the location of intermediate fields is of utmost concern, and our object here is to find a sublime connection between intermediate fields and subgroups of G.

Any intermediate field L gives rise to the subgroup $\mathrm{Gal}(E/L)$ of G, *cf.* Remark 4. Conversely for any subgroup $H \leq G$; clearly, E^H is an intermediate field. Thus we have a way of passing back and forth between the subgroups of G and the intermediate fields.

THEOREM 20. *(Fundamental Theorem of Galois Theory)*

(i) *Suppose E/F is a Galois extension, and $G = \mathrm{Gal}(E/F)$. Then there is a 1:1 correspondence*

$$\{\textit{subgroups of } G\} \leftrightarrow \{\textit{Intermediate fields (between } F \textit{ and } E)\}$$

given by

$$H \to E^H$$
$$\mathrm{Gal}(E/L) \leftarrow L.$$

In particular, $H_1 = H_2$ *iff* $E^{H_1} = E^{H_2}$.

(ii) *This correspondence satisfies the following extra properties, where* $H \leq G$:

(1) $|H| = [E : E^H], \quad [G : H] = [E^H : F]$;
(2) *It is order-reversing, i.e.,* $H_1 \subset H_2 \leftrightarrow E^{H_1} \supset E^{H_2}$;
(3) $H \triangleleft G$ *iff* E^H/F *is a normal field extension, in which case we have* $\mathrm{Gal}(E^H/F) \approx G/H$.

Proof. We shall show that the composite of these correspondences in each direction is the identity. E/E^H is Galois, and $\mathrm{Gal}(E/E^H) = H$, by Corollary 19. In the other direction, for any intermediate field L we have E/L Galois by Note 14, so $E^{\mathrm{Gal}(E/L)} = L$ by Theorem 15.

We turn to the additional assertions.

(1) $|H| = [E : E^H]$ by Theorem 7 applied to the Galois extension E/E^H; hence

$$[G : H] = \frac{|G|}{|H|} = \frac{[E : F]}{[E : E^H]} = [E^H : F].$$

(2) Immediate, from the definition of E^H and (1).

(3) We start with a general observation for any $\sigma, \tau \in G$ and $a \in E$:

$$\tau(a) = a \quad \text{iff} \quad \sigma\tau\sigma^{-1}(\sigma(a)) = \sigma(a).$$

Now take any subgroup $H < G$, and put $L = E^H$. Then

$$a \in L \quad \text{iff} \quad \tau(a) = a \text{ for all } \tau \text{ in } H, \quad \text{iff} \quad \sigma(a) \in E^{\sigma H \sigma^{-1}}.$$

In other words, for any σ in $\mathrm{Gal}(E/F)$, we have

$$E^{\sigma H \sigma^{-1}} = \sigma(L).$$

Now suppose that L/F is normal. If $a \in L$, then, for each σ in G, $\sigma(a)$ is a root of the minimal polynomial of a, implying $\sigma(a) \in L$; thus $\sigma(L) = L$. Hence

$$E^{\sigma H \sigma^{-1}} = \sigma(L) = L = E^H,$$

so $\sigma H \sigma^{-1} = H$, by (i). This holds for all σ in G, proving $H \triangleleft G$.

Conversely, suppose $H \triangleleft G$. Then $\sigma(L) = E^{\sigma H \sigma^{-1}} = E^H = L$ for all σ in G, so restriction of each automorphism to L yields a map

$$\Phi : \mathrm{Gal}(E/F) \to \mathrm{Gal}(L/F).$$

Clearly, $L^{\Phi(G)} \subseteq E^G = F$, so $\Phi(G) = \mathrm{Gal}(L/F)$, by Proposition 18, *i.e.*, Φ is onto. Furthermore, $\ker \Phi = \{\sigma \in G : \sigma|_L = 1|_L\} = \mathrm{Gal}(E/L)$, implying

$$\mathrm{Gal}(L/F) \approx \mathrm{Gal}(E/F)/\ker\Phi = \mathrm{Gal}(E/F)/\mathrm{Gal}(E/L).$$

In particular, $|\mathrm{Gal}(L/F)| = \frac{|\mathrm{Gal}(E/F)|}{|\mathrm{Gal}(E/L)|} = \frac{[E:F]}{[E:L]} = [L:F]$, proving L/F is Galois. □

We shall call this correspondence the *Galois correspondence*, and the remainder of this text involves its many applications to algebra.

Exercises

The Galois Group of the Compositum

The next few exercises involve the field compositum of subfields, *cf.* Proposition 21.18.

1. Suppose K, L are two subfields of E, with $K/K \cap L$ Galois having Galois group G. Then KL/L also is Galois with Galois group G.

(*Hint*: KL is the splitting field over L of the same separable polynomial as K over $K \cap L$. Let $H = \mathrm{Gal}(KL/L)$. Then there is an injection $\varphi: H \to G$, given by restriction to K, and $K^{\varphi(H)} = K \cap L$, proving $\varphi(H) = G$.)

2. With notation as in Exercise 1, show that $[KL : L]$ divides $[K : F]$.
3. Suppose that K_1, K_2 are subfields of E, and let $F = K_1 \cap K_2$. If K_i/F are Galois for $i = 1, 2$, then K_1K_2/F also is Galois, and $\mathrm{Gal}(K_1K_2/F) \approx \mathrm{Gal}(K_1K_2/K_1) \times \mathrm{Gal}(K_1K_2/K_2)$. (*Hint*: Write K_i as the splitting field of the separable polynomial f_i over F, and thus K_1K_2 also is the splitting field of a suitable separable polynomial. Let $G = \mathrm{Gal}(K_1K_2/F)$ and $H_i = \mathrm{Gal}(K_1K_2/K_i) \triangleleft G$. Note that $H_1 \cap H_2 = \mathrm{Gal}(K_1K_2/K_1K_2) = \{e\}$, and, by Exercise 1, $|H_1||H_2| = [K_1K_2 : K_1][K_1 : F] = |G|$; conclude with Proposition 6.13.)
4. What is the Galois group over $\mathbb{Q}$ of the splitting field of the polynomial $x^4 - 5x^2 + 6$? Of the polynomial $x^5 - 2$?
5. If $\mathrm{char}(F) = 0$ and $E = F[\rho]$, where ρ is a primitive nth root of 1 then $\mathrm{Gal}(E/F)$ is a subgroup of $\mathrm{Euler}(n)$. (*Hint*: $E = F\mathbb{Q}[\rho]$.)

More on Artin's Lemma

6. E/E^G is Galois and finite, for any finite group of automorphisms G of an arbitrary field E.
7. (Dedekind Independence Theorem.) Given any monoid M define a *(linear) character* of M into a field F to be a monoid homomorphism $\chi: M \to F \setminus \{0\}$, *i.e.*, $\chi(ab) = \chi(a)\chi(b)$. Prove that any distinct characters $\chi_1, \ldots, \chi_n$ are independent over F, in the sense that if $\alpha_1, \ldots, \alpha_n$ satisfy $\sum \alpha_i \chi_i(u) = 0$ for all u in M then each $\alpha_i = 0$. (*Hint*: This the same idea as in Artin's Lemma.)
8. Any distinct set of homomorphisms of fields $K \to E$ are independent over E in the sense of Exercise 7.
9. (Quadratic extensions in characteristic 2.) Assume $\mathrm{char}(F) = 2$, and K is a separable quadratic extension of F. Then $K = F[a]$, for suitable a whose minimal polynomial has the form $x^2 + x + \alpha$. (*Hint:* Replace x by βx for suitable β.) Observe that $a + 1$ is also a root of f, so K is Galois over F, with Galois group generated by the automorphism given by $a \mapsto a + 1$. (This result is generalized to Galois extensions of arbitrary prime degree, in exercise 23.9.)

   In case K is not separable, then $K = F[a]$ where the minimal polynomial has the form $x^2 + \alpha$. In this case, show that every element of $K \setminus F$ is inseparable.

We are about to embark on a series of fascinating applications in algebra, representing some of the high points in the history of mathematics. The underlying philosophy is to use the Galois correspondence to translate questions about intermediate fields to parallel questions in group theory, and then to use known results about groups (from Part I).

The applications here include the descriptions of the constructible regular n-gons, computation of the Galois groups of extensions of finite fields, and an algebraic proof of the "fundamental theorem of algebra" (which says that $\mathbb{C}$ is algebraically closed). In the next Chapter we shall also use Galois theory to prove the celebrated theorem of Ruffini-Abel that there is no formula to solve the general quintic equation.

We start with a rather straightforward application.

Finite Separable Field Extensions and the Normal Closure

PROPOSITION 1. *If E/F is finite Galois, then it has only a finite number of intermediate fields.*

Proof. The intermediate fields correspond to the subgroups of $\text{Gal}(E/F)$, which are finite in number (since $\text{Gal}(E/F)$ only has $2^{[E:F]}$ subsets). □

Note that a two-dimensional vector space V over $\mathbb{R}$ has infinitely many one-dimensional subspaces. (Indeed, identifying V with the real plane, its one-dimensional subspaces correspond to the lines passing through the origin, which are infinite in number). Thus, Proposition 1 is quite surprising. To push Proposition 1 farther, we resort to a useful technique, called the normal closure, which builds up a Galois extension from a separable extension. We borrow the idea from the proof of Theorem 25.15 (iii)$\Rightarrow$ (i). Clearly, one can write $K = F[a_1, \dots, a_t]$ for suitable a_i in K. Letting f_i be the (monic) minimal polynomial of a_i over F for $1 \leq i \leq t$, we take the product f of these f_i, discarding duplications, and let E be the splitting field of f over F. E is called the *normal closure* of K over F. Clearly, E/F is normal and thus Galois. (To justify the use of "the," one shows that (up to isomorphism) E does not depend on the choice of $a_1, \dots, a_t$; *cf.* Exercise 2.)

COROLLARY 2. *If K/F is finite and separable, then it has only finitely many intermediate fields.*

Proof. Let E be the normal closure of K. Then E/F is Galois, so its intermediate fields correspond to the subgroups of $\text{Gal}(E/F)$, which are finite in number. □

See Exercise 6 for a further application. The normal closure also enables us to tighten our results about separability and Galois extensions, *cf.* Exercises 7 and 8. Here is one tool for using this technique.

Remark 3. If E/F is Galois and an intermediate field K is invariant under $G = \text{Gal}(E/F)$, *i.e.*, $\sigma(K) = K$ for every σ in G, then K/F is Galois. (Indeed, as in the proof of Theorem 25.20(3) we see that $\text{Gal}(K/F) \approx \text{Gal}(E/F)/\text{Gal}(E/K)$, implying that $|\text{Gal}(K/F)| = [K : F]$, and thus K/F is Galois.)

PROPOSITION 4. *Suppose $K = F[a_1, \dots, a_t]$ with K/F separable. Let E be the normal closure of K over F. If $\text{Gal}(E/F) = \{\sigma_1, \dots, \sigma_n\}$, then*

$$E = F[\{\sigma_i(a_j) : 1 \le i \le n,\ 1 \le j \le t\}].$$

Proof. Let $E_0 = F[\sigma_i(a_j) : 1 \le i \le n,\ 1 \le j \le t]$. Then E_0 is invariant under G, so E_0/F is Galois (and thus normal), implying by Theorem 25.15 that $E = E_0$. □

Much of the theory of normal extensions and the normal closure can be accomplished without separability, *cf.* Exercises 2 through 6; however, since our applications all involve separable extensions we have taken the more convenient route, assuming separability throughout.

The Galois Group of a Polynomial

Summary 5. Before continuing with applications, let us review the basic setup, which utilizes various fundamental notions of abstract algebra. Suppose that $K = F[a_1, \dots, a_t]$ is a finite field extension of F. For convenience we assume F is perfect, which is the case when $\text{char}(F) = 0$ or when F is a finite field. Take f to be the product of the minimal monic polynomials of the a_i (discarding duplication), and let E be the splitting field of f over F. Then E/F is Galois, having some Galois group G of order $[E : F]$. G is also called the *Galois group of the polynomial* f, and is of great import in studying polynomials and their roots. Learning the group structure of G thereby becomes one of the main objectives of field theory. Some remarks along these lines, where $n = [K : F]$ and $k = \deg f$:

1. n divides $|G|$, by Remark 23.12;
2. $|G| \le n!$ by Theorem 23.11;
3. Recall by Proposition 23.6 that any automorphism of E/F permutes the roots of f; since the number of roots of f in E is k, we get a natural group injection $G \to S_k$, and, in particular, $|G|$ divides $k!$ Note when f is irreducible that $k = n$, in which case we conclude that $|G|$ divides $n!$

Constructible n-gons

As our first application of the theory, let us finish the classification of the constructible n-gons, started in Remark 22.10.

THEOREM 6. *A Galois field extension E/F is quadratically definable iff its Galois group is a 2-group.*

Proof. By the Galois correspondence, there is a chain of quadratic field extensions $F = F_0 \subset F_1 \subset F_2 \subset \cdots \subset F_t = E$, iff there is a corresponding chain of subgroups $H_0 = \mathrm{Gal}(E/F) \supset H_1 \supset H_2 \cdots \supset H_t = \{e\}$ with each $[H_i : H_{i+1}] = 2$. But any 2-group has such a chain of subgroups, by Proposition 12.8. □

THEOREM 7. *A regular n-gon is constructible iff n is a product of a power of 2, together (possibly) with distinct odd prime numbers each of the form $2^{2^t} + 1$.*

Proof. By Remark 22.10 the regular n-gon is constructible iff the primitive nth root ρ of 1 is constructible (as a complex number). By Example 25.12 $\mathbb{Q}[\rho]$ is Galois over $\mathbb{Q}$ with Galois group G of order $\varphi(n)$. So, by Theorem 6, ρ is constructible iff $|G|$ is a power of 2. Writing $n = p_1^{u_1} \ldots p_t^{u_t}$ with $p_1, \ldots, p_t$ distinct primes, we see

$$|G| = \varphi(n) = \varphi(p_1^{u_1}) \ldots \varphi(p_t^{u_t}),$$

so each $\varphi(p_j^{u_j})$ must be a power of 2.

Now $\varphi(2^u) = 2^{u-1}$ is always a power of 2. On the other hand, for $p \neq 2$, $\varphi(p^u) = (p-1)p^{u-1}$ is a power of 2 precisely when $u = 1$ and $p - 1$ is a power of 2, yielding the desired result in view of the easy Exercise 22.8. □

Incidentally, prime numbers of the form $n = 2^{2^t} + 1$ have an interesting history of their own; they are called *Fermat primes*, in honor of Fermat's only known mathematical mistake — he thought every number of this form is prime. In fact, n is prime for $t = 1, 2, 3, 4$, (n respectively is $3, 5, 17$, and $2^{16} + 1 = 65537$), but Euler proved $2^{32} + 1 = 641 \times 6700417$ is not prime, and no other Fermat prime is known. It is an open question whether or not there are infinitely many Fermat primes, and any new Fermat prime must be at least 10^{39456}, so Theorem 7 is somewhat tantalizing in this regard.

Finite Fields

Our next application is the computation of the Galois group of an arbitrary finite extension K/F of a finite field F. Of course, K is also a finite field, and K and F have the same characteristic, so $K = \mathrm{GF}(p^t)$ and $F = \mathrm{GF}(p^s)$ for a suitable prime number p and suitable s, t.

First assume $F = \mathrm{GF}(p) = \mathbb{Z}_p$. By Proposition 24.4, K is the splitting field of the separable polynomial $f = x^{p^t} - x$ over $\mathbb{Z}_p$, and so $K/\mathbb{Z}_p$ is Galois. Let us compute $G = \mathrm{Gal}(K/\mathbb{Z}_p)$. First of all, $|G| = [K : \mathbb{Z}_p] = t$. On the other hand, the Frobenius map $\sigma: K \to K$ (given by $a \mapsto a^p$) fixes $\mathbb{Z}_p$ and thus is an automorphism of order t, so $G = \langle \sigma \rangle$.

THEOREM 8. *Any finite extension K/F of finite fields is Galois; furthermore, $\mathrm{Gal}(K/F)$ is cyclic, generated by a suitable power of the Frobenius automorphism.*

Proof. We proved this for $F = \mathbb{Z}_p$. In general, $\mathbb{Z}_p \subseteq F \subseteq K$, so $\mathrm{Gal}(K/F)$ is a subgroup of the cyclic group $\mathrm{Gal}(K/\mathbb{Z}_p)$ and thus is cyclic. □

The Fundamental Theorem of Algebra

A field F is called *algebraically closed* if every polynomial over F splits into linear factors in $F[x]$. Algebraically closed fields are very useful, and form the foundation of algebraic geometry. Our final application of this section is to prove the fundamental theorem of algebra, that $\mathbb{C}$ is algebraically closed. The following observation comes in handy.

LEMMA 9. *A field F is algebraically closed iff there is no finite field extension $K \supset F$.*

Proof. ($\Rightarrow$) Otherwise take $a \in K \setminus F$. Then $[F[a] : F]$ is the degree of the minimal polynomial of a, which must be 1, so $a \in F$, contradiction.

($\Leftarrow$) If $f \in F[x]$ is irreducible then $K = F[x]/\langle f \rangle$ is a finite field extension of F, so $1 = [K : F] = \deg f$. □

Ironically, most proofs of the fundamental theorem of algebra come from analysis (such as the theory of analytic functions), but E. Artin found an elegant proof that is purely algebraic modulo the following basic topological fact:

LEMMA 10. *Any polynomial $f \in \mathbb{R}[x]$ of odd degree has a zero in $\mathbb{R}$.*

Proof. One may assume f is monic. Write $f = \sum_{i=0}^{n} a_i x^i$ with $a_n = 1$, and take $c > max_{\,0 \le i \le n}\{|a_i|\}$. Then

$$\begin{aligned} f(-nc) &= (-nc)^n + \sum_{i=0}^{n-1} a_i(-nc)^i < (-nc)^n + \sum_{i=0}^{n-1} c(nc)^i \\ &\le (-nc)^n + nc(nc)^{n-1} \le 0, \end{aligned}$$

so $f(-nc) < 0$, and likewise $f(nc) > 0$. Since f is a continuous function, which takes on a negative and a positive value, the intermediate value theorem shows that $f(a) = 0$ for some a in the interval $(-nc, +nc)$. □

COROLLARY 11. *Every irreducible nonlinear polynomial in $\mathbb{R}[x]$ has even degree. Consequently, every proper finite extension of $\mathbb{R}$ has even degree.*

Proof. The first assertion is clear. To prove the second assertion, suppose $K \supset \mathbb{R}$ had odd degree. Taking $a \in K \setminus \mathbb{R}$ yields $\deg a = [\mathbb{R}[a] : \mathbb{R}]$ which divides $[K : \mathbb{R}]$ and thus is odd, a contradiction. □

Now for the main result.

THEOREM 12. *(The Fundamental Theorem of Algebra.) $\mathbb{C}$ is algebraically closed.*

Proof. Assume on the contrary there exists a finite extension $K \supset \mathbb{C}$ (*cf.* Lemma 9). Take $a \in K \setminus \mathbb{C}$; let f be the minimal polynomial of a over $\mathbb{R}$, and let E be the splitting field of f over $\mathbb{C}$. Then E is the splitting field of $(x^2 + 1)f$ over $\mathbb{R}$, so $E/\mathbb{R}$ is Galois with some Galois group G.

Let S be a Sylow 2-subgroup of G. Then $[E^S : \mathbb{R}] = [G : S]$ is odd, and thus equals 1 by Corollary 11, *i.e.*, $E^S = \mathbb{R}$, so $S = G$ by Proposition 25.18. But then G is a 2-group, so its subgroup $\mathrm{Gal}(E/\mathbb{C})$ is also a 2-group. By Theorem 6, there is a quadratic field extension of $\mathbb{C}$ which, by Example 21.12(iv), can be obtained by adjoining the square root of a suitable element of $\mathbb{C}$. Thus, we can reach a contradiction by showing that $\sqrt{z} \in \mathbb{C}$ for each z in $\mathbb{C}$. This can be seen algebraically, *cf.* Exercise 12, but is obvious if we write $z = ae^{i\theta}$ in polar coordinates ($a \in \mathbb{R}^+$); then clearly $\sqrt{z} = \sqrt{a}e^{\frac{i\theta}{2}}$. □

COROLLARY 13. *$\mathbb{C}$ is the only proper algebraic extension of $\mathbb{R}$.*

Proof. Suppose E is an algebraic extension of $\mathbb{R}$. Then $E[\sqrt{-1}]$ is algebraic over $\mathbb{R}[\sqrt{-1}] = \mathbb{C}$, so $E[\sqrt{-1}] = \mathbb{C}$. Hence $\mathbb{R} \subseteq E \subseteq \mathbb{C}$; we are done since $[\mathbb{C} : \mathbb{R}] = 2$. □

Exercises

1. (Primitive root theorem revisited.) Show that any finite, separable field extension K/F has the form $K = F[a]$ for suitable a in K. (*Hint*: For F finite use Chapter 24; for F infinite use Steinitz's theorem (Exercise 23.7).)

The Normal Closure

2. The normal closure is uniquely defined. (*Hint*: Use the uniqueness of the splitting field of a given polynomial.)
3. Suppose E is the splitting field of a polynomial f over F. For any field $K \supseteq E$ and any homomorphism $\sigma: E \to K$ such that $\sigma(F) = F$, show that $\sigma(E) = E$.

4. Let m = number of F-injections from K to its normal closure E. Then $|\operatorname{Gal}(K/F)| \leq m \leq [K : F]$.
5. What is the normal closure of the field $\mathbb{Q}[\alpha^{1/n}]$ over $\mathbb{Q}$?
6. The normal closure of a quadratically defined field extension over F remains quadratically defined over F. (*Hint*: each new root can be obtained by continuing the chain of quadratic extensions.) Conclude that any constructible number is contained in a Galois extension of $\mathbb{Q}$, whose Galois group is a 2-group.

Separability Degree

Define the *separability degree* $[K : F]_s$ of a finite extension K/F to be the number of F-homomorphisms from K to its normal closure.

7. Using the idea of Theorem 23.14, show that $[K : F]_s \leq [K : F]$, with equality holding iff K/F is separable.
8. For any fields $F \subseteq K \subseteq E$ show $[E : F]_s = [E : K]_s[K : F]_s$. Conclude that E/F is separable iff E/K and K/F are separable.
9. An extension K/F is normal, iff every F-injection from K to its normal closure E is actually an F-automorphism of K, iff the number of F-automorphisms of K is $[K : F]_s$.

Finite Fields

10. For any finite field $K \supset \operatorname{GF}(p^t)$ show $\operatorname{Gal}(K/F) = \langle \sigma^t \rangle$, where σ is the Frobenius map.
11. If $a \in \operatorname{GF}(p^t)$ and f is the minimal polynomial of a, then the other roots of f are $a^p, a^{p^2}, a^{p^3}, \ldots$.

The Algebraic Closure

12. (Easy algebraic proof that $\mathbb{C}$ is closed under taking of square roots.) Given $c = a+bi$ for a, b in $\mathbb{R}$, solve $a+bi = (u+iv)^2 = u^2 - v^2 + 2uvi$ for suitable $u, v \in \mathbb{R}$. (*Hint*: If $b = 0$ then the assertion is obvious, so assume $b \neq 0$. Matching real and complex parts yields $a = u^2 - v^2$ and $b = 2uv$. Substitute $v = \frac{b}{2u}$ to obtain a quadratic equation in u^2 which has a positive solution.)
13. A field $\bar{F}$ is an *algebraic closure* of F if $\bar{F}$ is algebraically closed and is an algebraic extension of F. Prove that if $E \supset F$ is algebraically closed, then any finite extension of F is isomorphic to a subfield of E. Conclude that the algebraic closure of F in E is algebraically closed, and thus is an algebraic closure of F.
14. The algebraic closure of any field is infinite. (*Hint*: Lemma 9.) On the other hand, any infinite field has the same cardinality as its algebraic closure. (*Hint*: Compare with Exercise 21.6.)
15. What is the algebraic closure of $\mathbb{Q}$?

16. Viewing $GF(p^s) \subset GF(p^t)$ in the natural way when $s|t$, define the field $GF(p^\infty)$ to be the union of these fields, modulo an equivalence class identifying the appropriate elements. Prove this is algebraically closed, and thus is the algebraic closure of $\mathbb{Z}_p$.
17. Rather than rely on the fact that $\mathbb{C}$ is algebraically closed, one can prove, in general, that any field has an algebraic closure. The idea is to abstract the proof of Exercise 16: Let $\bar{F}$ be the field obtained by formally adjoining all roots of irreducible polynomials over F. This is obtained by taking the disjoint union and making the correct identifications via an equivalence relation. There are some delicate points in set theory that we are sloughing over.

In this chapter we present the crowning application of Galois theory: the determination, in terms of group theory, of the solvability of equations by formulas involving nth roots. To make life easier, we shall assume throughout that F is a perfect field, so that we do not need to worry about separability.

Root Extensions

Definition 0. A field K is a *root extension* of F (*of degree* n) if $K = F[a]$ for suitable $a \in K$ such that $a^n \in F$.

Note that this definition depends on the correct choice of a.

Example 1. If $\text{char}(F) \neq 2$, then any quadratic extension of F is a root extension of F of degree 2, by Example 21.12 (iv).

For degree > 2 the situation is more complicated, so we want to be able to iterate the extraction of roots, as in the next definition.

Definition 2. Suppose K/F is a field extension. A *root tower of height* $\leq n$, for K over F, is a chain of fields

$$K = K_0 \supset K_1 \supset \cdots \supset K_t = F \tag{1}$$

(for suitable t) such that K_i is a root extension of K_{i+1} of degree $\leq n$, for each i. A polynomial $f \in F[x]$ is *solvable by radicals of height* $\leq n$, if the splitting field E of f (over F) is contained in a field K that has a root tower of height $\leq n$.

Letting $\alpha = a^n \in F$ we can view a as $\sqrt[n]{\alpha}$, thereby justifying the terminology "root extension"; the term "radical" is a variant for "root," borrowed from the French.

By Example 1, every constructible number is contained in a root tower over $\mathbb{Q}$. On the other hand, it is easy to find a constructible number, such as $\sqrt{2} + \sqrt{3}$, which is not contained in any root extension of $\mathbb{Q}$.

Digression 3. There are two subtle points that should be noted.

(i) When one writes $\sqrt[n]{a}$, for $a \in \mathbb{R}$, one conventionally means the "principal" root. Thus, according to Definition 0, $\mathbb{Q}[i]$ is a root extension of $\mathbb{Q}$ (since $i^4 = 1$), although, strictly speaking, $\sqrt[4]{1}$ should be interpreted as 1, not i. On the other hand, $\sqrt{-1}$ can only be $\pm i$, so viewed in this way we see that i must be in the root extension $\mathbb{Q}(\sqrt{-1})$ of $\mathbb{Q}$. We shall define this notion precisely in Exercise 24. The situation becomes much more delicate when we consider higher roots of 1. Gauss proved that all the cyclotomic polynomials are solvable by radicals, even in this stricter sense. By using Definition 0, we bypass his deeper theory of cyclotomic extensions.

(ii) In the definition of solvability by radicals, one often can take $K = E$, for example, when F contains enough roots of 1, *cf.* Proposition 9 below. In general, however, one might not be able to take $K = E$, in which case one needs a technical lemma which shows that the normal closure of a separable extension with a root tower also has a root tower. The idea of the proof is quite clear, but the notation is cumbersome. In order to avoid complications, we set aside this technical lemma as Exercise 5, but it is needed to prove one direction of Theorem 11.

Example 4. Let $f = x^n - 1$. Its splitting field over $\mathbb{Q}$ is $\mathbb{Q}[\rho]$ (where ρ is a primitive nth root of 1), which is evidently a root extension of degree $\leq n$. Note that $\mathrm{Gal}(\mathbb{Q}[\rho]/\mathbb{Q}) = \mathrm{Euler}(n)$ is Abelian. (More generally, if $\mathrm{char}\, F = 0$ then $\mathrm{Gal}(F[\rho]/F)$ is Abelian, where ρ is a primitive nth root of 1; *cf.* Exercise 25.5.)

Digression 4'. Let us pause for a moment to consider the hypothesis that F contains a primitive nth root ρ of 1. Then the polynomial $f = x^n - 1$ has n distinct roots, namely $1, \rho, \ldots, \rho^{n-1}$, and thus is separable. In particular $0 \neq f' = nx^{n-1}$, so n is not divisible by $\mathrm{char}\, F$. Thus, our hypothesis on primitive roots of 1 implies the hidden hypothesis $\frac{1}{n} \in F$.

Example 5. Suppose $f = x^n - \alpha$ for α in F, and F contains a primitive nth root ρ of 1. Take $E = F[a]$, where a is a root of f, *i.e.*, $a^n = \alpha$. Then E is the splitting field of f, since the other roots $\rho^i a$ also lie in E, so f is solvable by radicals. Moreover, we claim $\mathrm{Gal}(E/F)$ is Abelian. Indeed any automorphism of E sends a to another root $\rho^i a$; if $\sigma, \tau \in \mathrm{Gal}(E/F)$, then σ, τ are given respectively by $a \mapsto \rho^i a$, $a \mapsto \rho^j a$ for suitable i, j, and thus

$$\sigma\tau(a) = \sigma(\rho^j a) = \rho^j \rho^i a = \rho^{i+j} a = \tau\sigma(a),$$

implying $\sigma\tau = \tau\sigma$. (In fact $\mathrm{Gal}(E/F)$ is cyclic, *cf.* Exercise 2.)

Example 6. Conversely, suppose K/F is Galois with $[K : F] = n$ prime, and suppose F contains a primitive nth root ρ of 1. Then $G = \mathrm{Gal}(K/F) = \langle\sigma\rangle$, for suitable σ in G, since $|G| = n$ is prime. We shall show that $K = F(\alpha^{1/n})$ for suitable $\alpha \neq 0$ in F. To see this, we must prove $\sigma(a) = \rho^i a$ for suitable a in K, $1 \leq i < n$. Then $a \notin F$, but $\sigma(a^n) = \rho^n a^n = a^n$, implying $a^n \in K^G = F$, as desired. Thus, we need prove

PROPOSITION 7. *Suppose σ is an automorphism of arbitrary order n in a field K, and suppose $F = K^\sigma$ contains a primitive nth root ρ of 1. Then there is some a in $K \setminus F$ such that $\sigma(a) = \rho^i a$ for suitable $1 \leq i < n$.*

Proof. Take $b \in K \setminus F$ and let

$$a_i = b + \rho^{-i}\sigma(b) + \rho^{-2i}\sigma^2(b) + \cdots + \rho^{-(n-1)i}\sigma^{n-1}(b), \quad 0 \le i < n. \quad (2)$$

Then $\sigma(a_i) = \rho^i a_i$, so $a_0 \in F$, and it suffices to show that $a_i \notin F$ for some $1 \le i < n$. Actually, we shall prove more generally that

$$\sigma^k(b) = \sum_{i=0}^{n-1} \frac{\rho^{ik}}{n} a_i \in \sum_{i=0}^{n-1} F a_i, \quad (3)$$

for each k, yielding the desired result (since otherwise $b \in F$, a contradiction).

To verify (3), note that (2) can be written more concisely as

$$a_i = \sum_{j=0}^{n-1} \rho^{-ij}\sigma^j(b).$$

Moreover, $\sum_{i=0}^{n-1} \rho^{iu} = \frac{\rho^{nu}-1}{\rho^u-1} = \frac{1-1}{\rho^u-1} = 0$ for any $1 \le u < n$, so we see

$$\sum_{i=0}^{n-1} \rho^{ik} a_i = \sum_{j=0}^{n-1} \sigma^j(b) \sum_{i=0}^{n-1} \rho^{i(k-j)}$$
$$= \sigma^k(b) \sum_{i=0}^{n-1} 1 = n\sigma^k(b),$$

as desired. □

This proof is full of interesting notions. First of all, the a_i in (2) are called the *Lagrange resolvents* of b, a key tool in computing root extensions, *cf.* Exercise 12. (Actually a more general proof of Proposition 7 can be obtained via Exercises 20 and 21.)

Solvable Galois Groups

The relation between solvability by radicals and the Galois group is our principal theme. Recall the definition of "solvable group" from Chapter 12. The next lemma is fundamental to the sequel.

LEMMA 8. *Suppose $F \subseteq E \subseteq K$, with K/F and E/F Galois. Then* $\text{Gal}(K/F)$ *is solvable, iff* $\text{Gal}(K/E)$ *and* $\text{Gal}(E/F)$ *are solvable.*

Proof. $\text{Gal}(E/F) \approx \text{Gal}(K/F)/\text{Gal}(K/E)$, so apply Theorem 12.9. □

PROPOSITION 9. *Suppose E is a splitting field of f over F, and F contains a primitive mth root of 1 for each $1 \le m \le n$, where $n = [E : F]$. The following are equivalent:*

(i) *f is solvable by radicals of height $\le n$;*

(ii) *E/F has a root tower);*

(iii) $\text{Gal}(E/F)$ *is solvable.*

Proof. ((ii) $\Rightarrow$ (i)) By definition.

((i) $\Rightarrow$ (iii)) We have a root tower

$$K = K_0 \supset K_1 \supset \cdots \supset K_t = F,$$

of height $\leq n$, with $K \supseteq E$; we need to show that $\mathrm{Gal}(K/F)$ is solvable, since then $\mathrm{Gal}(E/F)$ is solvable. We proceed by induction on t, noting the result is true by Example 5 if $t = 1$.

More generally, K_{t-1} is Galois over F, with Abelian Galois group. Now $\mathrm{Gal}(K/K_{t-1}) \triangleleft \mathrm{Gal}(K/F)$, and $\mathrm{Gal}(K/K_{t-1})$ is solvable, by induction on t. Hence $\mathrm{Gal}(K/F)$ is solvable, by Lemma 8.

((iii) $\Rightarrow$ (ii)) Let $G = \mathrm{Gal}(E/F)$. By Proposition 12.12, the solvability of G gives us a subnormal series $G = G_0 \supset G_1 \supset \cdots \supset \{e\}$, with each $[G_{i-1} : G_i]$ a suitable prime number $p_i \leq n$. Letting $K_i = E^{G_i}$ we see that each $[K_i : K_{i-1}] = p_i$, yielding a root tower of height $\leq n$, by Example 6. □

Even in the absence of primitive roots of 1, the same theory can be used, merely by adjoining roots of 1 at the onset. This is easy in characteristic 0, but in characteristic $\neq 0$ one must be careful - one cannot adjoin a primitive nth root of 1 if $\mathrm{char}\, F$ divides n, *cf* Digression 4′. Thus, we introduce the assumption $\mathrm{char}(F) \nmid n!$, *cf.* Exercise 1, which automatically holds in characteristic 0.

Example 10. Suppose $f = x^n - a$ for a in F, but with F not containing a primitive nth root ρ of 1. Then the splitting field of f is $E = F[\rho, a^{1/n}]$, so letting $K = F[\rho]$ we see that $E \supseteq K \supseteq F$ displays f as being solvable by radicals. Furthermore, $\mathrm{Gal}(E/F)$ is solvable, since $\mathrm{Gal}(K/F)$ and $\mathrm{Gal}(E/K)$ are Abelian.

THEOREM 11. *Suppose E is the splitting field of f over F, and $\mathrm{char}(F) \nmid n!$, where $n = [E : F]$. Then f is solvable by radicals of height $\leq n$, iff $\mathrm{Gal}(E/F)$ is solvable.*

Proof. Let E' be the splitting field of $(x^{n!} - 1)f$ over F; *i.e.*, $E' = E[\rho]$, where ρ is a primitive $n!$-root of 1. Then $\rho^{n!/m}$ is a primitive mth root of 1, for each $m < n$.

($\Rightarrow$) Suppose $E = E_0 \supset E_1 \supset \cdots \supset E_t = F$ is a root tower of height $\leq n$. Then

$$E' = E_0[\rho] \supset E_1[\rho] \supset \cdots \supset E_t[\rho] = F[\rho]$$

is a root tower of height $\leq n$ from E' to $F[\rho]$, which extends down to F via the root tower $F[\rho] \supset F[\rho^n] \supset F[\rho^{n(n-1)}] \supset \cdots \supset F$ of height $\leq n$.

Let L be the normal closure of E'. By Exercise 5, $L/F[\rho]$ has a root tower; by Proposition 9, $\mathrm{Gal}(L/F[\rho])$ is solvable. But $F[\rho]/F$ is Galois with Abelian Galois group, so $\mathrm{Gal}(L/F)$ is solvable, by Lemma 8; hence $\mathrm{Gal}(E/F)$ is solvable.

($\Leftarrow$) Let $G = \mathrm{Gal}(E/F)$. Then $\mathrm{Gal}(E'/F)/\,\mathrm{Gal}(E'/E) \approx G$; replacing E by E' we may assume $\rho \in E$. Then $\mathrm{Gal}(E/F[\rho]) \subseteq \mathrm{Gal}(E/F)$ is solvable, so Proposition 9 shows there is a root tower of height $\leq n$ from E to $F[\rho]$. Continuing the root tower from $F[\rho]$ to F (as above) shows f is solvable by radicals of height $\leq n$, over F. □

COROLLARY 12. *If* $\deg f \leq 4$, *then* f *is solvable by radicals.*

Proof. The splitting field of f has degree $\leq 4! = 24$, so its Galois group has order ≤ 24, and thus is solvable, by Corollary 12.11. □

Computing the Galois Group

In order to obtain an example of a polynomial f that is *not* solvable by radicals, we must find f whose Galois group is not solvable. Since S_n is not solvable for $n \geq 5$, let us find a polynomial whose Galois group is S_n. This is rather easy to do if we do not care about the base field, *cf.* Exercise 3, but our motivation at the outset was in studying polynomials over $\mathbb{Q}$, so it is only fair to look for an example over $\mathbb{Q}$. The explicit computation of the Galois group G of an arbitrary polynomial can be rather difficult, and is usually accomplished by viewing G explicitly as a group of permutations of the roots of f, as follows.

Remark 13. Recall (Summary 26.5) that if $E = F[a]$ then any automorphism of E/F permutes the roots $a_1 = a, \ldots, a_n$ of the minimal polynomial f of a, and thus we get a natural group injection $\mathrm{Gal}(E/F) \to S_n$, by which we shall view $\mathrm{Gal}(E/F)$ as a subgroup of S_n.

Digression 14. One can characterize certain subgroups explicitly by describing various combinations of the roots that they fix. Here is the main example of this technique.

Example 15. (The discriminant.) With notation as in Remark 13, view $G = \mathrm{Gal}(E/F)$ as a subgroup of S_n, and define

$$c = \{\prod (a_i - a_j) : 1 \leq j < i \leq n\}.$$

Any transposition in G sends c to $-c$. Thus, $d = c^2 \in F$, and is called the *discriminant* of f. Furthermore, the alternating group A_n fixes $F(c)$, so it follows that $\mathrm{Gal}(E/F(c)) = G \cap A_n$. In particular, d is a square in F iff $G \leq A_n$.

The discriminant can be computed via Exercises 20.12 and 20.13, and enables us to rederive the classical formulae for equations of degrees 3 and 4, *cf.* Exercise 11. Furthermore, the discriminant also provides a great deal of information about roots, *cf.* Exercises 8ff.

Let us return to the main theme. We assume $F \subseteq \mathbb{R}$. (In particular, we are in characteristic 0, with no problems about separability.) Then any $f \in F[x]$ factors (over $\mathbb{R}$) into irreducible polynomials $f_1 \dots f_t$, each of which has degree ≤ 2, by Corollary 26.13. Each root a_i of f_i can be viewed in $\mathbb{C}$ and is also a root of f. If $a_i \notin \mathbb{R}$, then, by the quadratic formula, its complex conjugate $\bar{a}_i$ is also a root of f_i and thus of f. Thus the nonreal roots of f (if they exist) pair off as complex conjugates, and complex conjugation in $\mathbb{C}$ induces an automorphism of order 2 on the splitting field of f over F.

To apply this to solvability we need one more result from group theory.

PROPOSITION 16. *Suppose p is prime. Then S_p is generated by any transposition τ and any cycle σ of order p.*

Proof. Reordering the indices, we may assume $\tau = (1\ 2)$. Replacing σ by a suitable power, we may assume $\sigma 1 = 2$; renumbering the other indices we may assume $\sigma = (1\ 2\ \dots p)$. But then we are done by the straightforward Exercise 5.8. □

COROLLARY 17. *If $f \in \mathbb{Q}[x]$ is irreducible of prime degree p and has precisely 2 nonreal roots in $\mathbb{C}$, then the Galois group of f is S_p.*

Proof. Since f splits over $\mathbb{C}$, we can view the splitting field E of f as a subfield of $\mathbb{C}$. By Remark 23.12, $p \mid [E : F]$, implying $G = \mathrm{Gal}(E/F)$ contains an automorphism σ of order p, by Cauchy's theorem. Viewing G in S_p, we see that σ corresponds to a cycle of order p. On the other hand, complex conjugation switches the two nonreal roots, leaving the real roots fixed, and thus yields an automorphism τ of E corresponding to a transposition in S_p. Hence σ, τ generate S_p, proving $\mathrm{Gal}(E/F) = S_p$. □

Example 18. Suppose $f = x^5 - pqx + p$, for $p, q \in \mathbb{N}$ and p prime. By Eisenstein's criterion, f is irreducible. On the other hand, we can determine precisely how many roots of f are real. One method is to observe that $f' = 5x^4 - pq$, which is positive for $|x|^4 > \frac{pq}{5}$ and negative otherwise. Since $f(0) = p > 0$, the intermediate value theorem implies that f has a real negative root. Moreover, since the graph of f has only two turning points, namely at $x_1 = -\sqrt[4]{\frac{pq}{5}}$ and $x_2 = +\sqrt[4]{\frac{pq}{5}}$, we see that f can have at most two more real roots; in fact, f has three real roots iff $f(x_1) > 0$ and $f(x_2) < 0$. For $i = 1, 2$,

$$f(x_i) = x_i(x_i^4 - pq) + p = x_i(\frac{pq}{5} - pq) + p = (-\frac{4}{5}qx_i + 1)p.$$

Clearly $f(x_1) > 0$, and for $f(x_2) < 0$ one needs $\frac{4}{5}qx_2 > 1$. Surely this holds for q is large enough, so in this case we conclude that f has precisely three real roots and two nonreal roots; therefore $\mathrm{Gal}(f) = S_5$, by Corollary 17. For a more precise analysis of the acceptable values of q, see Exercise 7.

(Another way of seeing that such f has precisely three real roots is via the discriminant, *cf.* Exercises 9 and 10.)

Besides providing negative criteria for solvability by radicals, Galois's theory enables one to obtain striking positive results, such as the following lovely theorem of Galois:

THEOREM 19. *Suppose $f \in F[x]$ is irreducible of prime degree p, and $a_1, \ldots, a_p$ are the roots of f in a splitting field E. Then f is solvable by radicals iff $E = F[a_i, a_j]$ for each $i \neq j$.*

The proof involves a fair amount of computation concerning transitive solvable subgroups of S_p, and is given in Exercises 16 through 19.

Exercises

1. If $\mathrm{char}\, F \nmid n$, then one can find a field extension E of F containing a primitive nth root of 1. (*Hint*: $f = x^n - 1$ is separable. Thus, any splitting field of f contains n distinct nth roots of 1, and one of these must be primitive, by a counting argument.)
2. In Example 5 show that $\mathrm{Gal}(E/F)$ actually is cyclic. (*Hint*: Identify $\mathrm{Gal}(E/F)$ with a subgroup of the cyclic multiplicative group of nth roots of 1.)

Prescribed Galois Groups

3. A quick example of a field extension whose Galois group is S_n. Let $x_1, \ldots, x_n$ be commuting indeterminates over $\mathbb{Q}$, and let E be the field of fractions of $\mathbb{Q}[x_1, \ldots, x_n]$. Every permutation σ in S_n defines the corresponding automorphism of $\mathbb{Q}[x_1, \ldots, x_n]$ given by $x_i \mapsto x_{\sigma i}$; this automorphism extends naturally to an automorphism of E, via $\sigma(\frac{c_1}{c_2}) = \frac{\sigma(c_1)}{\sigma(c_2)}$. Thus S_n can be viewed as a group of automorphisms of E; then E/E^{S_n} is Galois with Galois group S_n.
4. For any finite group G, find a Galois field extension having Galois group G. (*Hint*: View G as a subgroup of S_n, where $n = |G|$, and apply the Galois correspondence to Exercise 3.)

Root Towers

5. If K/F has a root tower of height $\leq n$ and E is the normal closure of K over F, then E/F has a root tower of height $\leq n$ that terminates with the original root tower of K/F. (*Hint*: Take a root tower

$K = K_0 \supset K_1 \supset \cdots \supset K_t = F$ (for suitable t) of height $\leq n$, *i.e.*, $K_i = K_{i+1}(a_i)$ with $a_i^{n_i} \in K_{i+1}$ and $n_i \leq n$, for each $i < t$. By Proposition 26.4, $E = F[\sigma_i(a_j) : 1 \leq i \leq n,\ 1 \leq j \leq t]$, where $\mathrm{Gal}(E/F) = \{\sigma_1 = 1,\ \sigma_2, \ldots, \sigma_n\}$. Put $L_0 = F$ and, for $1 \leq i \leq n$, $L_i = K\sigma_2(K)\ldots\sigma_i(K) = L_{i-1}\sigma_i(K)$, taken in E. Then $L_1 = K$ and $L_n = E$. By taking an analog of the original root tower at each stage, show that L_i/L_{i-1} has a root tower of height $\leq n$ for each i; putting them together then yields the desired root tower for E/F.)

6. Suppose K/F is Galois of degree p^t. Then for any series of extensions $K = K_0 \supset K_1 \supset \cdots \supset K_t = F$ such that each $p_i = [K_{i-1} : K_i]$ is prime, each K_{i-1}/K_i is necessarily Galois. (*Hint*: $\mathrm{Gal}(K/K_{i-1})$ has index p in $\mathrm{Gal}(K/K_i)$, so is a normal subgroup.)
7. Show, in general, (for $p, q > 0$ and n odd) that the polynomial $f = x^n - pqx + p$ always has one or three real roots, and has three real roots iff $p > \frac{n^n}{q^n(n-1)^{n-1}}$ (*Hint*: Carry out reasoning analogous to Example 18 to arrive at $f(x_2) = 0$ iff $x_2 > \frac{n}{q(n-1)}$; take the $(n-1)$-power, noting $x_2^{n-1} = \frac{pq}{n}$.)

Finding the Number of Real Roots via the Discriminant

Assume that $F \subseteq \mathbb{R}$ in Exercises 8 through 12.

8. Compute the discriminant d of $x^2 + bx + c$, as the familiar discriminant of the quadratic equation. Thus the number of real roots is determined by d.
9. Suppose $\deg f = n$, and f has m real roots. Then $n - m$ is an even number, which is divisible by 4 iff the discriminant of f is positive. (*Hint*: Let $t = \frac{n-m}{2}$, the number of pairs of nonreal roots. Compute the discriminant $d = \prod(a_i - a_j)^2$, by considering separately the various cases of whether or not a_i, a_j are real:

   If a_i, a_j are both real, then $(a_i - a_j)^2 > 0$.

   If a_i but not a_j is real, then $(a_i - a_j)(a_i - \bar{a}_j) = |a_i - a_j|^2 > 0$, and as noted above, both factors occur in the computation of d. Likewise if a_j but not a_i is real.

   If $a_i \neq a_j$ are both nonreal and $i \neq j$, then $(a_i - a_j)(\bar{a}_i - \bar{a}_j) = |(a_i - a_j)|^2 > 0$.

   Thus the sign of d is determined by $\prod(a_i - \bar{a}_i)^2$, taken over the t pairs of nonreal roots. But each $a_i - \bar{a}_i$ is purely imaginary, so its square is negative.)
10. If $\deg f \equiv 1 \pmod 4$ and $d < 0$, then f has at least three real roots and two nonreal roots.
11. Calculate the discriminant of the polynomial $x^n - pqx + p$, by means of Newton's formulas (*cf.* Exercises 20.12 and 20.13), and apply it

to exercise 9; compare with Exercise 7.

12. (Solving the cubic equation.) Suppose $f = x^3 + ux + v$ for $u, v \in \mathbb{Q}$. Using Exercises 20.12 and 20.13 compute the discriminant of f to be $4u^3 + 27v^2$; by adjoining the square root of this element one obtains a field F_1 over which the Galois group of f has order 3 and thus can be solved by adjoining a primitive cube root of 1 and taking a cube root. Carry out the explicit computation by means of Lagrange resolvents and the discriminant; the result is *Cardan's formula* for solving the cubic equation.
13. If $f \in F[x]$ has degree 4, and $a_1, a_2, a_3, a_4 \in E$ are roots, show that the subgroup fixing the element $(a_1 - a_3)(a_2 - a_4)$ is precisely the subgroup of permutations generated by $(1\ 2\ 3\ 4)$ and $(1\ 4)(2\ 3)$, which can be identified with D_4.

Galois Groups and Solvability

14. Suppose that $f \in F[x]$ is separable. Write $f = (x - a_1) \dots (x - a_n)$ in its splitting field E, with the a_i distinct. Then f is irreducible over F, iff $\mathrm{Gal}(E/F)$ acts transitively on $\{a_1, \dots, a_n\}$. (*Hint*: If f is irreducible, then $F[a_i] \approx F[a_j]$ over F; lift to $\mathrm{Gal}(E/F)$.)
15. The only possibilities for the Galois group G of an irreducible polynomial f of degree 4 include S_4, A_4, D_4, and the cyclic group of order 4.

    In Exercises 16 through 18, suppose p is a prime number.
16. Any normal subgroup $N \neq 1$ of a transitive subgroup of S_p is itself transitive. (*Hint*: The orbits have length dividing p.)
17. Let G be the subgroup of S_p consisting of all permutations that can be defined by a formula
$$j \mapsto aj + b \pmod{p}$$
    for suitable fixed a, b, where $1 \leq j \leq p$. Show that this actually is a subgroup of order $p(p-1)$ and is a subdirect product of the cyclic normal subgroup N of translations $j \mapsto j + b \pmod{p}$, by the cyclic subgroup of order $p - 1$ given by $j \mapsto aj \pmod{p}$.
18. Any transitive solvable subgroup H of S_p is conjugate to a subgroup of G containing N, with notation as in Exercise 17. (*Hint*: Using the derived subgroups, suppose $H^{(t-1)} \neq 0$ and $H^{(t)} = 0$. By Exercise 16, $H^{(t-1)}$ is a transitive Abelian subgroup and thus is cyclic, containing a cycle of length p which one may assume is $\sigma = (1\ 2\ \dots\ p)$, *i.e.*, $\sigma(j) = j+1 \pmod{p}$, for each j. Then $N \subseteq H$. For any τ in H, one has $\tau\sigma\tau^{-1} = \sigma^a$ for a suitable number a, so $\tau(j+1) = \sigma(j) + a$; letting $b = \tau(0)$, conclude $\tau \in G$.

19. Prove Theorem 19 by means of Exercise 18. (*Hint*: Exercise 18 implies that a transitive subgroup of S_n is solvable iff no element $\neq (1)$ fixes any two $i \neq j$. Show that $\mathrm{Gal}(E/F)$ is solvable iff $\mathrm{Gal}(E/F[a_i, a_j])$ is trivial for each i, j.)

The Norm and Trace

20. (Hilbert's Theorem 90.) Given a field extension K/F and $a \in K$ define the *norm* $N_{K/F}(a)$ to be $\prod_{\sigma \in \mathrm{Gal}(K/F)} \sigma(a)$. Note that $N_{K/F}$ is a linear character from K to F, in the sense of Exercise 25.7; also $N(\alpha a) = \alpha^n N(a)$, for α in F, where $n = |\mathrm{Gal}(K/F)|$. Show that if K/F is Galois with cyclic Galois group $\langle\sigma\rangle$ and if $N(a) = 1$ then $a = \frac{b}{\sigma(b)}$ for suitable b in K. (*Hint*: This is an elaboration of the idea of Proposition 7. Let χ_i denote the map $c \mapsto c^i$. Then $\{1, \chi_1, \dots \chi_{n-1}\}$ are independent linear characters, so there is c such that $b = \sum_i a\sigma(a) \dots \sigma^{i-1}(a)c^i \neq 0$.)
21. Reprove Proposition 7 as a Corollary to Exercise 20.
22. Given a field extension K/F and $a \in K$, define the *trace* $T_{K/F}(a)$ to be $\sum_{\sigma \in \mathrm{Gal}(K/F)} \sigma(a)$. Note that $T_{K/F}$ is a nonzero linear transformation of vector spaces. Show that if K/F is Galois with cyclic Galois group $\langle\sigma\rangle$ and $T_{K/F}(a) = 0$ then $a = \sigma(b) - b$ for suitable b in K. (*Hint*: This is similar to Exercise 20.)
23. Suppose that K/F is Galois of prime degree p, and $\mathrm{char}\, F = p$. Then $K = F[b]$ for some $b \in K$ such that $b^p - b \in F$. (*Hint*: Apply Exercise 22 to the element 1 to find b such that $\sigma(b) - b = 1$.)
24. Say K/F is a *strong root extension* if there is α in F and suitable n such that $K = F[\rho \sqrt[n]{a}]$ for every nth root ρ of 1. Define *strongly solvable by radicals* analogously to Definition 2, and prove that for each $n < 11$ the nth cyclotomic polynomial is strongly solvable by radicals. (*Hint*: Compute the Euler groups.) For $n = 11$, the problem becomes more difficult, but Gauss proved that all cyclotomic polynomials are strongly solvable by radicals; thus "strongly solvable by radicals" and "solvable by radicals" turn out to be the same.
25. Suppose ρ is a primitive mth root of 1, and $\eta_1, \dots, \eta_n$ are the nth roots of 1. Then $\sqrt[m]{\eta_j}$, $1 \leq j \leq n$, are the mnth roots of 1. (*Hint*: $x^{mn} - 1 = \prod_{j=1}^{n}(x^m - \eta_j)$.) Thus, in proving Gauss's theorem, one may assume n is prime.

Review Exercises for Part III

1. True or false: $\mathbb{C}$ is the algebraic closure of $\mathbb{Q}$.
2. What is the degree of $\sqrt[3]{2}+\sqrt{3}$ over $\mathbb{Q}$?
3. What is the degree of $\sqrt[3]{4}+\sqrt[4]{3}$ over $\mathbb{Q}$?
4. What is the degree of $\sqrt{i}$ over the following fields: $\mathbb{Q}$, $\mathbb{R}$, $\mathbb{C}$?
5. Prove that the algebraic closure of $\mathbb{Q}$ is not a finite extension of $\mathbb{Q}$.
6. Construct a field extension of $\mathbb{C}$ (which of course is not algebraic).
7. If ρ is a primitive tenth root of 1, then what is $[\mathbb{Q}[\rho] : \mathbb{Q}]$?
8. Prove algebraically that any angle can be bisected (into two equal parts).
9. Prove that the regular pentagon is constructible.
10. Prove that the regular 9-gon is not constructible.
11. Is the regular 11-gon constructible?
12. Is the regular 15-gon constructible?
13. For each of the following angles, show whether or not it can be trisected by means of compass and straight edge: $120°, 72°, 60°, 45°$.
14. Prove that an irreducible polynomial f is separable iff $f' \neq 0$.
15. For any p, give an example of an irreducible polynomial of degree p over $\mathbb{Z}_p$.
16. What is the splitting field of x^2+1 over $\mathbb{Z}_2$?
17. Suppose $f = x^6 - 3$. What is the splitting field E of f over F, and $[E : F]$, for the following fields: $F = \mathbb{Q}$, $F = \mathbb{Q}[i]$?
18. What is the splitting field of $x^4 - 4$ over $\mathbb{Q}$?
19. What is the splitting field of $x^5 + 1$ over $\mathbb{Q}$?
20. What is the splitting field of $x^5 - 2$ over $\mathbb{Q}$?
21. Prove that $\sqrt[4]{2}+\sqrt[4]{3}$ has degree 16 over $\mathbb{Q}$.
22. If possible, build fields of order 6,7,8,9 (or explain why they do not exist).
23. Factor the polynomial $x^{16} - x$ over $\mathbb{Z}_2$.
24. Determine all the irreducible polynomials of order 3 over $\mathbb{Z}_2$.
25. How many irreducible polynomials are there of order 4 over $\mathbb{Z}_2$?
26. Determine all the subfields of the field of order 16.
27. Define the cyclotomic polynomials, and show that they are monic polynomials over $\mathbb{Z}$.
28. Suppose K is a finite field containing $\mathbb{Z}_3$, and $a \in K$ has a minimal polynomial f of degree t over $\mathbb{Z}_3$. How explicitly can you factor f over K?
29. Prove that any finite separable extension has only a finite number of intermediate fields.

30. Prove from scratch that any field extension of degree 3 over $\mathbb{Q}$ is contained in a field obtained by adjoining a square root and a cube root.
31. Prove that any cubic equation can be solved over $\mathbb{Q}$, by taking two square roots and a cube root.
32. Prove that $\mathbb{C}$ is the only proper algebraic extension of $\mathbb{R}$.
33. Prove that the regular 17-gon is constructible.
34. What are the possible Galois groups for an irreducible polynomial of degree 3? of degree 4?
35. What is the Galois group of the polynomial $x^4 + 3x^2 + 2$ over $\mathbb{Q}$?
36. What is the Galois group over $\mathbb{Q}$ of the following polynomials: $x^4 - 1$, $x^4 - 2$, $x^4 - 4$?
37. What is the Galois group of the polynomial $x^{16} - x$ over $\mathbb{Z}_2$?
38. What is the Galois group of the polynomial $x^{16} - 1$ over $\mathbb{Z}_2$?
39. What is the Galois group of the polynomial $x^5 - 10x + 5$ over $\mathbb{Q}$?

Although our effort in field theory has been concentrated in studying algebraic numbers, "most" complex numbers are transcendental over $\mathbb{Q}$! (The argument was sketched in Exercises 21.2 through 21.5.) Strangely enough, the task of finding even one specific transcendental number is considerably more difficult. Many of the proofs of transcendence of particular numbers rely on the fundamental theorem of calculus: If f is a continuous function on the interval $[a, b]$ with continuous derivative f' on (a, b), then

$$f(b) - f(a) = (b - a)f'(c)$$

for some c in (a, b).

The basic idea in the application is as follows: Suppose a is algebraic over $\mathbb{Q}$. Then a satisfies a suitable polynomial $f \in \mathbb{Z}[x]$ of degree $t > 0$. Consider a closed interval I containing a, but no other root of f, in its interior. Then $|f'|$ is bounded on I by some number d; since $f(a) = 0$ we have

$$|f(b)| = |f(b) - f(a)| \leq d|b - a|$$

for any b in I. On the other hand, taking b to be rational, say $b = \frac{m}{n}$, we see that $0 \neq f(b) \in n^{-t}\mathbb{Z}$, implying $|f(b)| \geq n^{-t}$. Thus, we have a contradiction if given any $t, d \in \mathbb{N}^+$ we can find $b = \frac{m}{n}$ such that

$$n^{-t} > d|b - a|. \tag{1}$$

This method is applicable to numbers defined by series that converge very rapidly.

Example 1. (Liouville's Number.) $a = \sum_{u=1}^{\infty} 10^{-u!}$ is transcendental. Indeed, in the above notation take $b = \sum_{u=1}^{v} 10^{-u!}$, where v is to be determined in terms of t and d. Then $n = 10^{v!}$ and $|b - a| = \sum_{u=v+1}^{\infty} 10^{-u!} < 2 \cdot 10^{-(v+1)!}$. So for (1) to hold it is enough that

$$10^{-v!t} > 2d \cdot 10^{-(v+1)!}, \quad i.e., \quad 10^{v!(v+1-t)} > 2d.$$

This is certainly true when $v \geq 2d + t$. □

Transcendence of e

For more familiar numbers, such as $e = 1+1+\frac{1}{2}+\frac{1}{3!}+\ldots$, a more delicate procedure is called for. Having succeeded so far using calculus, let us begin by extracting the essence of a polynomial in terms of calculus. Write $f^{(i)}$ for the i-th derivative of f. Note that in what follows we are sloughing over problems of convergence, which could be bypassed altogether by dealing with formal power series (*cf.* Exercises 16.8ff).

Remark 2. An infinitely differentiable function $f: \mathbb{R} \to \mathbb{R}$ is a polynomial of degree n (over $\mathbb{R}$) iff $f^{(n+1)} = 0$. (Indeed, ($\Rightarrow$) is obvious, and ($\Leftarrow$) is clear by taking the antiderivative $n+1$ times.)

Define $f^{(0)} = f$. Then, given any polynomial $f(x) = \sum_{u=0}^{t} \alpha_u x^u$, we can define

$$\tilde{f} = \sum_{i=0}^{\infty} f^{(i)} = f + f^{(1)} + \cdots + f^{(t-1)} + f^{(t)}, \quad \text{and} \quad \tau_f = \tilde{f}(0).$$

The map $f \mapsto \tilde{f}$ defines an $\mathbb{R}$-linear transformation $\mathbb{R}[x] \to \mathbb{R}[x]$.

Remark 3. $\tau_f = \sum_{u=0}^{t} \alpha_u u!$, where $f = \sum_{u=0}^{t} \alpha_u x^u$. (Indeed, it is enough to check this for a monomial $f = x^u$, in which case $\tau_f = \tilde{f}(0) = u!$

Recalling $e^x = 1 + \frac{x}{1} + \frac{x^2}{2!} + \cdots + \frac{x^n}{n!} + \frac{x^{n+1}}{(n+1)!} + \ldots$, we next define

$$\delta_n(x) = \frac{x}{n+1} + \frac{x^2}{(n+1)(n+2)} + \ldots, \tag{2}$$

and note that $|\delta_n(x)| \leq e^{|x|}$.

Now consider $f = x^n$. Then $f^{(u)} = n(n-1)\ldots(n-u+1)x^{n-u}$, implying

$$\begin{aligned} \tau_f e^x = n!e^x &= n! + n!x + (n(n-1)\ldots 2)x^2 + \cdots + nx^{n-1} + x^n \\ &\quad + x^n\left(\frac{x}{n+1} + \frac{x^2}{(n+1)(n+2)} + \ldots\right) \\ &= \tilde{f} + x^n \delta_n(x). \end{aligned} \tag{3}$$

Thus, for general $f = \sum_{u=0}^{n} \alpha_u x^u$, we conclude

$$\tau_f e^x = \tilde{f} + \sum_u \alpha_u x^u \delta_u(x). \tag{4}$$

Furthermore, (2) implies $|\sum_{u=0}^{n} \alpha_u c^u \delta_u(c)| \leq |\sum_u \alpha_u c^u| e^{|c|} = |f(c)| e^{|c|}$ for any $c \in \mathbb{R}$. In order to use (4) to its fullest we need a way to estimate τ_f for certain polynomials f.

LEMMA 4. *Suppose $h \in \mathbb{Z}[x]$, and define*

$$f(x) = \frac{x^{n-1}}{(n-1)!}h(x), \quad g(x) = \frac{x^n}{(n-1)!}h(x).$$

Then τ_f, τ_g are integers;

(i) $\tau_f \equiv h(0) \pmod{n}$; *and*

(ii) $\tau_g \equiv 0 \pmod{n}$.

Proof. Write $h = \sum_{u=0}^{t} \alpha_u x^u$, $\alpha_u \in \mathbb{Z}$. Then

$$\begin{aligned}\tau_f &= \sum_{u=0}^{t} \alpha_u \frac{(n+u-1)!}{(n-1)!} \\ &= \alpha_0 + \sum_{u=1}^{t} \alpha_u n(n+1)\dots(n+u-1);\end{aligned}$$

by inspection $h(0) = \alpha_0 \equiv \tau_f \pmod{n}$.

$$\tau_g = \sum \alpha_u \frac{(n+u)!}{(n-1)!} = \sum \alpha_u n(n+1)\dots(n+u) \equiv 0 \pmod{n}. \quad \square$$

THEOREM 5. *e is transcendental over $\mathbb{Q}$.*

Proof. On the contrary, suppose e is algebraic over $\mathbb{Q}$. Then there are β_v in $\mathbb{Z}$, with $\beta_0 \neq 0$ and $\sum_{v=0}^{m} \beta_v e^v = 0$. For any polynomial $f = \sum \alpha_u x^u$ in $\mathbb{Q}[x]$ we have (by (4))

$$0 = \tau_f \sum_{v=0}^{m} \beta_v e^v = \sum_{v=0}^{m} \beta_v \tau_f e^v = \sum_{v=0}^{m} \beta_v (\tilde{f}(v) + \sum_u \alpha_u v^u \delta_u(v)) = S_1 + S_2 + S_3,$$

where $S_1 = \beta_0 \tau_f$, $S_2 = \sum_{v=1}^{m} \beta_v \tilde{f}(v)$, and $S_3 = \sum_{v=1}^{m} \beta_v \sum_u \alpha_u v^u \delta_u(v)$ (so that $|S_3| \leq \sum_v |\beta_v||f(v)|e^{|v|}$).

We want to find a polynomial f for which S_1 is a nonzero integer mod p, $S_2 \equiv 0 \pmod{p}$, and $|S_3| < 1$; this would yield a contradiction, since their sum could not be 0. We take the "Hermite polynomial"

$$f = \frac{x^{p-1}}{(p-1)!}(x-1)^p(x-2)^p \dots (x-m)^p, \tag{5}$$

where p is a sufficiently large prime number.

$|S_3| \to 0$ as $p \to \infty$; indeed, $|f(v)| \to 0$ as $p \to \infty$, whereas β_v and $e^{|v|}$ are independent of p.

Furthermore, by Lemma 4, S_1 is a nonzero integer $\not\equiv 0 \pmod{p}$ whenever $p > \beta_0 m$.

Finally, to show $S_2 \equiv 0 \pmod{p}$, put $f_v(x) = f(x+v) = \frac{x^p}{(p-1)!} g_v(x)$ for a suitable polynomial g_v, for $1 \le v \le m$. Each $\tilde{f}(v) = \tilde{f}_v(0) = \tau_{f_v} \equiv 0 \pmod{p}$, by Lemma 4(ii). □

Although rather intricate in execution, the basic idea of the proof is quite simple, once one manages to find the Hermite polynomial f with the correct properties.

Transcendence of π

The proof that π is transcendental follows the same lines, but involves symmetric polynomials at a key step, so Exercises 20.11 and 23.10 are prerequisites. We want to apply the same considerations that concern the exponential function, so we relate π to e via the formula $e^{\pi i} = -1$. Thus, it is natural to prove the equivalent statement:

THEOREM 6. *πi is transcendental.*

Proof. Otherwise πi satisfies some polynomial $g = \sum_{u=0}^{m} \beta_u x^u$, for $\beta_u \in \mathbb{Z}$, $\beta_0 \neq 0$. Letting $a_1 = \pi i, \ \dots, a_m$ be the roots of g in a suitable splitting field, we have $1 + e^{a_1} = 0$, and thus

$$\begin{aligned} 0 &= (1+e^{a_1})\dots(1+e^{a_m}) \\ &= 1 + e^{a_1} + e^{a_2} + \dots + e^{a_m} + e^{a_1}e^{a_2} + e^{a_1}e^{a_3} + \dots \end{aligned} \tag{6}$$

Noting $e^{a_i}e^{a_j} = e^{a_i+a_j}$, we can rewrite the right-hand side of (6) as

$$1 + \sum_{v=1}^{2^m-1} e^{b_v},$$

where each b_v is a sum of various a_i. Suitably rearranging the b_v, we have some $n < 2^m$ such that $b_v \neq 0$ for $1 \le v \le n$ and $b_v = 0$ for all $v > n$. Taking $m' = 2^m - n$, (6) becomes

$$0 = m' + \sum_{v=1}^{n} e^{b_v}.$$

In view of (4), any polynomial $f = \sum \alpha_u x^u \in \mathbb{Q}[x]$ satisfies

$$0 = \tau_f m' + \sum_{v=1}^{n} \tau_f e^{b_v} = \tau_f m' + \sum_{v=1}^{n} (\tilde{f}(b_v) + \sum_u \alpha_u b_v^u \delta_u(b_v)) = S_1 + S_2 + S_3$$

where $S_1 = \tau_f m'$, $S_2 = \sum_{v=1}^{n} \tilde{f}(b_v)$, $|S_3| < \sum_{v=1}^{n} |f(b_v)|e^{|b_v|}$. This time we take

$$f = \frac{\beta_0^{np} x^{p-1}}{(p-1)!}(x - b_1)^p \dots (x - b_n)^p. \tag{7}$$

Then f can be written as $\frac{x^{p-1}}{(p-1)!} \sum_{i=0}^{np} q_i x^i$, where the q_i are expressions symmetric in the $\beta_0 b_1, \dots, \beta_0 b_n$ and thus are symmetric in the $\beta_0 a_1, \dots, \beta_0 a_m$ and so are integers by Exercise 23.10.

Now we conclude as in Theorem 5. By Lemma 4, S_1 is a nonzero integer. Take $p > S_1$. As before, S_2 is an integer $\equiv 0 \pmod p$ and $S_3 \to 0$ as $p \to \infty$, contrary to $S_1 + S_2 + S_3 = 0$. □

COROLLARY 7. *π is not constructible, and, in particular, one cannot square the circle by means of compass and straight edge.*

These notes have dealt almost exclusively with commutative rings, in part because of the greater ease in dealing with one-sided ideals than with two-sided ideals. However, noncommutative rings do arise naturally in many mathematical contexts, and should also be considered. In harmony with one of the themes of these notes, we shall apply certain noncommutative rings (namely skew fields) to elementary number theory, although there are several deep connections to geometry and other subjects that are beyond the scope of these notes. Our brief excursion into skew fields is motivated by the following theorem:

LAGRANGE'S FOUR SQUARE THEOREM. *Every natural number is a sum of four squares.*

The reader will notice at once the similarity to Fermat's theorem concerning primes of the form $a^2 + b^2$, which equals $(a + bi)(a - bi)$ in $\mathbb{Z}[i]$. So we would like to find a similar way to factor $a^2 + b^2 + c^2 + d^2$ Of course we need too many square roots of -1 to work in $\mathbb{C}$ (or indeed in any field), and instead must seek our solution in a skew field. (Recall the definition from Chapter 13.) In order to proceed, we need noncommutative generalizations of some of the notions described in earlier Chapters. But first we need to know more about the nature of commutativity. We say elements a, b *commute* if $ab = ba$.

Remark 0. Suppose $a \in R$ is invertible. If a commutes with b in R, then a^{-1} also commutes with b. (Indeed, $a^{-1}b = a^{-1}baa^{-1} = a^{-1}aba^{-1} = ba^{-1}$.)

Although arbitrary elements need not commute, certain elements (such as $0, 1$) commute with every element of the ring. Define the *center* $\mathrm{Cent}(R)$ of a ring R to be $\{c \in R : cr = rc \text{ for all } r \text{ in } R\}$. Obviously, $\mathrm{Cent}(R)$ is a commutative ring, which is a field if R is a skew field, by Remark 0. Thus any skew field can be viewed as a vector space over its center.

Generalizing Chapter 21, we say an F-*algebra* is a ring R containing an isomorphic copy of F in its center C, *i.e.*, there is an injection $F \to C$. (There is a more abstract definition of algebra, given in Exercise 2, but this definition serves our purposes.) Perhaps the most familiar noncommutative algebra is $M_n(F)$, the algebra of $n \times n$ matrices over a field F, where the injection $F \to M_n(F)$ sends α to the scalar matrix αI.

The Quaternion Algebra

Definition 1. (Hamilton's Algebra of Quaternions.) Let $\mathbb{H}$ denote the two-dimensional vector space over $\mathbb{C}$, with base $\{1, j\}$; thus $\mathbb{H} = \mathbb{C} + \mathbb{C}j =$

$\mathbb{R} + \mathbb{R}i + \mathbb{R}j + \mathbb{R}k$, where $k = ij$. We define multiplication on $\mathbb{H}$ via the rules

$$\begin{aligned} i^2 = j^2 = k^2 &= -1; \\ ij = -ji &= k; \\ jk = -kj &= i; \\ ki = -ik &= j. \end{aligned}$$

(Note: The last three lines are encompassed by rule $ijk = -1$. For example, $kj = -(-1)kj = -ijkkj = -i$.)

This multiplication extends via distributivity to all of $\mathbb{H}$, by

$$(a_1 + b_1 i + c_1 j + d_1 k)(a_2 + b_2 i + c_2 j + d_2 k) = a_3 + b_3 i + c_3 j + d_3 k$$

for $a_u, b_u, c_u, d_u \in \mathbb{R}$, where

$$\begin{aligned} a_3 &= a_1 a_2 - b_1 b_2 - c_1 c_2 - d_1 d_2; \\ b_3 &= a_1 b_2 + b_1 a_2 + c_1 d_2 - d_1 c_2; \\ c_3 &= a_1 c_2 - b_1 d_2 + c_1 a_2 + d_1 b_2; \\ d_3 &= a_1 d_2 + b_1 c_2 - c_1 b_2 + d_1 a_2. \end{aligned}$$

We have no guarantee *a priori* that $\mathbb{H}$ is an $\mathbb{R}$-algebra, although it is not difficult to verify the axioms directly. However, it is more elegant to view $\mathbb{H}$ as a subalgebra of an algebra we already know, namely $M_2(\mathbb{C})$.

PROPOSITION 2. *$\mathbb{H}$ is a skew field.*

Proof. There is a map $\varphi\colon \mathbb{H} \to M_2(\mathbb{C})$ given by $y + zj \mapsto \begin{pmatrix} y & z \\ -\bar{z} & \bar{y} \end{pmatrix}$. Writing $\hat{i}, \hat{j}, \hat{k}$ for the respective images of i, j, k, we see that $\hat{i} = \begin{pmatrix} i & 0 \\ 0 & -i \end{pmatrix}$, $\hat{j} = \begin{pmatrix} 0 & 1 \\ -1 & 0 \end{pmatrix}$, and $\hat{k} = \begin{pmatrix} 0 & i \\ i & 0 \end{pmatrix}$ and we have $\hat{i}^2 = \hat{j}^2 = \hat{k}^2 = \hat{i}\hat{j}\hat{k} = -1$. Thus φ respects the defining relations for $\mathbb{H}$ and thereby preserves multiplication. $\ker \varphi = 0$ by inspection, so we conclude $\mathbb{H}$ is a ring, and φ is a ring injection.

Furthermore, let $d = \det \begin{pmatrix} y & z \\ -\bar{z} & \bar{y} \end{pmatrix} = y\bar{y} + z\bar{z}$, a positive real number whenever y or $z \neq 0$; then

$$\begin{pmatrix} y & z \\ -\bar{z} & \bar{y} \end{pmatrix}^{-1} = d^{-1} \begin{pmatrix} \bar{y} & -z \\ \bar{z} & y \end{pmatrix},$$

implying $(y + zj)^{-1} = d^{-1}(\bar{y} - \bar{z}j)$. Hence $\mathbb{H}$ is a skew field. □

If it seems that we have pulled φ out of a hat, see Exercise 4 for a more systematic approach.

Digression. $\mathbb{H}$ has dimension 4 as a vector space over $\mathbb{R}$, whence the name "quaternions"; this algebra is closely related to the group of quaternions, which we studied earlier, *cf.* Exercise 18. For many years Hamilton had tried in vain to find a skew field of dimension 3 over $\mathbb{R}$, although today it is rather easy to see that there is none, *cf.* Exercise 15. Indeed, there is a theorem (which we shall not prove here) that the dimension of a skew field over its center must be a square number.

Remark 3. In the proof of Proposition 2 we defined a multiplicative map $N: \mathbb{H} \setminus \{0\} \to \mathbb{R}^+$ given by

$$N(y+zj) = \det \begin{pmatrix} y & z \\ -\bar{z} & \bar{y} \end{pmatrix} = y\bar{y} + z\bar{z}.$$

(N is multiplicative because det is multiplicative.) We can rewrite N as $N(a+bi+cj+dk) = a^2+b^2+c^2+d^2$, which is a sum of four squares.

Proof of Lagrange's Four Square Theorem

This discussion bears directly on Lagrange's Four Square Theorem, in the following manner.

COROLLARY 4. *If n_1 and n_2 are sums of four square integers, then so is $n_1 n_2$.*

Proof. Write $n_u = a_u^2 + b_u^2 + c_u^2 + d_u^2 = N(a_u + b_u i + c_u j + d_u k)$ for $u = 1, 2$. Then $n_1 n_2 = N(q)$, where

$$q = (a_1 + b_1 i + c_1 j + d_1 k)(a_2 + b_2 i + c_2 j + d_2 k). \quad \square$$

COROLLARY 5. *To prove Lagrange's Four Square Theorem it suffices to assume the natural number n is prime.*

Proof. By induction applied to Corollary 4. $\square$

LEMMA 6. *For any prime number p, there is a number $n < \frac{p}{2}$ such that np is a sum of three square numbers prime to p.*

Proof. One may assume p is odd. In the field $\mathbb{Z}_p$ the sets

$$\{-[a]^2 : -\frac{p}{2} < a < \frac{p}{2}\} \quad \text{and} \quad \{[b]^2 + 1 : -\frac{p}{2} < b < \frac{p}{2}\}$$

each take on at least $\frac{p+1}{2}$ distinct values, since any duplication comes at most in pairs. (In any field the equation $x^2 - t = 0$ has at most two solutions.) But $\mathbb{Z}_p$ has p elements, so $-[a]^2 = [b]^2 + 1$ for suitable a, b, *i.e.*, p divides $a^2 + b^2 + 1$. Since $a^2 + b^2 + 1 < 2(\frac{p-1}{2})^2 + 1$ one sees $n < \frac{p}{2}$. □

The proof of Lagrange's Four Square Theorem can be concluded by induction on the smallest n such that np is a sum of four squares and is not too difficult, *cf.* Exercises 6 and 7. However, it is enlightening to recast the proof in terms of the structure of algebras, as was done in proving Fermat's theorem. We shall sketch the proof, leaving the verifications as exercises.

We need a quaternion version of the Gaussian integers $\mathbb{Z}[i]$. The obvious candidate is $\mathbb{Z} + \mathbb{Z}i + \mathbb{Z}j + \mathbb{Z}k$, where $i^2 = j^2 = k^2 = 1 = -ijk$. The first step then is to define a Euclidean algorithm analogous to the Gaussian integers, but the natural definition doesn't quite work for this ring. However, one can define instead the *ring of integral quaternions* $R = \{a + bi + cj + dk : 2a, 2b, 2c, 2d \in \mathbb{Z}\}$, and this indeed satisfies the Euclidean algorithm (Exercise 8). Consequently, as in the proof of Proposition 16.11 one concludes that every left ideal is principal. Now we obtain a noncommutative arithmetic on R by defining an element r to be irreducible whenever Rr is a maximal left ideal and defining $\gcd(r_1, r_2)$ to be that r_3 such that $Rr_1 + Rr_2 = Rr_3$. One can conclude that an integral quaternion q is irreducible iff $N(q)$ is a prime number in $\mathbb{Z}$ (*cf.* Exercise 10), from which it follows at once that any prime p in $\mathbb{Z}$ can be written as $N(q)$, where q is an irreducible integral quaternion dividing p. Hence, $4p = N(2q)$ is a sum of four squares in $\mathbb{Z}$, so we conclude with Exercise 6.

Let us return to quaternion algebras. We could build a quaternion algebra Q starting from any field F; namely take a four-dimensional vector space $F + Fi + Fj + Fk$ with multiplication defined by the relations $i^2 = j^2 = k^2 = -1 = ijk$. One could hope for Q to be a skew field, but this is not the case when $F = \mathbb{Z}_p$. (If $a^2 + b^2 + c^2 = np$ in Lemma 6, then $ai + bj + ck$ is a zero-divisor in $\mathbb{Z}_p$.) More generally, every finite skew field is commutative, but this result needs some preparation.

Polynomials over Skew Fields

In this discussion, D denotes a skew field having center F. Note that $F[x] \subseteq \text{Cent}(D[x])$, so in particular $xd = dx$ for all d in D, where we view $D \subset D[x]$ naturally as the constant polynomials. If f, g are polynomials, we say g divides f if $f = hg$ for some h in $D[x]$. Also, given $f = \Sigma d_i x^i$ in $D[x]$ write $f(d)$ for $\Sigma d_i d^i$, *i.e.*, "right substitution" for d. (We must be careful how we substitute, since d need not commute with the d_i.)

Remark 7. Given $f \in D[x]$ and d in D, we have

$$f(x) = q(x)(x - d) + f(d)$$

for some q in $D[x]$. In particular, $f(d) = 0$ iff $x - d$ divides $f(x)$.

The proof follows from the Euclidean algorithm for polynomials (Proposition 16.7). Explicitly, if $f = \Sigma_{i=0}^{n} d_i x^i$, then let $g(x) = f - d_n x^{n-1}(x - d)$, of degree $< n$; one sees by induction on n that

$$g = q_1(x)(x - d) + g(d) = q_1(x)(x - d) + f(d),$$

implying $f = g + d_n x^{n-1}(x - d) = (q_1 + d_n x^{n-1})(x - d) + f(d)$.

REMARK 8. *(i) Given $f = hg$, let us put $\bar{f} = h(x)g(d)$; writing $h = \Sigma d_i x^i$, we see $f(x) = \Sigma d_i g(x) x^i$ and $\bar{f} = \Sigma d_i g(d) x^i$, so $f(d) = \Sigma d_i g(d) d^i = \bar{f}(d)$.*

(ii) If $0 \neq a \in D$ and $g|f$ then aga^{-1} divides afa^{-1}. (Indeed, if $f = hg$ then $afa^{-1} = ahga^{-1} = aha^{-1}aga^{-1}$.)

These simple computations provide our main tool:

PROPOSITION 9. *With notation as in Remark 8, if d is a root of f but not of g, then $g(d)dg(d)^{-1}$ is a root of h. (Note that here we should assume D is a skew field, to insure that $g(d)$ is invertible.)*

Proof. By Remark 8, $\bar{f}(d) = f(d) = 0$. Hence, $x - d$ divides $\bar{f} = h(x)g(d)$; hence $x - g(d)dg(d)^{-1} = g(d)(x - d)g(d)^{-1}$ divides $g(d)(h(x)g(d))g(d)^{-1} = g(d)h(x)$ and thus divides h. □

By a *conjugate* of d in D we merely mean an element of the form ada^{-1}, for suitable $a \neq 0$ in D. Note that if $f(d) = 0$ for $f \in F[x]$ then, for any $a \neq 0$ in D, ada^{-1} is a root of $afa^{-1} = f$, implying every conjugate of d is also a root of f. The big difference between the commutative and noncommutative cases is that d is its only conjugate when D is commutative, whereas there can be an infinite number of distinct conjugates when $d \in D \setminus F$. In fact, there are enough conjugates for the following theorem.

THEOREM 10. *If $f(x) \in F[x]$ is monic irreducible and has a root d_1 in D, then $f = (x - d_n) \dots (x - d_1)$ in $D[x]$, where each d_i is a conjugate of d_1.*

Proof. Let $g = x - d_1$, and write $f = hg$. Any element $d = ad_1a^{-1} \neq d_1$ is a root of f, but not of g, so applying Proposition 9 yields a root $d_2 = (d - d_1)d(d - d_1)^{-1}$ of h; we continue by induction on degree. To make sure the procedure does not break down, we require the following observation.

Remark 11. Assumptions as in Theorem 10, if $0 \neq h \in D[x]$ with $\deg h \leq \deg f$ and every conjugate of d_1 is a root of h, then $h = f$. (Indeed, take a monic counterexample h of minimal degree $< \deg f$; then for any $a \neq 0$ of D, each conjugate of d is a root of the polynomial aha^{-1}, and thus of

$h - aha^{-1}$, which has lower degree since both h and aha^{-1} are monic; hence $h = aha^{-1}$ for all $a \neq 0$ in D. Thus $h \in F[x]$, since each coefficient is in F. But f is the minimal polynomial of d_1 in D, contradiction.) □

Actually, Theorem 10 is only half a theorem. There is a sort of uniqueness to the factorization, as follows:

THEOREM 12. *If* $f = (x - d_n) \dots (x - d_1) \in D[x]$ *and* $f(d) = 0$, *then* d *is a conjugate of some* d_i; *in fact, writing* $f_i = (x - d_i)(x - d_{i-1}) \dots (x - d_1)$ *and* $f_0 = 1$, *one has* $d = f_i(d)^{-1} d_{i+1} f_i(d)$ *for some* i.

Proof. If $f_1(d) = 0$ then $d = d_1$. Hence, we may assume $f_1(d) \neq 0$. Take i with $f_i(d) \neq 0$ and $f_{i+1}(d) = 0$. Letting $f = f_{i+1}$ and $g = f_i$ in Proposition 9 we see $f = (x - d_{i+1})g$, so $g(d)dg(d)^{-1}$ is a root of $x - d_{i+1}$, *i.e.*, is d_{i+1} itself. Hence, $d = g(d)^{-1} d_{i+1} g(d)$. □

Structure Theorems for Skew Fields

THEOREM 13. *(Skolem-Noether Theorem) Suppose* $a \in D$ *is algebraic over* F. *Any other root* $b \in D$ *of the minimal polynomial* $f(x)$ *of* a *is conjugate to* a.

Proof. Write $f = (x - d_n) \dots (x - d_1)$ by means of Theorem 10, so that each d_i is conjugate to a, and apply Theorem 12 (since f also is the minimal polynomial of b). □

Remark 14. Another way of expressing this result is as follows: If K/F is separable and $L \supset F$ is another subfield of D that is F-isomorphic to K, then the isomorphism is given by conjugation by some element d of D; in particular, $dKd^{-1} = L$. (Indeed, by Exercise 23.3 we can write $K = F[a]$; letting $b \in L$ be the image under the isomorphism, we see that a and b have the same minimal polynomial, so $b = dad^{-1}$ for suitable d in D.)

It is possible to prove this result also for nonseparable extensions, but the proof will not be given here.

Next, define the *centralizer* $C_R(S)$ of an arbitrary subset S of a ring R to be $\{r \in R : rs = sr \text{ for all } s \text{ in } S\}$. It is easy to check that $C_R(S)$ is a subring of R, and in view of Remark 0, if D is a skew field then $C_D(S)$ is a skew field. We are ready for a deep result of Wedderburn.

THEOREM 15. *(Wedderburn) Every finite skew field* D *is commutative.*

Proof. Let $F = \text{Cent}(D)$, a finite field that thus has order p^t for a suitable prime p and some t; let $K \supset F$ be a commutative subring of D having maximal order. Then K is a finite integral domain and thus a field, and being a vector space over F, has order $m = p^u$ for some multiple u of t. Supposing D noncommutative, we see $K \neq F$, since $F[d]$ is commutative for any d in D. By Theorem 26.8, K/F is cyclic Galois, implying by the Galois correspondence that there is an intermediate field $F \subseteq L \subset K$ such that $[K : L] = p$.

Let $D' = C_D(L)$. By Theorem 13, the nontrivial automorphism of K/L is given by conjugation by an element d of D; clearly, $d \in D'$, implying D' is not commutative. Let $n = |D'|$. *A fortiori*, K is a maximal commutative subring of D'; since $|L| = p^{u-1}$, we see $L = \text{Cent}(D')$. Any commutative subring K' of D' properly containing L must also have order precisely $p^u = m$. (Indeed, K' is a field extension of L, so $[K' : L] \geq p$, implying $|K'| \geq m \geq |K|$ so by hypothesis $|K'| = m$.)

Now viewing $H = K \setminus \{0\}$ as a subgroup of the group $G = D' \setminus \{0\}$, we see that the number of subgroups conjugate to H is at most $\frac{|G|}{|H|} = \frac{n-1}{m-1}$. But each subgroup contains the element 1, so the number of conjugates of elements of H in D' is at most $\frac{(m-2)(n-1)}{m-1} + 1 < n - 1$, implying some element a of D' is not conjugate to any element of K.

Let $K' = F[a]$. As noted above, K' is a field of order m. But any extension of L of order m is a splitting field of the polynomial $x^m - x$ over $\mathbb{Z}_p$ and thus over L, and hence they all are isomorphic over L; thus K and K' are conjugate in D' by Theorem 13, contrary to the assumption that a is not conjugate to any element of K. □

Are there skew fields other than the quaternions? It took algebraists another 50 years to produce a skew field other than Hamilton's. In fact, Frobenius proved $\mathbb{H}$ is the only noncommutative skew field whose elements all are algebraic over $\mathbb{R}$, *cf.* Exercise 15.

Another glance at Example 2 reveals the key role played by matrices, and, indeed, rings of matrices are the focus of noncommutative algebra. But that starts another tale that must be told elsewhere.

Exercises

1. $M_n(F)$ is an F-algebra, where we identify F with the ring of scalar matrices (*cf.* Exercise 13.1).
2. Define an *F-algebra* to be a F-vector space R that also is a ring satisfying the following property, for all α in F and r_i in R (where multiplication is taken to be the ring multiplication or the scalar multiplication of the vector space, according to its context):

$$\alpha(r_1 r_2) = (\alpha r_1) r_2 = r_1(\alpha r_2).$$

Define the ring homomorphism $\varphi: F \to R$ given by $\alpha \mapsto \alpha \cdot 1$. Then

$F \cdot 1$ is a subfield of Cent(R), so we have arrived at the definition in the text.

This definition is very useful in generalizing to algebras over arbitrary commutative rings, not necessarily fields.

3. Show that the regular representation (Exercise 21.15) also works for noncommutative algebras.
4. Derive Proposition 2 via the regular representation, viewing $\mathbb{H}$ as the vector space $\mathbb{C}+\mathbb{C}j$ and considering the matrices corresponding to right multiplication by i, j, k respectively.
5. The map $^-: \mathbb{H} \to \mathbb{H}$ given by $a + bi + cj + dk \mapsto a - bi - cj - dk$ is an anti-automorphism (*i.e.*, reversing the order of multiplication) of order 2. (Hint: $N(q) = q\bar{q}$.)

Proof of Lagrange's Four Square Theorem

The following exercises provide two proofs of Lagrange's Four Square theorem.

6. If $2n = a^2 + b^2 + c^2 + d^2$, then rearranging the summands so that $a \equiv b \pmod 2$ and $c \equiv d \pmod 2$ one can write n as the sum $\left(\frac{a+b}{2}\right)^2 + \left(\frac{a-b}{2}\right)^2 + \left(\frac{c+d}{2}\right)^2 + \left(\frac{c-d}{2}\right)^2$.
7. (Direct computational proof of Lagrange's Four Square Theorem.) Suppose p is not a sum of four squares. Take $n > 1$ minimal, such that $np = a^2 + b^2 + c^2 + d^2$ is a sum of four squares; n is odd by Exercise 6. Take the respective residues a', b', c', d' of a, b, c, d $\pmod n$, each of absolute value $< \frac{n}{2}$. Then there exists m such that $nm = (a')^2 + (b')^2 + (c')^2 + (d')^2 < n^2$; hence, $m < n$. Since $N(a+bi+cj+dk) = np$, note that $(a+bi+cj+dk)(a'-b'i-c'j-d'k)$ is a quaternion each of whose coefficients is congruent to $0 \pmod n$; dividing through by n yields a quaternion whose norm is $< np$, contradiction.
8. Verify the Euclidean algorithm for the ring of integral quaternions. (Hint: Use a similar trick to that used for the Gaussian integers.)
9. Any integer $m > 1$ is reducible as a quaternion. (Hint: Assume that m is an odd prime. Then m divides $1 + a^2 + b^2$ for suitable integers a, b prime to m. Let $q = 1 + ai + bj$ and $d = \gcd(m, q) = r_1 m + r_2 q$. $N(d - r_1 m) = N(r_2 q) = N(r_2)(1 + a^2 + b^2)$ is a multiple of m, implying $m | N(d)$. In particular, d is not invertible, and $Rm \subset Rd$.)
10. An integral quaternion r is irreducible iff $N(r)$ is prime. (Hint: Take a prime divisor p of $N(r)$ in $\mathbb{Z}$, and let $d = \gcd(p, r)$. Then $p = N(d) = N(r)$.)
11. (Quaternion algebras over an arbitrary field) Given a field F of characteristic $\neq 2$ and α, β in F, define the *(generalized) quaternion*

algebra R to be the 4-dimensional algebra $F+Fa+Fb+Fab$, whose multiplication satisfies the rules $a^2=\alpha$, $b^2=\beta$, and $ba=-ab$. Prove $\text{Cent}(R)=F$, and R has no nonzero ideals.

12. Any skew field D of dimension 4 over its center F is a quaternion algebra. (*Hint*: For d in $D\setminus F$, let $K=F[d]$. Then $[K:F]=2$, so $K=F[a]$ for suitable a satisfying $a^2\in F$. The nontrivial automorphism σ of K is given by $a\mapsto -a$, so there is some b such that $bab^{-1}=-a$, by Skolem-Noether. Then $ba=-ab$, so b^2 commutes with a. Hence $K[b^2]$ is a field; counting dimensions show $b^2\in K$. But then $\sigma(b^2)=bb^2b^{-1}=b^2$, so $b^2\in K^{\langle\sigma\rangle}=F$.)
13. If K is a maximal commutative subring of R then $C_R(K)=K$.
14. If $S,T\subseteq R$, then $C_R(S\cup T)=C_R(S)\cap C_R(T)$.
15. (Frobenius' Theorem.) The only skew fields D algebraic over $\mathbb{R}$ are $\mathbb{R},\mathbb{C}$, and $\mathbb{H}$. (Hint: Suppose $D\neq\mathbb{R},\mathbb{C}$. Take a maximal commutative subring C of D. Then $C\approx\mathbb{C}$, so $\text{Cent}(D)=\mathbb{R}$. Write $C=\mathbb{R}[i]$, where $i^2=-1$. Claim: Any element a in $D\setminus C$ is contained in a copy of $\mathbb{H}$ that also contains C. Indeed, $\mathbb{R}[a]\approx\mathbb{C}$, and thus $\mathbb{R}[a]=\mathbb{R}[j]$ with $j^2=-1$. Let $c=ij+ji$. Then c commutes with both i and j, and thus by Exercises 13 and 14 is in $\mathbb{R}$. This proves that $D'=\mathbb{R}+\mathbb{R}i+\mathbb{R}j+\mathbb{R}ij$ is a subring of D. Now let $b=ij-ji\in D'$. Then $ib=-bi$; replacing j by b enables us to assume $ij=-ji$. Now $j^2i=ij^2$, so $j^2\in\mathbb{R}[i]\cap\mathbb{R}[j]=\mathbb{R}$. But $j\notin\mathbb{R}$, implying $j^2<0$; writing $j^2=-\alpha$ and replacing j by $\frac{j}{\sqrt{\alpha}}$, one has $j^2=-1$, so $D'\approx\mathbb{H}$, as claimed.

    Any d in D is contained in a subring $D''=\mathbb{R}+\mathbb{R}i+\mathbb{R}j'+\mathbb{R}ij'\approx\mathbb{H}$, where $ij'=-j'i$ and $j'^2=-1$; hence $j'j^{-1}$ commutes with i, implying $j'j^{-1}\in\mathbb{R}[i]$, or $j'\in\mathbb{R}[i]j\subset D'$, proving $d\in D'$.)
16. Suppose R is a ring containing a field F, and $n=[R:F]<\infty$. Then R is algebraic over F, in the sense that any r in R satisfies a nontrivial equation $\sum_{i=0}^{n}\alpha_i r^i=0$ for α_i in F.
17. Given any group G and field F, define the *group algebra* $F[G]$ to be the F-algebra which as vector space over F has base G, with multiplication of the base elements given by multiplication in G. Show that $Z(G)\subseteq\text{Cent}(F[G])$.
18. Let Q denote the quaternion group (with relations $a^4=1=b^4$; $bab^{-1}=a^{-1}$; $a^2=b^2$). Show that $\mathbb{H}$ contains a quaternion group (generated by i,j). Moreover, there is an algebra homomorphism $\mathbb{R}[Q]\to\mathbb{H}$ given by $a\mapsto i$, $b\mapsto j$; the kernel is $\langle a^2+1\rangle$.
19. Generalizing Exercise 17, define the *monoid algebra* $F[M]$ for any monoid M. Show that $F[x]$ can be written as a monoid algebra.

INDEX

Printed in the United States
by Baker & Taylor Publisher Services